ADVANCES IN PROSTAGLANDIN, THROMBOXANE, AND LEUKOTRIENE RESEARCH
VOLUME 9

Leukotrienes and Other Lipoxygenase Products

Advances in Prostaglandin,
Thromboxane, and Leukotriene Research

Series Editors: Bengt Samuelsson and Rodolfo Paoletti

Advances in Prostaglandin, Thromboxane, and Leukotriene Research Series
Volume 9
(Formerly Advances in Prostaglandin and Thromboxane Research Series)

Leukotrienes and Other Lipoxygenase Products

Editors

Bengt Samuelsson
Professor
Department of Physiological Chemistry
Karolinska Institute
Stockholm, Sweden

Rodolfo Paoletti
Professor and Director
Institute of Pharmacology
and Pharmacognosy
University of Milan
Milan, Italy

Raven Press ■ New York

Raven Press, 1140 Avenue of the Americas, New York, New York 10036

Made in the United States of America

International Standard Book Number 0–89004–741–3
Library of Congress Catalog Number 81–40607

Great care has been taken to maintain the accuracy of the information contained in the volume. However, Raven Press cannot be held responsible for errors or for any consequences arising from the use of the information contained herein.

Materials appearing in this book prepared by individuals as part of their official duties as U.S. Government employees are not covered by the above-mentioned copyright.

Preface

This volume of *Advances in Prostaglandin, Thromboxane, and Leukotriene Research* contains most of the papers presented at an international conference on Leukotrienes and Other Lipoxygenase Products held in Florence, Italy on June 10–12, 1981. We would like to take this opportunity to express our gratitude to Dr. E. Folco and the staff of the Fondazione Giovanni Lorenzini, as well as to Drs. F. Berti, G. C. Folco, and E. Granström for their important contributions before and during the conference.

We are grateful to the authors for their prompt submission of the papers and to Dr. Alan Edelson of Raven Press for arranging rapid publication.

The leukotrienes constitute a new group of biologically active compounds derived from arachidonic acid and other polyunsaturated fatty acids. The transformations are initiated by oxygenation at C-5 in a lipoxygenase catalyzed reaction. The discovery of the leukotriene system was first described during the Fourth International Prostaglandin Conference in Washington D.C. in May 1979. This field has subsequently grown rapidly in several directions including biochemistry, pharmacology and physiology, and one can begin to see clinical implications of the new knowledge. Furthermore, it is being recognized that the leukotrienes might play a role not only in immediate hypersensitivity reactions but also in host defense mechanisms in general.

The Editors

Contents

Biosynthesis

Biological Effects

Contributors

Fadia E. Ali
Research and Development Division
Smith Kline and French Laboratories
Philadelphia, Pennsylvania 19101

R. E. C. Altounyan
Fisons Limited
Pharmaceutical Division
Research and Development Laboratories
Loughborough, Leicestershire
LE11 OQY, England

K. Frank Austen
Department of Medicine
Harvard Medical School, and
Department of Rheumatology and
Immunology, Brigham and Women's
  Hospital
Boston, Massachusetts 02115

Michael K. Bach
Hypersensitivity Diseases Research and
  Experimental Chemistry
The Upjohn Company
Kalamazoo, Michigan 49001

J. M. Bailey
Department of Biochemistry
George Washington University Medical
  School
2300 Eye Street, N.W.
Washington, D.C. 20037

S. R. Baker
Lilly Research Centre Limited
Erl Wood Manor, Windlesham
Surrey, GU20 6PH, England

J. R. Bantick
Research and Development Laboratories
Fisons Limited
Pharmaceutical Division
Loughborough, Leicestershire
LE11 OQY, England

C. M. R. Bax
Departments of Prostaglandin Research
  and Pharmacology

The Wellcome Research Laboratories
Beckenham, Kent BR3 3BS England

Barry Berkowitz
Research and Development Division
Smith Kline and French Laboratories
Philadelphia, Pennsylvania 19101

F. Berti
Institute of Pharmacology and
  Pharmacognosy
University of Milan
20129 Milan, Italy

Parimal Bhattacherjee
The Wellcome Research
Laboratories
Beckenham, Kent BR3 3BS
England

Jakob Björk
Department of Experimental Medicine
Pharmacia AB
S-751 04 Uppsala, Sweden

J. R. Boot
Lilly Research Centre Limited
Erl Wood Manor, Windlesham
Surrey, GU20 6PH, England

P. Borgeat
Departement d'Endocrinologie
  Moléculaire
Le Centre Hospitalier de
l'Université Laval
Ste-Foy, Quebec, Canada GIV 4G2

Alan R. Brash
Departments of Pharmacology and
  Medicine
Vanderbilt University School of Medicine
Nashville, Tennessee 37232

John R. Brashler
Hypersensitivity Diseases Research
The Upjohn Company
Kalamazoo, Michigan 49001

G. Brunelli
Institute of Pharmacology and
Pharmacognosy
University of Milan
20129 Milan, Italy

R. W. Bryant
Department of Biochemistry
George Washington University Medical
School
2300 Eye Street, N.W.
Washington, D.C. 20037

James A. Burke
Department of Pharmacology
Cornell University
Medical College
1300 York Avenue
New York, New York 10021

L. Casey
Casualty Care Research Program Center
Naval Medical Research Institute
Bethesda, Maryland 20014

Hans-Erik Claesson
Department of Physiological Chemistry
Karolinska Institute
S-104 01 Stockholm, Sweden

J. Clarke
Hyperbaric Medicine Program Center
Naval Medical Research Institute
Bethesda, Maryland 20014

F. Coceani
Research Institute and
Department of Pediatrics
The Hospital for Sick Children
555 University Avenue
Toronto, Ontario, Canada M5G 1X8

M. Cole
Fisons Limited
Pharmaceutical Division
Research and Development Laboratories
Loughborough, Leicestershire
LE11 OQY, England

E. J. Corey
Department of Chemistry

Harvard University
Cambridge, Massachusetts 02138

O. Cromwell
Department of Clinical Immunology
Cardiothoracic Institute
Fulham Road
London, SW3 6HP, England

Sven-Erik Dahlén
Department of Physiology
Karolinska Institute
S-104 01 Stockholm, Sweden

W. Dawson
Lilly Research Centre Limited
Erl Wood Manor, Windlesham
Surrey, GU20 6PH, England

Denise M. DiSantis
Department of Pharmacology
Washington University Medical School
660 South Euclid Avenue
St. Louis, Missouri 63110

M. Dixon
Fisons Limited
Pharmaceutical Division
Research and Development Laboratories
Loughborough, Leicestershire
LE11 OQY, England

Jeffrey M. Drazen
Department of Physiology
Harvard School of Public Health,
and Departments of Medicine
Harvard Medical School
and Brigham and Women's Hospital
Boston, Massachusetts 02115

Kenneth E. Eakins
The Wellcome Research Laboratories
Beckenham, Kent BR3 3BS
England

E. V. Elliott
Fisons Limited
Pharmaceutical Division
Research and Development Laboratories
Loughborough, Leicestershire
LE11 OQY, England

Lars Engstedt
Department of Medicine IV
Karolinska Institute
Södersjukhuset
S-10064 Stockholm, Sweden

Joanne C. Figueiredo
Department of Medicine
Harvard Medical School, and
Department of Rheumatology and
Immunology, Brigham and Women's
Hospital
Boston, Massachusetts 02115

J. Fletcher
Casualty Care Research Program Center
Naval Medical Research Institute
Bethesda, Maryland 20014

Roderick J. Flower
Department of Prostaglandin Research
The Wellcome Research Laboratories
Beckenham, Kent BR3 3BS, England

G. C. Folco
Institute of Pharmacology and
Pharmacognosy
University of Milan
20129 Milan, Italy

Robert Friedman
Division of Rheumatology
Department of Medicine
New York University School of Medicine
New York, New York 10016

B. Fruteau de Laclos
Departement d'Endocrinologie
Moléculaire
Le Centre Hospitalier de l'Universite
Laval
Ste-Foy, Quebec
Canada G1V 4G2

A. M. Ghelani
Research and Development Laboratories
Fisons Limited
Pharmaceutical Division
Loughborough, Leicestershire
LE11 OQY, England

John G. Gleason
Research and Development Division
Smith Kline and French Laboratories
Philadelphia, Pennsylvania 19101

E. J. Goetzl
Howard Hughes Medical Institute
Laboratories
Harvard Medical School
Boston, Massachusetts 02115

D. W. Goldman
Howard Hughes Medical Institute
Laboratories
Harvard Medical School
and Departments of Medicine,
Harvard Medical School and
The Brigham and Women's Hospital
Boston, Massachusetts 02115

Ingiäld Hafström
Department of Medicine III
Karolinska Institute
Södersjukhuset
S-10064 Stockholm, Sweden

F. Hamilton
Research Institute and
Department of Pediatrics
The Hospital for Sick Children
555 University Avenue
Toronto, Ontario, Canada MG5 1X8

Sven Hammarström
Department of Physiological Chemistry
Karolinska Institute
S-104 01 Stockholm, Sweden

Brian Hammond
The Wellcome Research Laboratories
Beckenham, Kent BR3 3BS
England

Göran Hansson
Department of Physiological Chemistry
Karolinska Institute
S-104 01 Stockholm, Sweden

Per Hedqvist
Department of Physiology
Karolinska Institute
S-104 01 Stockholm, Sweden

G. A. Higgs
*Departments of Prostaglandin Research
and Pharmacology
The Wellcome Research Laboratories
Beckenham, Kent BR3 3BS, England*

M. C. Holroyde
*Fisons Limited
Pharmaceutical Division
Research and Development Laboratories
Loughborough, Leicestershire
LE11 OQY, England*

Tony E. Hugli
*The Scripps Clinic and Research
Foundation
La Jolla, California 92037*

D. Huhtanen
*Research Institute and
Department of Pediatrics
The Hospital for Sick Children
555 University Avenue
Toronto, Ontario, Canada M5G 1X8*

Barbara A. Jakschik
*Department of Pharmacology
Washington University Medical School
660 South Euclid Avenue
St. Louis, Missouri 63110*

W. B. Jamieson
*Lilly Research Centre Limited
Erl Wood Manor, Windlesham
Surrey, GU20 6PH, England*

W. Jubiz
*Veterans Administration
Medical Center
University of Utah
Salt Lake City, Utah 84148*

Anne Kagey-Sobotka
*Clinical Immunology Division
Department of Medicine
The Johns Hopkins University
School of Medicine at The Good
Samaritan Hospital
Baltimore, Maryland 21239*

Michael A. Kaliner
*Allergic Diseases Section
National Institute for Allergy and
Infectious Diseases
Bethesda, Maryland 20205*

Howard B. Kaplan
*Division of Rheumatology
Department of Medicine
New York University School of Medicine
New York, New York 10016*

A. B. Kay
*Department of Clinical Immunology
Cardiothoracic Institute
Fulham Road, London
SW3 6HP, England*

W. König
*Lehrstuhl für Medizinische Mikrobiologie
und Immunologie
Arbeitsgruppe für
Infektabwehrsmechanismen
Ruhr-Universität Bochum
Postfach, 4630 Bochum, West Germany*

Helen M. Korchak
*Division of Rheumatology
Department of Medicine
New York University School of Medicine
New York, New York 10016*

Robert D. Krell
*Research and Development Division
Smith Kline and French Laboratories
Philadelphia, Pennsylvania 19101*

H. W. Kunau
*Institut für Physiologische Chemie
Arbeitsgruppe Bioorganische Chemie
Ruhr-Universität Bochum
Postfach, 4630 Bochum, West Germany*

T. B. Lee
*Research and Development Laboratories
Fison Limited
Pharmaceutical Division
Loughborough, Leicestershire
LE11 OQY, England*

L. G. Letts
*Department of Pharmacology
Institute of Basic Medical Sciences
Royal College of Surgeons of England
Lincoln's Inn Fields
London WC2A 3PN, England*

Roberto Levi
*Department of Pharmacology
Cornell University Medical College
1300 York Avenue
New York, New York 10021*

Robert A. Lewis
Department of Medicine
Harvard Medical School
and Department of Rheumatology and
*   Immunology*
Brigham and Women's Hospital
Boston, Massachusetts 02115

Lawrence M. Lichtenstein
Clinical Immunology Division
Department of Medicine
The Johns Hopkins University School of
*   Medicine at The Good Samaritan*
*   Hospital*
Baltimore, Maryland 21239

Jan Åke Lindgren
Department of Physiological Chemistry
Karolinska Institute
S-104 01 Stockholm, Sweden

C. E. Low
Department of Biochemistry
George Washington University Medical
*   School*
2300 Eye Street, N.W.
Washington, D.C. 20037

U. Lundberg
Department of Physiological Chemistry
Karolinska Institute
S-104 01 Stockholm, Sweden

Richard L. Maas
Departments of Pharmacology and
*   Medicine*
Vanderbilt University School of Medicine
Nashville, Tennessee 37232

Curt L. Malmsten
Department of Physiological Chemistry
Karolinska Institute
S-104 01 Stockholm, Sweden

Gianni Marone
Clinical Immunology Division
Department of Medicine
The Johns Hopkins University School of
*   Medicine at The Good Samaritan*
*   Hospital*
Baltimore, Maryland 21239

K. Mizuno
Department of Biochemistry
Tokushima University School of Medicine
Tokushima, Japan

S. Moncada
Departments of Prostaglandin Research
*   and Pharmacology*
The Wellcome Research Laboratories
Beckenham, Kent BR3 3BS, England

H. R. Morris
Department of Biochemistry
Imperial College
South Kensington, London
SW7 England

Douglas R. Morton
Hypersensitivity Diseases Research and
*   Experimental Chemistry*
The Upjohn Company
Kalamazoo, Michigan 49001

P. H. Naccache
Departments of Pathology and Physiology
University of Connecticut Health Center
Farmington, Connecticut 06032

S. Narumiya
Department of Medical Chemistry
Kyoto University
Kyoto 606, Japan

Phillip Needleman
Department of Pharmacology
Washington University Medical School
St. Louis, Missouri 63110

John A. Oates
Departments of Pharmacology and
*   Medicine*
Vanderbilt University School of Medicine
Nashville, Tennessee 37232

B. R. C. O'Driscoll
Department of Clinical Immunology
Cardiothoracic Institute
Fulham Road, London
SW3 6HP England

P. M. Olley
Research Institute and
Department of Pediatrics
The Hospital for Sick Children
555 University Avenue
Toronto MG5 1X8, Canada

C. Omini
Institute of Pharmacology and
*   Pharmacognosy*
University of Milan
20129 Milan, Italy

D. J. Osborne
Lilly Research Centre Limited
Erl Wood Manor, Windlesham
Surrey GU20 6PH, England

C. R. Pace-Asciak
Research Institute
Hospital for Sick Children
555 University Avenue
Toronto, Ontario, Canada M5G 1X8

Jan Palmblad
Department of Medicine IV
Karolinska Institute
Södersjukhuset
S-10064 Stockholm, Sweden

M. A. Palmer
Department of Pharmacology
Institute of Basic Medical Sciences
Royal College of Surgeons of England
Lincoln's Inn Fields
London WC2A 3PN, England

Charles W. Parker
Department of Internal Medicine
Washington University Medical School
660 South Euclid Avenue
St. Louis, Missouri 63110

Stephen Peters
Clinical Immunology Division
Department of Medicine
The Johns Hopkins University School of
  Medicine at The Good Samaritan
  Hospital
Baltimore, Maryland 21239

S. Picard
Department d'Endocrinologie
  Moléculaire
Le Centre Hospitalier de l'Universite
  Laval
Ste-Foy, Quebec, Canada G1V 4G2

W. C. Pickett
Lederle Laboratories
Pearl River, New York 10965

Priscilla J. Piper
Department of Pharmacology
Institute of Basic Medical Sciences
Royal College of Surgeons of England
Lincoln's Inn Fields
London WC2A 3PN, England

M. B. Pupillo
Department of Biochemistry
George Washington University Medical
  School
2300 Eye Street, N.W.
Washington, D.C. 20037

Olof Rådmark
Department of Physiological Chemistry
Karolinska Institute
S-104 01 Stockholm, Sweden

P. Ramwell
Georgetown University Medical Center
Washington, D.C. 20007

G. Rossoni
Institute of Pharmacology and
  Pharmacognosy
University of Milan
20129 Milan, Italy

John A. Salmon
Department of Prostaglandin Research
The Wellcome Research Laboratories
Beckenham, Kent BR3 3BS
England

Marwa N. Samhoun
Department of Pharmacology
Institute of Basic Medical Sciences
Royal College of Surgeons of England
Lincoln's Inn Fields
London WC2A 3PN, England

Alvin R. Sams
Department of Pharmacology
Washington University Medical School
660 South Euclid Avenue
St. Louis, Missouri 63110

Bengt Samuelsson
Department of Physiological Chemistry
Karolinska Institute
S-104 01 Stockholm, Sweden

S. K. Sankarappa
Department of Physiological Chemistry
Ohio State University
Columbus, Ohio 43210

Robert Schleimer
Clinical Immunology Division
Department of Medicine
The Johns Hopkins University School of
 Medicine at The Good Samaritan
 Hospital
Baltimore, Maryland 21239

Charles Serhan
Divison of Rheumatology
Department of Medicine
New York University School of Medicine
New York, New York 10016

R. I. Sha'afi
Departments of Pathology and Physiology
University of Connecticut Health Center
Farmington, Connecticut 06032

P. Sheard
Research and Development Laboratories
Fisons Limited
Pharmaceutical Division
Loughborough, Leicestershire
LE11 OQY, England

E. Sideris
Research Institute and Department of
 Pediatrics
The Hospital for Sick Children
555 University Avenue
Toronto, M5G 1X8, Canada

Marvin Siegel
Clinical Immunology Division
Department of Medicine
The Johns Hopkins University School of
 Medicine at The Good Samaritan
 Hospital
Baltimore, Maryland 21239

P. Sirois
Unite de Recherches Pulmonaire
Center Hospitalier Universitaire
Sherbrooke, Quebec, Canada J1H 5N4

M. J. H. Smith
Department of Chemical Pathology
King's College Hospital Medical School
Denmark Hill
London, SE5 8RX, England

James E. Smolen
Division of Rheumatology
Department of Medicine
New York University School of Medicine
New York, New York 10016

Nicholas A. Soter
Department of Dermatology
Harvard Medical School, and
Department of Rheumatology and
 Immunology, Brigham and Women's
 Hospital
Boston, Massachusetts 02115

Howard Sprecher
Department of Physiological Chemistry
Ohio State University
Columbus, Ohio 43210

Linda K. Steel
Allergic Diseases Section
National Institute for Allergy and
 Infectious Diseases
Bethesda, Maryland 20205

W. J. F. Sweatman
Lilly Research Centre Limited
Erl Wood Manor, Windlesham
Surrey, GU20 6PH, England

G. W. Taylor
Department of Biochemistry
Imperial College
South Kensington, London
SW7 England

J. P. Tippins
Department of Pharmacology
Institute of Basic Medical Sciences
Royal College of Surgeons of England
Lincoln's Inn Fields
London WC2A 3PN, England

Ann-Mari Udén
Department of Medicine IV
Karolinska Institute
Södersjukhuset
S-10064 Stockholm, Sweden

P. Vallerand
Department d'Endocrinologie
Moléculaire
Le Centre Hospitalier de l'Universite
Laval
Ste-Foy, Quebec, Canada G1V 4G2

J. Y. Vanderhoek
Department of Biochemistry
George Washington University Medical
School
2300 Eye Street, N.W.
Washington, D.C. 20037

John R. Vane
Department of Prostaglandin Research
The Wellcome Research Laboratories
Beckenham, Kent BR3 3BS, England

T. Viganò
Institute of Pharmacology and
Pharmacognosy
University of Milan
20129 Milan, Italy

M. J. Walport
Department of Clinical Immunology
Cardiothoracic Institute
Fulham Road, London
SW3 6HP, England

Barry M. Weichman
Research and Development Division
Smith Kline and French Laboratories
Philadelphia, Pennsylvania 19101

Gerald Weissmann
Division of Rheumatology
Department of Medicine
New York School of Medicine
New York, New York 10016

S. Yamamoto
Department of Biochemistry
Tokushima University School of Medicine
Tokushima, Japan

K. Yokochi
Research Institute and
Department of Pediatrics
The Hospital for Sick Children
555 University Avenue
Toronto, Ontario
Canada M5G 1X8

Leukotrienes and Other Lipoxygenase Products,
edited by B. Samuelsson and R. Paoletti.
Raven Press, New York © 1982.

The Leukotrienes: An Introduction

Bengt Samuelsson

Department of Physiological Chemistry, Karolinska Institute,
S-104 01 Stockholm, Sweden

Polyunsaturated fatty acids play a role as precursors of biologically active compounds that can act as mediators or modulators of various cell functions. Thus three main groups of derivatives—the prostaglandins, the thromboxanes, and the recently discovered leukotrienes—are formed by oxygenation and further transformation of various polyunsaturated fatty acids, of which arachidonic acid plays the most significant role (58,60). The leukotrienes are of particular interest since the cysteine-containing leukotrienes are responsible for the biological activity earlier referred to as slow-reacting substance of anaphylaxis (SRS-A). These derivatives are potent bronchoconstrictors and seem to play a pathophysiological role in immediate hypersensitivity reactions. There is also evidence that both these compounds and a leukotriene with pronounced effects on leukocyte migration might function as mediators in inflammation. This chapter summarizes some of the developments of this area of research.

DISCOVERY OF LEUKOTRIENES

Anti-inflammatory drugs such as aspirin inhibit the enzyme (cyclo-oxygenase) responsible for conversion of arachidonic acid into prostaglandins (66). However, anti-inflammatory corticosteroids prevent formation of prostaglandins by a different mechanism; viz., they inhibit the release of arachidonic acid from the phospholipids (23,33). In view of the pronounced differences in the anti-inflammatory effects of steroids and aspirin-type drugs, it seemed conceivable to us that arachidonic acid might generate additional proinflammatory derivatives which were not dependent on the cyclo-oxygenase for their formation. The production of such products should be inhibited by anti-inflammatory steroids unless free arachidonic acid was added. To test this hypothesis, the metabolism of arachidonic acid in polymorphonuclear leukocytes (PMNLs) was studied.

It was found that with PMNLs from the peritoneal cavity of rabbits 5(S)-hydroxy-6,8,11,14-eicosatetraenoic acid (5-HETE) was the major arachidonic acid-derived product (6). Additional work showed that more polar products were also formed. These were identified as: (a) 5(S),12(R)-dihydroxy-6,8,10,14-eicosatetraenoic acid (major product; leukotriene B_4, see below); (b) two additional 5(S),12-dihydroxy-6,8,10-*trans*,14-*cis*-eicosatetraenoic acids, epimeric at C-12; and (c) two isomeric

5,6-dihydroxy-7,9,11,14-eicosatetraenoic acids (7,8) (Fig. 1). Stereochemical analysis, demonstrating formation of two acids with an all-*trans* conjugated triene, epimeric at C-12, and one major isomer (12R) with different configuration of the triene raised the question of the mechanism of formation (8). Using isotopic oxygen, it could be demonstrated that the oxygen of the alcohol group at C-5 originated in molecular oxygen, whereas the oxygen of the alcohol group at C-12 was derived from water (9) (Fig. 2). Based on these observations, it was postulated that an unstable intermediate was generated from arachidonic acid by the leukocytes. This intermediate would undergo nucleophilic attack by water, alcohols, and other nucleophiles.

Rabbit peritoneal PMNLs were therefore incubated for 30 sec with arachidonic acid before adding 10 volumes of methanol, 10 volumes of ethanol, or 0.2 volumes of 1 N HCl. The products were analyzed by reverse-phase high-pressure liquid chromatography (RP-HPLC). The chromatogram in Fig. 3 shows the pattern of products obtained (polar metabolites only). The material formed upon trapping with methanol (or ethanol) consisted of two additional compounds present in equal amounts. Their ultraviolet spectra were identical to those of compounds I and II, indicating the presence of three conjugated double bonds. Infrared spectrometry further indicated that the conjugated double bonds had *trans* geometry. Gas chro-

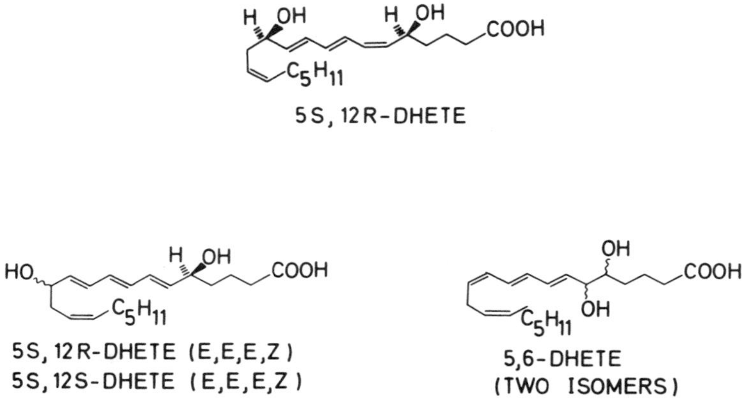

5S, 12R-DHETE

5S, 12R-DHETE (E,E,E,Z)
5S, 12S-DHETE (E,E,E,Z)

5,6-DHETE
(TWO ISOMERS)

FIG. 1.   Structures of dihydroxylated metabolites of arachidonic acid in rabbit PMNLs.

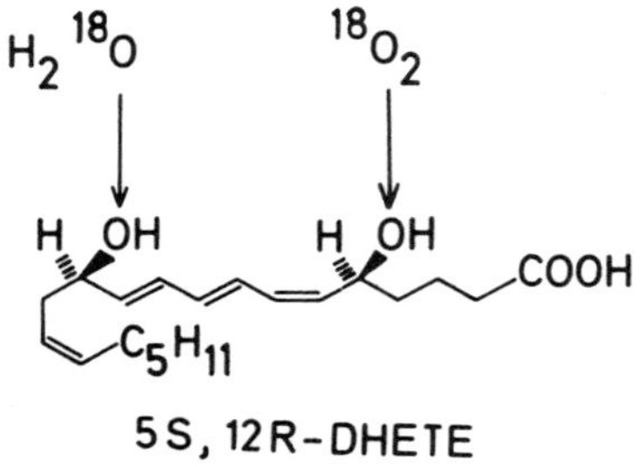

5S, 12R-DHETE

FIG. 2.   Origin of oxygen in hydroxyl groups of 5(S),12(R)-dihydroxy-6,14-*cis*-8,10-*trans*-eicosatetraenoic acid.

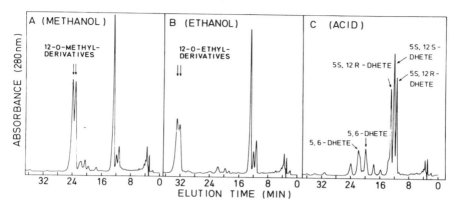

FIG. 3. RP-HPLC chromatograms of the products obtained upon addition of (**A**) 10 volumes methanol; (**B**) 10 volumes of ethanol; and (**C**) 0.2 volumes of 1 N HCl to suspensions of PMNL incubated for 30 sec with arachidonic acid.

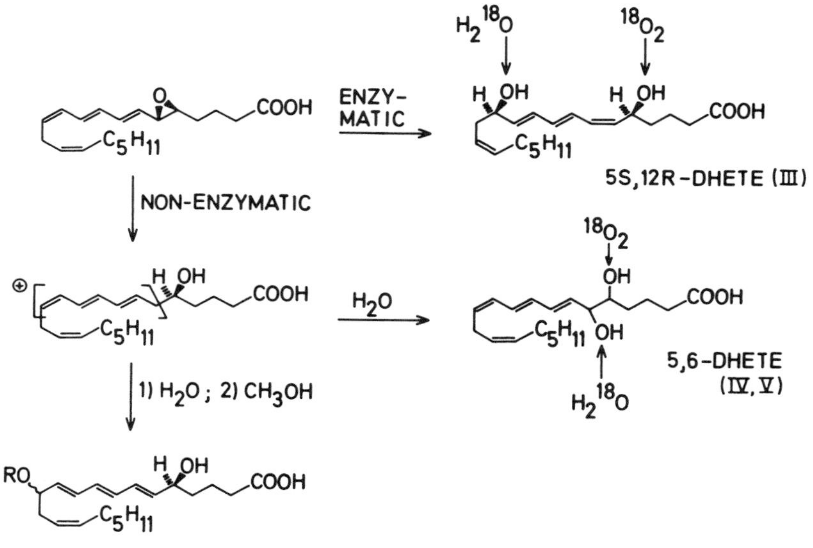

1) 5,12-DHETE (I, II), 2) 12-O-METHYL-DERIV.

FIG. 4. Formation of dihydroxy derivatives from an unstable intermediate. Origin of oxygen and trapping experiments.

matographic-mass spectrometric (GC-MS) analyses of several derivatives of the two compounds showed that they were isomeric and carried hydroxyl groups at C-5 and methoxy groups at C-12. Steric analyses showed that the alcohol groups had (S) configuration. Although the configurations at C-12 were not determined, it is clear that the compounds are the C-12 epimers of 5(S)-hydroxy,12-methoxy-6,8,10,14 (E.E.E.Z)-eicosatetraenoic acid (Fig. 4).

Analogous derivatives were identified when ethanol or ethylene glycol were used for trapping. These data show that a metabolite of arachidonic acid in leukocytes can undergo a facile nucleophilic reaction with alcohols. Interestingly, RP-HPLC analysis of samples obtained from trapping experiments performed under various conditions always indicated inverse relationships between the amount of compounds I and II formed and their 12-O-alkyl derivatives. This suggested that compounds I and II were formed nonenzymatically from the same intermediate that gave rise to the 12-O-alkyl derivatives.

To determine the stability of the intermediate, rabbit PMNLs were incubated with arachidonic acid for 45 sec before adding 1 volume of acetone (to stop enzymatic activity). At various time intervals, aliquots of the mixture were transferred to flasks containing 15 volumes of methanol. The relative amounts of metabolites were estimated by RP-HPLC. Figure 5 shows the decay of the intermediate, measured as the 12-O-methyl derivative, at pH 7.4 and 37°C [half-life $(t_{1/2}) = 3$ to 4 min]. Simultaneously, the concentrations of compounds I, II, IV, and V increased with time. The concentrations of compounds III and 5-hydroxy-6,8,11,14-eicosatetraenoic acid remained constant. This suggested that compounds I, II, IV, and V are formed nonenzymatically by hydrolysis of a common unstable intermediate, whereas compound III arises by enzymatic hydrolysis of the same intermediate. Similar experiments performed at acidic and alkaline pH indicated that the intermediate was acid-labile and somewhat stabilized at alkaline pH.

Based on the experimental data described above, the structure 5,6-oxido-7,9,11,14-eicosatetraenoic acid was proposed for the intermediate (9). Hydrolysis of epoxides is acid-catalyzed, and opening of allylic epoxides is favored at allylic positions (C-6 in this case). This agrees with the retention of [18]O at C-5 in compounds IV and V. The formation of compounds I through V from the epoxide intermediate is shown in Fig. 4. Except for compound III, these are formed by chemical hydrolysis of

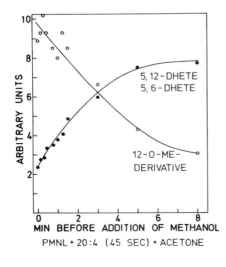

FIG. 5. Time course of the formation of compounds I, II, IV, or V (●) and the disappearance of the unstable intermediate measured as 12-O-methyl compounds I-II (○) in a mixture of water/acetone (1:1, v/v) at pH 7.4 and 37°C. Prostaglandin B_2 was added as an internal standard for quantitation by RP-HPLC.

the epoxide through a mechanism involving a carbonium ion. The latter added hydroxyl anion preferentially at C-6 and C-12 to yield four isomeric products which contain the stable conjugated triene structure. Compound III is formed enzymatically from the intermediate as it is not racemic at C-12 and because it is formed only by nondenatured cell preparations.

A proposed pathway for the formation of the epoxide from arachidonic acid is shown in Fig. 6. It involves initial formation of 5-hydroperoxy-6,8,11,14-eicosatetraenoic acid (5-HPETE), which is the precursor of the 5-hydroxy acid. The epoxide is formed from 5-HPETE by abstraction of a proton at C-10 and elimination of hydroxyl anion from the hydroperoxy group. The proposed structure (9) of the intermediate, 5,6-oxido-7,9,11,14-eicosatetraenoic acid (leukotriene A₄, see below), has been confirmed by chemical synthesis and the stereochemistry elucidated (54). The configuration of the double bonds of the conjugated triene in leukotriene B₄ was also determined using a synthetic approach (14). The allylic epoxide intermediate was recently isolated from human PMNLs (55). It can thus exist in free form in cells and tissues.

The concept that the unstable epoxide plays a role in the transformation of arachidonic acid in leukocytes was a prerequisite for the work leading to elucidation of the structure and biogenesis of slow reacting substance of anaphylaxis (SRS-A)

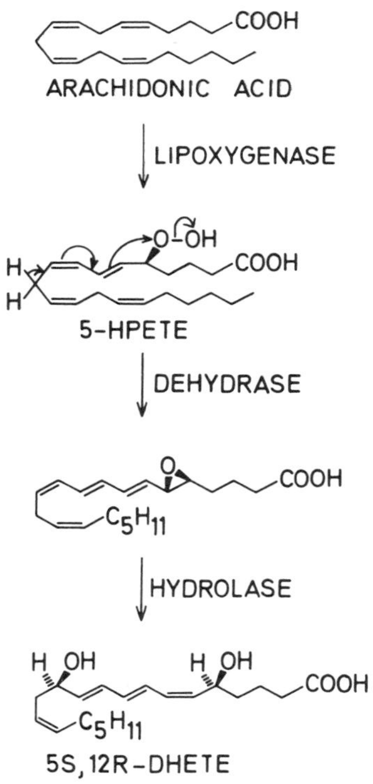

FIG. 6. Mechanism of formation of unstable intermediate.

(9). Thus quantitative studies on the formation of the 5-HETE and 5,12-dihydroxy-eicosatetraenoic acids (5,12-DHETE) by human PMNLs demonstrated that the ionophore A23187 stimulates the synthesis of 5-HETE and 5,12-DHETE (10). This was of particular interest as previous studies had shown that the ionophore also stimulates release of SRS from leukocytes (13). Furthermore, the ultraviolet (UV) absorption of the dihydroxy acids (λ_{max} about 270 nm) discovered in the leukocytes was similar to that reported for SRS-A (43,48). The effects of the ionophore, the UV absorbance data, and other considerations led us to develop the hypothesis that there was a biogenetic link between the unstable allylic epoxide intermediate and SRS-A (9).

LEUKOTRIENES AND SRS-A

The term slow reacting substance (SRS) was introduced in 1938 by Feldberg and Kellaway for a smooth-muscle-contracting factor that appears in the perfusate of guinea pig lung following treatment with cobra venom (19). Subsequent studies suggested that SRS is an important mediator in asthma and other types of immediate hypersensitivity reactions (1,11,37,47). Immunologically released SRS is usually referred to as SRS-A. It is considered to be released together with other mediators (e.g., histamine and chemotactic factors) after interaction between immunoglobulin E (IgE) molecules, bound to membrane receptors, and antigens as pollen, etc.

Earlier structural work on SRS was severely limited by the difficulty of obtaining sufficient quantities of pure preparations of SRS. However, it had been characterized as a polar lipid (12,48,65) possibly containing sulfur and having UV absorption (43,48,49). Studies with labeled arachidonic acid indicated that it was incorporated into SRS (2,35).

When we investigated different methods for producing SRS, we found that murine mastocytoma cells treated with ionophore A23187 and L-cysteine generated SRS. This method proved superior to previously described systems with respect to formation of spasmogenic material antagonized by the SRS antagonist FPL55712 and incorporation of isotopically labeled precursors (see below) (46). The procedure used for isolation involved precipitation of protein with ethanol, alkaline hydrolysis, separation on Amberlite XAD-8 and silicic acid, and two steps of RP-HPLC. The material obtained by this procedure was essentially pure, showed an absorbance maximum at 280 nm, and produced a typical contraction of guinea pig ileum which was reversed by FPL55712 (46). The UV spectrum resembled those of the dihydroxy acids described above, however the maximum was shifted 10 nm, to a higher wavelength. This was consistent with a sulfur substituent α to a conjugated triene. Experiments with labeled precursors showed that arachidonic acid and cysteine were incorporated into the products.

Desulfurization of SRS by Raney nickel gave 5-hydroxyarachidonic acid, which indicated that the arachidonic acid derivative and cysteine were linked by a thioether bond (Fig. 7). The alcohol group at C-5 in the fatty acid reinforced the hypothesis of a biogenetic relationship between the arachidonic acid metabolites we had found in leukocytes and SRS.

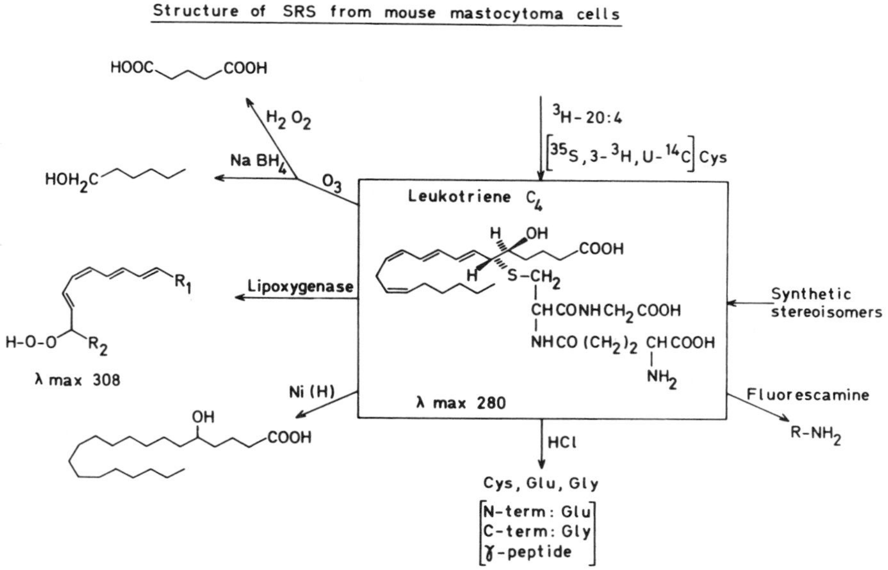

FIG. 7. Some transformations in structural studies on SRS.

The double bonds in SRS biosynthesized from tritium-labeled arachidonic acid were localized by reductive ozonolysis. This yielded [^{3}H]1-hexanol, demonstrating retention of the Δ^{14} double bond of arachidonic acid. A unique method was used for locating the conjugated triene. Previous studies in our laboratory had shown that arachidonic acid and related fatty acids containing two methylene interrupted *cis* double bonds at the ω6 and ω9 positions are oxygenated by soybean lipoxygenase to form ω6 oxygenated derivatives with isomerization of the ω6 double bond to ω7. Treatment of SRS with the lipoxygenase resulted in isomerization of the Δ^{14} double bond into conjugation with the conjugated triene (forming a tetraene) as there was a bathochromic shift of 30 nm. This indicated the presence of a Δ^{11}-*cis* double bond and additional double bonds at Δ^7 and Δ^9 in SRS. The structural studies at this stage showed that the SRS was a derivative of 5-hydroxy-7,9,11,14-eicosatetraenoic acid with a cysteine-containing substituent in thioether linkage at C-6. Derivatization of cysteine was suggested by failure to isolate alanine after desulfurization. The cysteine-containing substituent was therefore referred to as RSH in the reports of this work (46,59,61). Additional studies involving amino acid analyses of acid-hydrolyzed SRS demonstrated that, in addition to cysteine, 1 mole of glycine and 1 mole of glutamic acid were present per mole of SRS. The structure of the peptide as determined by end group (dansyl method and hydrazinolysis) and sequence analyses (dansyl-Edman procedure) was γ-glutamylcysteinylglycine (glutathione). The experiments described above showed that the structure of SRS from murine mastocytoma cells is 5-hydroxy-6-*S*-glutathionyl-7,9,11,14-eicosatetraenoic acid, leukotriene (LT) C_4 (see below) (24). The structure was confirmed by comparison with synthetic material. This represented the first structure determination of an SRS-A (24). The preparation and some properties of corresponding cysteinyl-glycine derivative (LTD_4) and cysteinyl derivative (LTE_4) were also reported at the same time (24). These compounds were later isolated from natural sources (see below). The proposed stereochemistry of LTC_4 was confirmed and unambiguously assigned by total synthesis including preparation of stereoisomers of LTC_4 (25). The synthetic work was carried out by Corey and co-workers (14). LTC_4 is thus 5(S)-hydroxy,6(R)-S-glutathionyl-7,9-*trans*-11,14-*cis*-eicosatetraenoic acid. The previously proposed biogenetic relationship between LTA_4 and LTC_4 was recently confirmed by the actual conversion of synthetic LTA_4 into LTC_4 in human PMNLs (56) (Fig. 8).

Subsequent studies with a different cell type, RBL-1 cells, demonstrated that the major slow reacting substance was less polar than LTC_4 (50). That the fatty acid part of this compound and LTC_4 were identical was indicated by their UV spectra, the product obtained after Raney nickel desulfurization and the spectral change observed after treatment with soybean lipoxygenase. Amino acid analyses, however, showed that the less polar product lacked glutamic acid. Edman degradation indicated that glycine was C-terminal. Incubation of LTC_4 with γ-glutamyl trans-peptidase yielded additional proof for the structure. The product, 5(S)-hydroxy,6(R)-S-cysteinyl-glycine-7,9-*trans*-11,14-*cis*-eicosatetraenoic acid (LTD_4) was identical

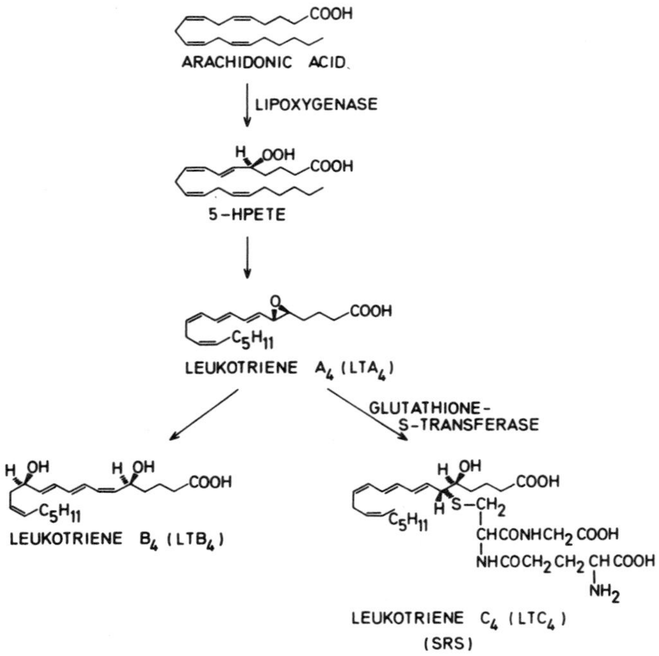

FIG. 8. Formation of leukotrienes.

with the less polar product from RBL-1 cells (Fig. 9). LTD_4 is more potent than LTC_4 in the guinea pig ileum bioassay (50).

Following the structure determination of SRS from mastocytoma cells (24,46) and synthetic preparation of LTC_4, LTD_4, and LTE_4 (24), all of these cysteine-containing leukotrienes have been found in a variety of biological systems using a comparison with synthetic material or partial characterization by chemical or physical methods for identification. These studies are summarized in Table 1. SRS-A is thus a mixture of the cysteine-containing leukotrienes, i.e., the parent compound LTC_4 and metabolites LTD_4 and LTE_4. The relative proportion of these leukotrienes depends on the procedure used to prepare the SRS-A.

The metabolism of LTC_3 has been investigated using tritium-labeled material of high specific activity (27). The results showed that guinea pig lung homogenates

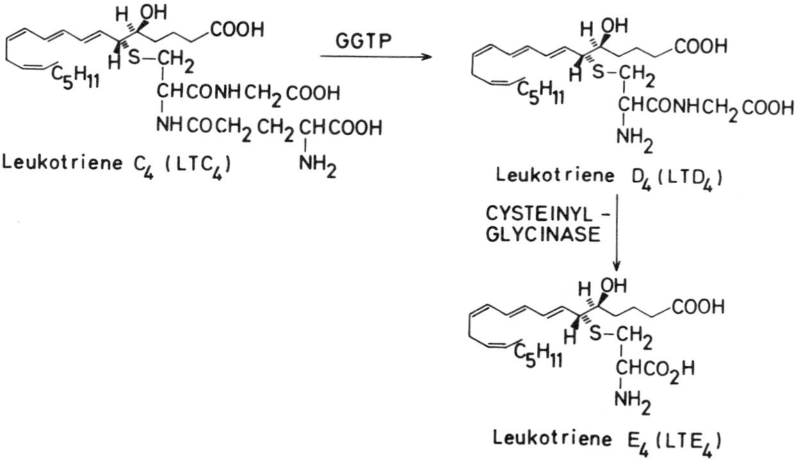

FIG. 9. Structures found in preparations of SRS-A.

TABLE 1. *Identification of leukotrienes from different sources*

Source	LTA_4	LTB_4	LTC_4	LTD_4	LTE_4	Refs.
Rabbit peritoneal leukocytes	+	+				7,9
Human peripheral leukocytes	+	+	+			10,30,55
Mouse mastocytoma cells	+		+			24,26,46
Rat basophilic leukemia cells				+	+	44,50,53
Rat peritoneal monocytes			+	+		3,4
Rat peritoneal cells (anaphylactic)			+	+	+	38
Rat peripheral leukocytes		+				21
Rat pleural neutrophils		+				63
Rat macrophages		+				17
Mouse macrophages			+			57
Human lung			+	+		39
Guinea pig lung				+		45
Cat paws				+	+	34

rapidly converted LTC_3 to LTD_3. Liver and kidney homogenates did not catabolize LTC_3 appreciably. This was apparently due to high tissue concentrations of glutathione which prevented LTD_3 formation because LTD_3 was rapidly metabolized by liver and kidney homogenates through hydrolysis of the peptide bond to give 5-hydroxy-6-S-cysteinyl-7,9,11-eicosatrienoic acid (LTE_3). In accordance with these findings, labeled LTC_3 administered into the right atrium of male monkey was rapidly transformed into LTD_3 and LTE_3 (29).

NOMENCLATURE

The biological significance of the biosynthetic pathways described and the cumbersome systematic names of the compounds involved suggested the introduction of a trivial name for these entities (59). The term "leukotriene" was chosen because the compounds were first detected in leukocytes and the common structural feature is a conjugated triene. Various members of the group have been designated alphabetically (Fig. 10): Leukotrienes A are 5,6-oxido-7,9-*trans*-11-*cis*; leukotrienes B are 5(S),12(R)-dihydroxy-6-*cis*-8,10-*trans*; leukotrienes C are 5(S)-hydroxy-6(R)-S-γ-glutamyl-cysteinylglycyl-7,9-*trans*-11-*cis*; leukotrienes D are 5(S)-hydroxy-6(R)-S-cysteinylglycyl-7,9-*trans*-11-*cis*; and leukotrienes E are 5(S)-hydroxy-6(R)-S-cysteinyl-7,9-*trans*-11-*cis*-eicosapolyenoic acids. As various precursor acids can be converted to leukotrienes containing three to five double bonds, a subscript denoting this number is used (62). Leukotriene A_4 is thus the epoxy derivative of arachidonic acid which can be further transformed to leukotrienes B_4, C_4, D_4, and E_4 as detailed in Figs. 8 and 9.

It was recently demonstrated that leukotrienes can also be formed after initial oxygenation at C-15 and C-8. Thus the unstable 14,15-dihydroxy-eicosatetraenoic acid (15-HPETE) is transformed into 14,15-dihydroxy derivatives (14,15-LTB_4)

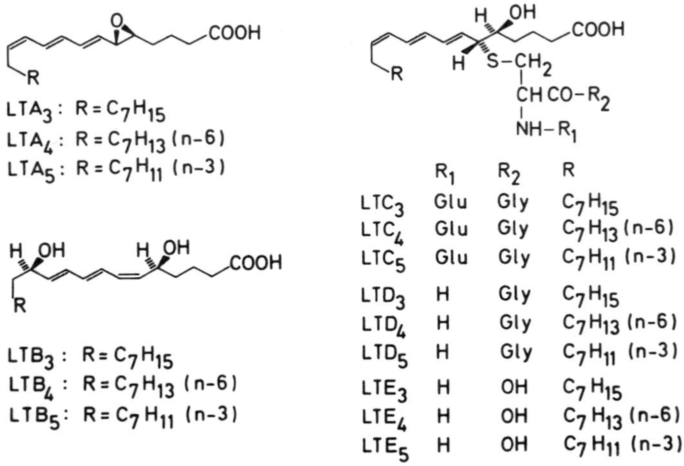

LTA3 : R = C7H15
LTA4 : R = C7H13 (n-6)
LTA5 : R = C7H11 (n-3)

LTB3 : R = C7H15
LTB4 : R = C7H13 (n-6)
LTB5 : R = C7H11 (n-3)

	R_1	R_2	R
LTC_3	Glu	Gly	C_7H_{15}
LTC_4	Glu	Gly	C_7H_{13} (n-6)
LTC_5	Glu	Gly	C_7H_{11} (n-3)
LTD_3	H	Gly	C_7H_{15}
LTD_4	H	Gly	C_7H_{13} (n-6)
LTD_5	H	Gly	C_7H_{11} (n-3)
LTE_3	H	OH	C_7H_{15}
LTE_4	H	OH	C_7H_{13} (n-6)
LTE_5	H	OH	C_7H_{11} (n-3)

FIG. 10. Nomenclature for leukotrienes.

and 8,15-dihydroxy derivatives (8,15-LTB$_4$) (36,41) (Fig. 11). Furthermore, bis-homo-γ-linolenic acid, which is oxygenated to give an 8-hydroperoxy derivative (6), can be converted to an 8,9-isomer of LTC$_3$ (28).

An isomer of LTB$_4$, 5S,12S-dihydroxy-6-*trans*-8-*cis*-10-*trans*-14-*cis*-eicosa-tetraenoic acid, was recently isolated as a metabolite of arachidonic acid in human leukocytes. This derivative is not formed via the leukotriene pathway and is therefore referred to as 5S,12S-DHETE (40) (Fig. 12).

BIOLOGICAL STUDIES

The suggested importance of SRS-A in asthma and anaphylactic reactions and the finding that SRS-A belonged to the leukotrienes stimulated the interest in studies of the biological effects of these substances.

LTC$_4$ and LTD$_4$ (0.1 to 1.0 nM) caused concentration-dependent contractions of guinea pig ileum (31). This is the classical preparation for determining biological activity of SRS-A in relation to histamine. It was found that on a molar basis histamine was 200 times less active than LTC$_4$, suggesting that 1 unit of SRS-A (i.e., 6 ng histamine hydrochloride) corresponds to approximately 0.2 pmole LTC$_4$. LTC$_4$ and LTD$_4$ also increased vascular permeability in guinea pig skin and had smooth-muscle-stimulating properties which were practically identical to those previously observed for SRS-A (18,31). Thus low concentrations of LTC$_4$ and LTD$_4$

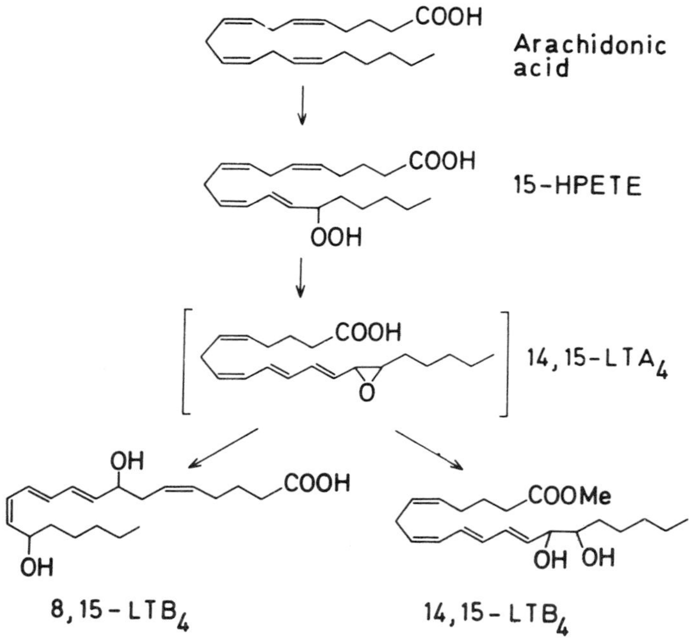

FIG. 11. Leukotrienes formed by initial oxygenation at C-15.

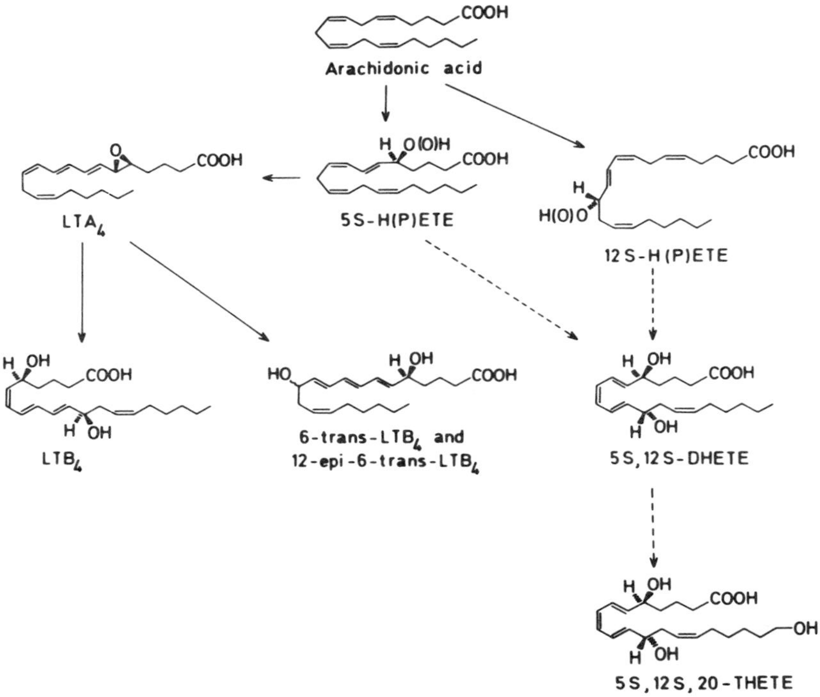

FIG. 12. Formation of dihydroxy acids.

caused contraction of human and guinea pig bronchi and virgin guinea pig uterus but not estrous rat uterus or smooth muscle preparations from the rabbit (15,31).

Additional studies with the pure leukotrienes, discussed in detail in this volume, have provided more detailed information about the effects on the pulmonary and cardiovascular systems (64). Furthermore, the cysteine-containing leukotrienes have specific effects on the microcirculation, and the dihydroxy derivative LTB_4 influences leukocyte migration by causing leukocyte adhesion to the endothelium in postcapillary venules and by potent chemotactic effects (16,21,22,42,51,52). These results (Fig. 13) indicate that the leukotrienes might be important mediators in such host defense mechanisms as immediate hypersensitivity reactions and acute inflammatory reactions.

The effects of some cyclo-oxygenase products and the leukotrienes are complementary (Fig. 14). Thus synergism between the leukotrienes causing plasma leakage and the vasodilators prostaglandins E_2 (PGE_2) and prostacyclin (PGI_2) might be of importance in the formation of edema. Furthermore, synergistic effects between the leukotrienes with bronchoconstrictor properties and thromboxane A_2 (TXA_2) are also conceivable. LTC_4 and LTD_4 cause release of TXA_2 in guinea pig lung (20). As TXA_2 is a potent constrictor of airways, its release might contribute to

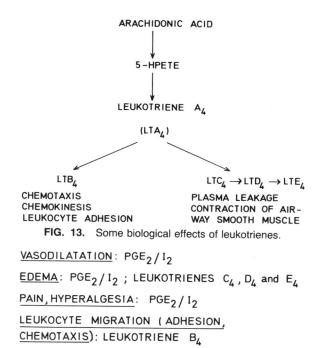

FIG. 13. Some biological effects of leukotrienes.

VASODILATATION: PGE_2/I_2

EDEMA: PGE_2/I_2 ; LEUKOTRIENES C_4, D_4 and E_4

PAIN, HYPERALGESIA: PGE_2/I_2

LEUKOCYTE MIGRATION (ADHESION, CHEMOTAXIS): LEUKOTRIENE B_4

FIG. 14. Role of arachidonic acid-derived products in inflammation.

the bronchospasm in allergic manifestations. The biochemical interrelationship between the cyclo-oxygenase pathway (yielding prostaglandins and thromboxanes) and the leukotriene pathway is illustrated in Fig. 15.

Anti-inflammatory steroids prevent the release of the precursor acid, arachidonic acid, whereas such cyclo-oxygenase inhibitors as aspirin block the transformation of this acid into prostaglandins and thromboxanes. It was recently proposed that anti-inflammatory steroids act by stimulating the synthesis of an inhibitor of phospholipase A_2 (5,32). By inhibiting the release of arachidonic acid, steroids prevent formation of not only prostaglandins and thromboxanes but also leukotrienes and other oxygenated derivatives. The inhibition of leukotriene formation might be responsible for some of the therapeutic effects of steroids which are not shared by aspirin-type drugs.

The increased knowledge about the leukotriene system seems to indicate new possibilities for the development of novel and more specific therapeutic agents, particularly in diseases related to immediate hypersensitivity reactions and inflammation. Such drugs might be based on antagonism of end products or inhibition of enzymes involved in the generation and further transformation of the key intermediate leukotriene A_4. A dual effect on the leukotriene pathway and the cyclo-oxygenase pathway might also be of value.

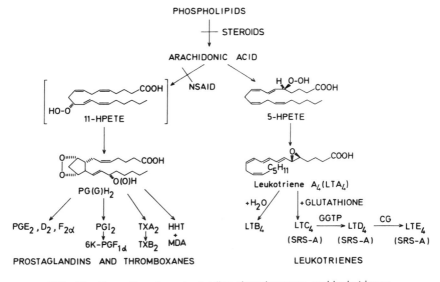

FIG. 15. Formation of prostaglandins, thromboxanes, and leukotrienes.

ACKNOWLEDGMENT

The work from the author's laboratory was supported by the Swedish Medical Research Council (project 03X-217).

REFERENCES

1. Austen, K. F. (1978): Hemostasis of effector systems which can be recruited for immunological reactions. *J. Immunol.*, 121:793–805.
2. Bach, M. K., Brashler, J. R., and Gorman, R. R. (1977): On the structure of SRS-A: evidence of biosynthesis from arachidonic acid. *Prostaglandins*, 14:21–38.
3. Bach, M. K., Brashler, J. R., Hammarström, S., and Samuelsson, B. (1980): Identification of leukotriene C-1 as a major component of slow reacting substance from rat mononuclear cells. *J. Immunol.*, 125:115–117.
4. Bach, M. K., Brashler, J. R., Hammarström, S., and Samuelsson, B. (1980): Identification of a component of rat mononuclear cell SRS as leukotriene D. *Biochem. Biophys. Res. Commun.*, 93:1121–1126.
5. Blackwell, G. J., Carnuccion, R., Di Rosa, M., Flower, R. J., Parente, L., and Persico, P. (1980): Macrocortin: a polypeptide causing the anti-phopholipase effect of glucocorticoids. *Nature*, 287:147–149.
6. Borgeat, P., Hamberg, M., and Samuelsson, B. (1976): Transformation of arachidonic acid and homo-γ-linolenic acid by rabbit polymorphonuclear leukocytes. *J. Biol. Chem.*, 251:7816–7820.
7. Borgeat, P., and Samuelsson, B. (1979): Transformation of arachidonic acid by rabbit polymorphonuclear leukocytes. *J. Biol. Chem.*, 254:2643–2646.
8. Borgeat, P., and Samuelsson, B. (1979): Metabolism of arachidonic acid in polymorphonuclear leukocytes: structural analysis of novel hydroxylated compounds. *J. Biol. Chem.*, 254:7865–7869.
9. Borgeat, P., and Samuelsson, B. (1979): Arachidonic acid metabolism in polymorphonuclear leukocytes: unstable intermediate in formation of dihydroxy acids. *Proc. Natl. Acad. Sci. USA*, 76:3213–3217.
10. Borgeat, P., and Samuelsson, B. (1979): Arachidonic acid metabolism in polymorphonuclear leukocytes: effects of ionophore A23187. *Proc. Natl. Acad. Sci. USA*, 76:2148–2152.

11. Brocklehurst, W. E. (1953): Occurrence of an unidentified substance during anaphylactic shock in cavy lung. *J. Physiol. (Lond.)*, 120:16P-17P.
12. Brocklehurst, W. E. (1962): Slow reacting substance and related compounds. *Prog. Allergy*, 6:539–558.
13. Conroy, M. C., Orange, R. P., and Lichtenstein, L. M. (1976): Release of slow-reacting substance of anaphylaxis (SRS-A) from human leukocytes by the calcium ionophore A23187. *J. Immunol.*, 116:1677–1681.
14. Corey, E. J., Marfat, A., Goto, G., and Brion, F. (1980): Leukotriene B: total synthesis and assignment of stereochemistry. *J. Am. Chem. Soc.*, 102:7984–7985.
15. Dahlén, S.-E., Hedqvist, P., Hammarström, S., and Samuelsson, B. (1980): Leukotrienes are potent constrictors of human bronchi. *Nature*, 288:484–486.
16. Dahlén,S.-E., Björk, J., Hedqvist, P., Arfors, K.-E., Hammarström, S., Lindgren, J.-Å., and Samuelsson, B. (1981): Leukotrienes promote plasma leakage and leukocyte adhesion in postcapillary venules: in vivo effects with relevance to the acute inflammatory response. *Proc. Natl. Acad. Sci. USA*, 78:3887–3891.
17. Doig, M. V., and Ford-Hutchinson, A. W. (1980): The production and characterization of products of the lipoxygenase enzyme system released by rat peritoneal macrophages. *Prostaglandins*, 20:1007–1019.
18. Drazen, J. M., Austen, F. K., Lewis, R. A., Clark, D. A., Goto, G., Marfat, A., and Corey, E. J. (1980): Comparative airway and vascular activities of leukotrienes C-1 and D in vivo and in vitro. *Proc. Natl. Acad. Sci. USA*, 77:4354–4358.
19. Feldberg, W., and Kellaway, C. H. (1938): Liberation of histamine and formation of lysolecithin-like substances by cobra venom. *J. Physiol. (Lond.)*, 94:187–226.
20. Folco, G., Hansson, G., and Granström, E. (1981): Leukotriene C_4 stimulates TXA_2 formation in isolated sensitized guinea pig lungs. *Biochem. Pharmacol.*, 30:2491–2493.
21. Ford-Hutchinson, A. W., Bray, M. A., Doig, M. V., Shipley, M. E., and Smith, M. J. H. (1980): Leukotriene B, a potent chemokinetic and aggregating substance released from polymorphonuclear leukocytes. *Nature*, 286:264–265.
22. Goetzl, E. J., and Pickett, W. C. (1980): The human PMN leukocyte chemotactic activity of complex hydroxy-eicosatetraenoic acids (HETEs). *J. Immunol.*, 125:1789–1791.
23. Gryglewski, R., Panczeko, B., Korbut, R., Grodzinska, L., and Ocetkiewica, A. (1975): Corticosteroids inhibit prostaglandin release from perfused mesenteric blood vessels of rabbit and from perfused lungs of sensitized guinea pig. *Prostaglandins*, 10:343–355.
24. Hammarström, S., Murphy, R. C., Samuelsson, B., Clark, D. A., Mioskowski, C., and Corey, E. J. (1979): Structure of leukotriene C: identification of the amino acid part. *Biochem. Biophys. Res. Commun.*, 91:1266–1272.
25. Hammarström, S., Samuelsson, B., Clark, D. A., Goto, G., Marfat, A., Mioskowski, C., and Corey, E. J. (1980): Stereochemistry of leukotriene C-1. *Biochem. Biophys. Res. Commun.*, 92:946–953.
26. Hammarström, S., and Samuelsson, B. (1980): Detection of leukotriene A_4 as an intermediate in the biosynthesis of leukotriene C_4 and D_4. *FEBS Lett.*, 122:83–86.
27. Hammarström, S. (1981): Metabolism of leukotriene C_3 in the guinea pig: identification of metabolites formed by lung, liver and kidney. *J. Biol. Chem.*, 256:9573–9578.
28. Hammarström, S. (1981): Conversion of dihomo-γ-linolenic acid to an isomer of leukotriene C_3, oxygenated at C-8. *J. Biol. Chem.*, 256:7712–7714.
29. Hammarström, S., Bernström, K., Örning, L., Dahlén, S.-E., Hedqvist, P., Smedegård, G., and Revenäs, B. (1981): Rapid in vivo metabolism of leukotriene C_3 in the monkey. *Biochem. Biophys. Res. Commun.*, 101:1109–1115.
30. Hansson, G., and Rådmark, O. (1980): Leukotriene C_4: isolation from human polymorphonuclear leukocytes. *FEBS Lett.*, 122:87–90.
31. Hedqvist, P., Dahlén, S.-E., Gustafsson, L., Hammarström, S., and Samuelsson, B. (1980): Biological profile of leukotrienes C_4 and D_4. *Acta Physiol. Scand.*, 110:331–333.
32. Hirata, F., Schiffmann, E., Venkatasubramanian, K., Salomon, D., and Axelrod, J. (1980): A phospholipase A_2 inhibitory protein in rabbit neutrophils induced by glucocorticoids. *Proc. Natl. Acad. Sci. USA*, 77:2533–2536.
33. Hong, S.-C. L., and Levine, L. (1976): Inhibition of arachidonic acid release from cells as the biochemical action of anti-inflammatory corticosteroids. *Proc. Natl. Acad. Sci. USA*, 73:1730–1734.

34. Houglum, J., Pai, J.-K., Atrache, V., Sok, D.-E., and Sih, C. J. (1980): Identification of the slow reacting substance from cat paws. *Proc. Natl. Acad. Sci. USA*, 77:5688–5692.
35. Jakschik, B. A., Falkenhein, S., and Parker, C. W. (1977): Precursor role of arachidonic acid in release of slow reacting substance from rat basophilic leukemia cells. *Proc. Natl. Acad. Sci. USA*, 74:4577–4581.
36. Jubiz, W., Rådmark, O., Lindgren J.-Å., Malmsten, C., and Samuelsson, B. (1981): Novel leukotrienes: products formed by initial oxygenation of arachidonic acid at C-15. *Biochem. Biophys. Res. Commun.*, 99:976–986.
37. Kellaway, C. H., and Trethewie, E. R. (1940): The liberation of a slow-reacting smooth muscle-stimulating substance in anaphylaxis. *Q. J. Exp. Physiol.*, 30:121–145.
38. Lewis, R. A., Drazen, J. M., Austen, K. F., Clark, D. A., and Corey, E. J. (1980): Identification of the C(6)-S-conjugate of leukotriene A with cysteine as a naturally occurring slow reaction substance of anaphylaxis (SRS-A): importance of the 11-cis geometry for biological activity. *Biochem. Biophys. Res. Commun.*, 96:271–277.
39. Lewis, R. Å., Austen, K. F., Drazen, J. M., Clark, D. A., Marfat, A., and Corey, E. J. (1980): Slow reacting substances of anaphylaxis: identification of leukotrienes C-1 and D from human and rat sources. *Proc. Natl. Acad. Sci. USA*, 77:3710–3714.
40. Lindgren, J.-A., Hansson, G., and Samuelsson, B. (1981): Formation of novel hydroxylated eicosatetraenoic acids in preparations of human polymorphonuclear leukocytes. *FEBS Lett.*, 128:329–335.
41. Lundberg, U., Rådmark, O., Malmsten, C., and Samuelsson, B. (1981): Transformation of 15-hydroperoxy-5,9,11,13-eicosatetraenoic acid into novel leukotrienes. *FEBS Lett.*, 126:127–132.
42. Malmsten, C., Palmblad, J., Udén, A.-M., Rådmark, O., Engstedt, L., and Samuelsson, B. (1980): Leukotriene B$_4$: a highly potent and stereospecific factor stimulating migration of polymorphonuclear leukocytes. *Acta Physiol. Scand.*, 110:449–451.
43. Morris, H. R., Taylor, G. W., Piper, P. J., Sirois, P., and Tippins, J. R. (1978): Slow-reacting substance of anaphylaxis: purification and characterization. *FEBS Lett.*, 87:203–206.
44. Morris, H. R., Taylor, G. W., Piper, P. J., Samhoun, M. N., and Tippins, J. R. (1980): Slow reacting substances (SRSs): the structure identification of SRSs from rat basophilic leukemia (RBL-1) cells. *Prostaglandins*, 19:185–201.
45. Morris, H. R., Taylor, G. W., Piper, P. J., and Tippins, J. R. (1980): Structure of slow reacting substance of anaphylaxis from guinea pig lung. *Nature*, 285:104–106.
46. Murphy, R., Hammarström, S., and Samuelsson, B. (1979): Leukotriene C: a slow-reacting substance from murine mastocytoma cells. *Proc. Natl. Acad. Sci. USA*, 76:4275–4279.
47. Orange, R. P., and Austen, K. F. (1969): Slow-reacting substance of anaphylaxis. *Adv. Immunol.*, 10:104–144.
48. Orange, R. P., Murphy, R. C., Karnovsky, M. L., and Austen, K. F. (1973): The physicochemical characteristics and purification of SRS-A. *J. Immunol.*, 110:760–770.
49. Orange, R. P., Murphy, R. C., and Austen, K. F. (1973): Inactivation of slow-reacting substance of anaphylaxis (SRS-A) by arylsulfatases. *J. Immunol.*, 113:316–322.
50. Örning, L., Hammarström, S., and Samuelsson, B. (1980): Leukotriene D: a slow reacting substance from rat basophilic leukemia cells. *Proc. Natl. Acad. Sci. USA*, 77:2014–2017.
51. Palmblad, B., Malmsten, C., Udén, A.-M., Rådmark, O., Engstedt, L., and Samuelsson, B. (1981): Leukotriene B$_4$ is a potent and stereospecific stimulator of neutrophil chemotaxis and adherence. *Blood (in press)*.
52. Palmer, R. M. J., Stepheny, R. J., Higgs, G. A., and Eakins, K.-E. (1980): Chemokinetic activity of arachidonic acid lipoxygenase products on leukocytes of different species. *Prostaglandins*, 20:411–418.
53. Parker, C. W., Falkenhein, S. F., and Huber, M. M. (1980): Sequential conversion of the glutathionyl side chain of slow reacting substance (SRS) to cysteinyl-glycine and cysteine in rat basophilic leukemia cells stimulated with A23187. *Prostaglandins*, 20:863–886.
54. Rådmark, O., Malmsten, C., Samuelsson, B., Clark, D. A., Giichi, G., Marfat, A., and Corey, E. J. (1980): Leukotriene A: stereochemistry and enzymatic conversion to leukotriene B. *Biochem. Biophys. Res. Commun.*, 92:954–961.
55. Rådmark, O., Malmsten, C., Samuelsson, B., Goto, G., Marfat, A., and Corey, E. J. (1980): Leukotriene A: isolation from human polymorphonuclear leukocytes. *J. Biol. Chem.*, 255:11828–11831.
56. Rådmark, O., Malmsten, C., and Saumelsson, B. (1980): Leukotriene A$_4$: enzymatic conversion to leukotriene C$_4$. *Biochem. Biophys. Res. Commun.*, 96:1679–1687.

57. Rouzer, C. A., Scott, W. H., Cohn, Z. A., Blackburn, P., and Manning, J. M. (1980): Mouse peritoneal macrophages release leukotriene C in response to a phagocytotic stimulus. *Proc. Natl. Acad. Sci. USA*, 77:4928–4932.
58. Samuelsson, B., Goldyne, M., Granström, E., Hamberg, M., Hammarström, S., and Malmsten, C. (1978): Prostaglandins and thromboxanes. *Annu. Rev. Biochem.*, 47:997–1029.
59. Samuelsson, B., Borgeat, P., Hammarström, S., and Murphy, R. C. (1979): Introduction of a nomenclature: leukotrienes. *Prostaglandins*, 17:785–787.
60. Samuelsson, B., Hammarström, S., Murphy, R. C., and Borgeat, P. (1980): Leukotrienes and slow reacting substance of anaphylaxis (SRS-A). *Allergy* 35:375–381.
61. Samuelsson, B., Borgeat, P., Hammarström, S., and Murphy, R. C. (1981): Leukotrienes: a new group of biologically active compounds. *Adv. Prostaglandin Thromboxane Res.*, 6:1–18.
62. Samuelsson, B., and Hammarström, S. (1980): Nomenclature for leukotrienes. *Prostaglandins*, 19:645–648.
63. Siegel, M. I., McConnell, R. T., Bonser, R. W., and Cautrecasas, P. (1981): The production of 5-HETE and leukotriene B in rat neutrophils from carrageenan pleural exudates. *Prostaglandins*, 21:123–132.
64. Smedegård, G., Revenäs, B., Hedqvist, P., Dahlén, S.-E., Hammarström, S., and Samuelsson, B. (1982): Cardiovascular and pulmonary effects of leukotriene C_4 in the monkey (Macaca iris). *Nature*, 295, January.
65. Strandberg, K., and Uvnäs, B. (1971): Purification and properties of the slow reacting substance formed in the cat paw perfused with compound 48/80. *Acta Physiol. Scand.*, 82:358–374.
66. Vane, J. R. (1971): Inhibition of prostaglandin synthesis as a mechanism of action for aspirin-like drugs. *Nature [New Biol.]*, 231:232–235.

Leukotrienes and Other Lipoxygenase Products,
edited by B. Samuelsson and R. Paoletti.
Raven Press, New York © 1982.

Selective Inhibitors of Platelet Arachidonic Acid Metabolism: Aggregation Independent of Lipoxygenase

Alvin R. Sams, *Howard Sprecher, *S. K. Sankarappa, and Philip Needleman

*Department of Pharmacology, Washington University Medical School, St. Louis, Missouri 63110; and *Department of Physiological Chemistry, Ohio State University, Columbus, Ohio 43210*

Platelets contain two arachidonic-acid-metabolizing enzymes. Cyclo-oxygenase converts arachidonate via the prostaglandin (PG) endoperoxide to thromboxane A_2 (TXA_2) and 12-hydroxy-5,8,10-heptadecatrienoic acid (HHT), whereas lipoxygenase metabolizes the fatty acid to 12-hydroperoxy-5,8,10,14-eicosatetraenoic acid (HPETE) (4). The aqueous decay products of these arachidonate metabolic pathways are thromboxane B_2 (TXB_2) and 12-hydroxy-5,8,10,14-eicosatetraenoic acid (HETE). The critical participation of TXA_2 and the PG endoperoxides has frequently been demonstrated. Indeed, exogenous addition of the endoperoxides or preformed TXA_2 will aggregate aspirin-treated platelets (5). What has remained elusive is the role of the lipoxygenase pathway in platelet function. Inhibitors of the lipoxygenase pathway [e.g., the acetylenic acid 5,8,11,14-eicosatetraynoic acid (ETYA) (6) or BW755C (8)] also inhibited cyclo-oxygenase.

Recently several compounds have been described which preferentially inhibit platelet lipoxygenase over the cyclo-oxygenase. Hammarström (7) found a 10-fold differential in favor of inhibition of platelet lipoxygenase over cyclo-oxygenase with 5,8,11-eicosatriyonic acid. Vanderhoek et al. (13) reported that 15-hydroxy-5,8,11,13-eicosatetraenoic acid selectively inhibited platelet lipoxygenase. Wilhelm et al. (14) and Sun et al. (12) had similar results. They also observed that the acetylenic analogs 4,7,10,13-ETYA and 5,8,11,14-heneicosatetraynoic acid inhibit platelet lipoxygenase at concentrations that did not block TXA_2 synthesis. However, the effects of these compounds on platelet aggregation were not extensively analyzed. Such experiments seem particularly pertinent since Dutilh et al. (1,2) reported that low concentrations of ETYA block platelet lipoxygenase and not cyclo-oxygenase, and simultaneously reverse arachidonate-induced aggregation. They concluded that the lipoxygenase is essential for blood platelet aggregation.

In the current investigation we performed a comparative study of the structural requirements of numerous acetylenic analogs to preferentially inhibit either platelet cyclo-oxygenase or lipoxygenase and to determine rank order effectiveness of these compounds to inhibit human platelet aggregation. From these studies emerged a series of selective enzyme inhibitors that permit study of the biochemical and pharmacological effects of the various arachidonic acid metabolites.

MATERIALS

The acetylenic fatty acids were prepared by total organic synthesis as previously described (10,11). Prostaglandin standards were gifts of The Upjohn Co., and 5,8,11,14-eicosatetraynoic acid ETYA was from Roche. [^{14}C]Arachidonic acid (55 Ci/mole) was purchased from New England Nuclear, and unlabeled arachidonic acid from Nu-Chek Prep. Thin layer chromatography (TLC) was performed on silicic acid plates from Brinkman (Westbury, N. Y.).

METHODS

Washed human platelet suspensions were prepared as previously described (9). Aggregation was monitored on a dual-channel Payton aggregometer. The acetylenic acid analogs were preincubated with 0.3 ml of the washed platelet suspension in the aggregometer cuvette at 37°C for 3 min prior to adding the aggregation agent (arachidonic acid, 1 to 3 μmole/l).

The radiochemical experiments were also performed with the washed human platelet suspensions in siliconized aggregometer cuvettes. The acetylenic acid analogs (0 to 100 μmole/l) were preincubated with 0.3 ml washed platelets for 3 min at 37°C prior to the incubation with [^{14}C]arachidonic acid (300,000 cpm, 10 μM) for 15 min. The samples were acidified with formic acid (pH 3.5), extracted with ethyl acetate, concentrated, and applied with unlabeled standard to silicic acid thin-layer plates. The arachidonate metabolites were separated with the system "C" solvent system (chloroform:methanol:acetic acid:water, 90:8:1:0.8). Autoradiograms of the thin layer plates were obtained in Kodak X-Omat X-ray film (48 hr at room temperature). The appropriate lanes were cut from the thin-layer chromatogram and counted in scintillation fluor (4a20, RPI Corp.) in a liquid scintillation counter.

Complete dose-response curves were constructed. IC$_{50}$ designates the concentration of acetylenic compound required to cause a 50% reduction in response compared to the control. The IC$_{50}$ for cyclo-oxygenase was calculated as the concentration of ETYA analog needed to cause a 50% reduction in the sum of the TXB$_2$ and HHT zones from the thin-layer plates. The IC$_{50}$ for lipoxygenase was calculated from the HETE peak. Each experiment was repeated three to four times, and mean values are presented. The standard errors were <10 to 15% of the mean. In each aggregation or radiochemistry experiment ETYA was run as a control antagonist.

RESULTS

Effect of Acetylenic Acids on Platelet Arachidonate Metabolism

5,8,11,14-Eicosatetraynoic acid (ETYA) served as the primary control in the metabolic experiments. This acetylenic fatty acid simultaneously inhibited platelet cyclo-oxygenase and lipoxygenase activity in a concentration-dependent fashion, with an IC_{50} of 4.2 μM (Fig. 1). Autoradiographic data from a radiochemical experiment is presented in Fig. 2. The control illustrates the separation of metabolites following the incubation of intact washed human platelets with [^{14}C]arachidonic acid. The primary cyclo-oxygenase products are TXB_2 and HHT; HETE is the lipoxygenase product; and trace amounts of other metabolites can be seen (Fig. 2). Preincubation of the platelets with acetylenic acid 22:3 (ω6) simultaneously inhibits both enzymes, with the resulting disappearance of the TXB_2, HHT, and HETE zones. On the other hand, 20:2 (ω7) (Fig. 1) proved to be a selective cyclo-oxygenase inhibitor without apparent effect on the lipoxygenase activity at acetylenic acid concentrations up to 30 μM. Quantitation of the individual zones by scraping from the thin-layer plate and counting demonstrate that as the cyclo-oxygenase is inhibited [e.g., by the acetylenic acid 18:3 (ω4)] more of the arachidonic acid is shuttled into the lipoxygenase product HETE (Fig. 3). Alternatively, a number of the fatty acid analogs prove to be selective lipoxygenase inhibitors. The autoradiogram in Fig. 4 demonstrates that 21:4 (ω7) abolishes the HETE band without altering the cyclo-oxygenase products. In fact, careful quantitation shows that more arachidonate is shuttled into the TXB_2 and HHT bands with enhanced lipoxygenase inhibition by 21:4 (ω7) (Fig. 5).

With the validation that the acetylenic analogs possess a high degree of selectivity for inhibition of the metabolic enzymes, we next evaluated the effects of these compounds on platelet aggregation. The data rather strikingly indicate a bimodal distribution of acetylenic acids based on their rank order potency to inhibit platelet

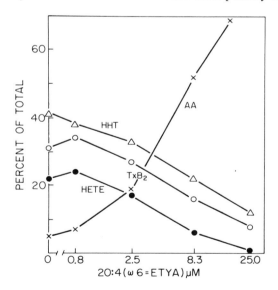

FIG. 1. Concentration-dependent inhibition of arachidonic acid metabolism in human washed platelet suspensions by ETYA. Authentic standards were applied on both sides of the thin-layer chromatogram and were visualized by iodine staining.

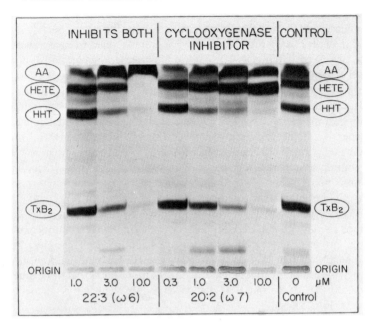

FIG. 2. Autoradiograph of the effects of the acetylenic fatty acids 22:3 (ω6) and 20:2 (ω7) on platelet arachidonate metabolism. The 22:3 pretreatment abolished the HETE, TXB$_2$, and HHT peaks, whereas the dynoic fatty acid 20:2 inhibited only the TXB$_2$ and HHT peaks and enhanced the HETE peak.

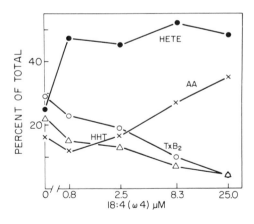

FIG. 3. Concentration-dependent inhibitors of platelet cyclo-oxygenase and simultaneous enhancement of the lipoxygenase by 18:4 (ω4).

aggregation (Table 1). The acetylenic acid analogs that selectively inhibited either cyclo-oxygenase alone or cyclo-oxygenase and lipoxygenase at concentrations of <10 μM were potent inhibitors (at concentrations of <2.5 μM) of arachidonate-induced aggregation of washed human platelet suspensions (Table 1). In contrast, the acetylenic acid analogs that were potent lipoxygenase inhibitors, or that inhibited

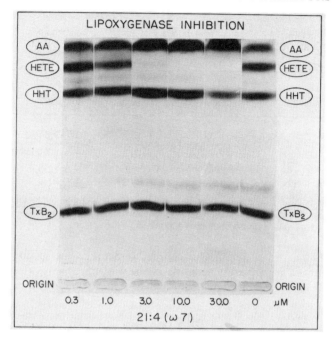

FIG. 4. Autoradiograph of the effectiveness of the 21:4 (ω7) acetylenic analog as a lipoxygenase inhibitor.

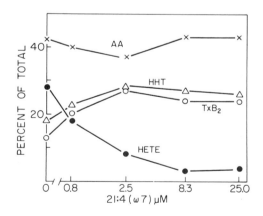

FIG. 5. Concentration-dependent inhibition of platelet lipoxygenases by the 21:4 (ω7) analog.

neither the cyclo-oxygenase nor the lipoxygenase, were much less potent in suppressing aggregation, requiring >5 μM to produce 50% inhibition. There was no detectable difference in potency in suppressing arachidonate-induced platelet aggregation between the inactive acetylenic acids (i.e., inhibited neither enzyme) and those that were potent inhibitors of lipoxygenase alone. The effectiveness as enzyme and aggregation inhibitors of all the acetylenic acids tested is outlined in Table 2.

TABLE 1.   *Comparative potency of acetylenic acid analogs as inhibitors of platelet arachidonate enzymatic metabolism and aggregation*

Enzyme inhibition	Enzyme IC_{50} (μmoles/l)	Aggregation IC_{50} (μmoles/l)
Cyclo-oxygenase only	2.5–8.3	0.6–2.3
Cyclo-oxygenase and lip-oxygenase	4.2–7.5	1.4–2.3
Lipoxygenase only	1.7–8.3	5.1–16.0
Neither cyclo-oxygenase nor lipoxygenase	> 83	4.8–21.0

See *Methods* for experimental details.

The fatty acid analogs are grouped according to their enzyme specificity. Clearly, there is a direct correlation between the potency of the analogs to inhibit cyclo-oxygenase and to suppress arachidonate-induced platelet aggregation.

DISCUSSION

Our data clearly differ from those of Dutilh et al. (1,2) who demonstrated that ETYA was a more potent inhibitor of lipoxygenase than cyclo-oxygenase and that the former enzyme influences platelet aggregation. We found that in our system ETYA was equipotent as a lipoxygenase and as a cyclo-oxygenase inhibitor. Similarly, Hammarström (7) and Sun et al. (12) found no apparent selectivity of ETYA in suppressing platelet arachidonate metabolism. Our data demonstrating the direct relationship between cyclo-oxygenase and aggregation seem to correlate for a large number of analogs.

The number of analogs studied permits a partial analysis of structure activity relationships. If acetylenic fatty acids that possess their initial triple bond at position 5 (i.e., $\Delta 5$) are compared, some conclusions can be drawn (Table 3). The triynoic analogs (18–20 carbons) that possess three triple bonds and lack $\Delta 14$ are ineffective as cyclo-oxygenase inhibitors and are impotent as inhibitors of platelet aggregation. Goetz et al. (3) previously studied some 20 carbon triynoic acid analogs and observed that the rank order of potency for the inhibition of bovine seminal vesicle cyclo-oxygenase and the inhibition of platelet aggregation was as follows: 20:3 ($\omega 6$) > 20:3 ($\omega 7$) > 20:3 ($\omega 9$). We found that 20:3 ($\omega 8$) and 20:3 ($\omega 6$) were potent cyclo-oxygenase inhibitors and potent inhibitors of platelet aggregation, whereas 20:3 ($\omega 9$) inhibited only platelet lipoxygenase and was a weak antithrombotic agent (Table 2). Comparison of the triynoic acetylenic derivatives with triple bonds at positions 8,11,14 (i.e., $\Delta 8$) clearly demonstrate that unsaturation at $\Delta 14$ is critical for inhibition of cyclo-oxygenase activity, whereas the lack of the triple bond at $\Delta 5$ is not essential to suppress TXA_2 biosynthesis (Table 3). This relationship seems

TABLE 2. *Rank order potency of acetylenic acid analogs as inhibitors of arachidonic acid metabolism and aggregation in washed human platelets*

Acetylenic acid	ω	Δ	Enzymatic IC_{50} (μmoles/l)	Aggregation IC_{50} (μmoles/l)
Inhibits cyclo-oxygenase only				
Triynoics				
19:3	5	8	2.5	0.6
18:3	6	6	3.3	1.3
20:3	8	6	3.3	1.4
20:3	6	8	4.2	1.4
18:3	4	8	6.7	2.3
17:3	3	8	15.0	4.2
Tetraynoics				
19:4	5	5	2.5	0.9
18:4	4	5	3.3	1.0
19:4	4	6	3.3	0.9
20:4	4	7	5.0	1.4
20:4	3	8	8.3	2.2
Diynoic				
20:2	7	10	6.7	1.8
Inhibits cyclo-oxygenase and lipoxygenase				
20:4	6	5	4.2	1.4
21:4	4	8	4.2	1.7
22:3	6	10	4.2	2.1
22:4	6	7	5.0	2.3
22:4	7	6	7.5	2.3
Inhibits lipoxygenase only				
21:4	7	5	1.7	5.1
21:3	7	8	4.2	5.4
20:3	9	5	8.3	6.3
20:4	7	4	2.5	16.0
Inhibits neither cyclo-oxygenase nor lipoxygenase				
20:2	9	8	0	6.3
21:3	8	7	0	4.8
19:3	8	5	0	5.4
18:3	7	5	0	4.8
16:3	6	4	0	21.0

ω designates the position of the last double bond counted from the terminal (ω) carbon on the fatty acid. Δ designates the position of the first double bond counted from the carboxyl end of the fatty acid. Thus 19:3 ($\omega5$, $\Delta8$) is 8,11,14-nonadecatrynoic acid. The enzymatic IC_{50} is the concentration of acetylenic acid needed to cause a 50% reduction in the sum of the TXB_2 and HHT for cyclo-oxygenase inhibition, or 50% fall in HETE levels of lipoxygenase activity.

fairly independent of chain length (i.e., from 17 to 20 carbon acetylenics); however when the chain length was extended to 20 carbons the analog no longer inhibited the cyclo-oxygenase but was preferentially effective against the 12-lipoxygenase and therefore a poor antithrombotic agent. The tetraynoic analogs that possess triple bonds at the 5,8,11,14 positions demonstrate the impact of the length of the ω-

TABLE 3. *Inhibitory effect of acetylenic acids with fixed positions of the triple bonds but varying chain length*

Compound	Enzyme inhibition (μmole/l)			Aggregation IC$_{50}$ (μmole/l)
	C.O.	L.O.	IC$_{50}$	
Δ_5 Acetylenics				
18:3 (ω7)	No	No	—	4.8
19:3 (ω8)	No	No	—	5.4
20:3 (ω9)	No	Yes	8.3	6.3
18:4 (ω4)	Yes	No	3.3	1.0
19:4 (ω5)	Yes	No	2.5	0.9
20:4 (ω6)	Yes	Yes	4.2	1.40
21:4 (ω7)	No	Yes	1.7	5.10
Δ_8 Acetylenics				
17:3 (ω3)	Yes	No	3.3	1.8
18:3 (ω4)	Yes	No	6.7	2.3
19:3 (ω5)	Yes	No	2.5	0.6
20:3 (ω6)	Yes	No	4.2	1.4
21:3 (ω7)	No	Yes	4.2	5.4
20:4 (ω3)	Yes	No	8.3	2.2
21:4 (ω4)	Yes	Yes	4.2	1.7

C.O. = cyclo-oxygenase, L.O. = lipoxygenase.

TABLE 4. *Effect of 20-carbon acetylenics on platelet arachidonate metabolizing enzymes and aggregation*

Eicosatetraynoics	Enzyme inhibition (μmole/l)			Aggregation IC$_{50}$ (μmole/l)
	C.O.	L.O.	IC$_{50}$	
20:4 (ω7)	No	Yes	2.5	16.0
20:4 (ω6)	Yes	Yes	1.7	1.4
20:4 (ω5)	Yes	No	5.0	1.3
20:4 (ω4)	Yes	No	5.0	1.7
20:4 (ω3)	Yes	No	8.3	2.2

carbon chain. Thus with 18 to 20 carbons they are potent inhibitors of cyclo-oxygenase and platelet aggregation (Table 3). At 20 carbons the Δ5 acetylenic fatty acid inhibits cyclo-oxygenase and lipoxygenase, and further extension of the ω-chain to 21 carbons results in a compound which inhibits only lipoxygenase. Variation of the position of the triple bonds also profoundly influences the inhibitory effects of these analogs. With the chain length fixed at 20 carbons and comparing only tetraynoic acid compounds, it appears that the ω3, ω4, and ω5 compounds inhibit cyclo-oxygenase only and are potent inhibitors of aggregation (Table 4). The ω6 compound (ETYA) inhibits both, but the ω7 inhibits lipoxygenase only.

The apparent lack of participation of the 12-lipoxygenase in platelet aggregation leaves the function of this enzyme unresolved in this tissue. An exciting possibility was recently demonstrated by Borgeat at the 1981 Winter Prostaglandin Meeting (April 1981). His work indicated that leukocytes metabolized arachidonic acid to 5-HETE. However, when he mixed leukocytes and platelets, 5,12-diHETE was formed, suggesting transfer of the 5-HETE to platelets, where the 12-lipoxygenase catalyzed the formation of 5,12-diHETE. Thus, two cell types were required to participate in the synthesis of a potent biological compound. Such cell-cell communication and interaction would have profound implications.

The availability of agents with such a wide spectrum of activity provides a family of tools that will permit discrimination of metabolic pathways. What remains to be determined is the utility of these analogs in other cell types. Elsewhere in this volume, the effectiveness of these analogs in inhibiting cyclo-oxygenase, 5-lipoxygenase, and leukotriene sythesis is evaluated by Jakschik et al. in rat basophilic leukemic cells.

SUMMARY

The use of acetylenic acid analogs differing in chain length or position of the triple bonds permitted the systematic study of structure activity relationships for the arachidonate metabolizing enzymes (i.e., cyclo-oxygenase and lipoxygenase) in platelets and the relationship of these enzymes to aggregation. We were able to demonstrate analogs that were differentially selective in altering platelet arachidonic acid metabolism. We found analogs that preferentially: (a) inhibited cyclo-oxygenase only, (b) inhibited the 12-lipoxygenase only, (c) inhibited the cyclo-oxygenase and lipoxygenase, and (d) inhibited neither enzyme in platelets. There was a direct correlation between the rank order of potency of the acetylenic analogs to inhibit platelet cyclo-oxygenase and to suppress aggregation. Certain structural features of the triynoic acetylenic analogs were critical in influencing platelet function; thus the presence of a triple bond at position 14 as well as the lack of the triple bond at position 5 resulted in analogs which inhibited both cyclo-oxygenase and platelet aggregation. On the other hand, analogs that inhibited only platelet 12-lipoxygenase were very weak inhibitors of platelet aggregation. These inhibitors provide potentially powerful tools for dissociating the two arachidonate metabolic pathways. If other tissues are as readily manipulated as platelets, the analogs may be especially useful for gaining insight into the contribution of lipoxygenase products to biological function.

ACKNOWLEDGMENTS

This work was supported by PN grants from NIH HL-14397 and HL-20787, and HS grants NO1HV82930, AM-20387, and AM-18844.

REFERENCES

1. Dutilh, C. E., Haddeman, E., Don, J. A., and Ten Hoor, F. (1981): The role of arachidonate lipoxygenase and fatty acid during irreversible blood platelet aggregation in vitro. *Prostaglandin Med.*, 6:111–126.
2. Dutilh, C. E., Haddeman, E., Jouvenaz, G. H., Ten Hoor, F., and Nugteren, D. H. (1978): Study of the two pathways for arachidonate oxygenation in blood platelets. *Lipids*, 14:241–246.
3. Goetz, J. M., Sprecher, H., Cornwell, D. G., and Panganamala, R. V. (1976): Inhibition of prostaglandin biosynthesis of triyonic acids. *Prostaglandins*, 12:187–192.
4. Hamberg, M., and Samuelsson, B. (1974): Prostaglandin endoperoxides: novel transformations of arachidonic acid in human platelets. *Proc. Natl. Acad. Sci. USA*, 71:3400–3404.
5. Hamberg, M., and Samuelsson, B. (1975): A new group of biologically active compounds derived from prostaglandin endoperoxides. *Proc. Natl. Acad. Sci. USA*, 72:2994–2998.
6. Hamberg, M., Svensson, J., and Samuelsson, B. (1974): Prostaglandin endoperoxides: a new concept in the mode of action and release of prostaglandins. *Proc. Natl. Acad. Sci. USA*, 71:3824–3828.
7. Hammarström, S. (1977): Selective inhibition of platelet n-8 lipoxygenase by 5,8,11-eicosatriyonic acid. *Biochim. Biophys. Acta*, 487:517–519.
8. Higgs, G. A., Flower, R. J., and Vane, J. R. (1979): A new approach to anti-inflammatory drugs. *Biochem. Pharmacol.*, 28:1959–1961.
9. Minkes, M., Stanford, N., Chi, M.-Y., Roth, G. J., Raz, A., Needleman, P., and Majerus, P. W. (1977): Cyclic adenosine-3',5'-monophosphate inhibits the availability of arachidonate to prostaglandin synthetase in human platelets. *J. Clin. Invest.*, 59:449–454.
10. Osbond, J. M., Philpott, P. G., and Wickens, J. C. (1961): Essential fatty acids. I. Synthesis of linolenic, α-linolenic, arachidonic and docosa-4,7,10,13,16-pentaenoic acid. *J. Chem. Soc.*, 2779.
11. Sprecher, H. (1978): In: *Prog. Chem. Fats Lipids*, edited by R. T. Holman, p. 219. Pergamon Press, London.
12. Sun, F. F., McGuire, J. C., Morton, D. R., Pike, J. E., Sprecher, H., and Kunau, W. H. (1981): Inhibition of platelet arachidonic acid 12-lipoxygenase by acetylenic acid compounds. *Prostaglandins*, 21:333–343.
13. Vanderhoek, J. Y., Bryant, R. W., and Bailey, J. M. (1980): 15-Hydroxy-5,8,11,13-eicosatetraenoic acid. *J. Biol. Chem.*, 255:5996–5998.
14. Wilhelm, T. E., Sankarappa, S. K., Van Rollins, M., and Sprecher, H. (1981): Selective inhibitors of platelet lipoxygenase: 4,7,10,13-icosatetraynoic acid and 5,8,11,14-henicosatetraynoic acid. *Prostaglandins*, 21:323–332.

Leukotrienes and Other Lipoxygenase Products,
edited by B. Samuelsson and R. Paoletti.
Raven Press, New York © 1982.

Novel Leukotrienes and Lipoxygenase Products from Arachidonic Acid

Richard L. Maas, Alan R. Brash, and John A. Oates

Departments of Pharmacology and Medicine, Vanderbilt University School of Medicine, Nashville, Tennessee 37232

Since their discovery and initial structural characterization, the leukotrienes have attracted considerable attention because of their powerful biological activities (18). As an initial step toward eventual clinical studies in the area, we studied the formation of leukotrienes and related lipoxygenase products in three representative systems: elicited rat peritoneal mononuclear cells, porcine leukocytes, and human leukocytes. This chapter describes the identification of, and some biosynthetic studies on, several new lipoxygenase and leukotriene products, including ω-hydroxylated derivatives of leukotriene (LT) B_4, a novel dihydroxy acid, 5,15-DHETE, formed by double lipoxygenation of arachidonic acid, and several 8,15- and 14,15-dihydroxy leukotrienes which, of note, originate from 15(S)-lipoxygenation via a new epoxide intermediate, 14,15(S)-oxido-5,8,10,12-eicosatetraenoic acid.

ω-HYDROXYLATION OF LTB_4 BY RAT MONONUCLEAR CELLS

Thioglycolate-induced rat peritoneal mononuclear cells were prepared and harvested as described (16). The cells were routinely stimulated with the divalent cation ionophore A23187 (38 μM) and in some experiments with arachidonic acid (100 μM), conditions known to markedly stimulate leukotriene biosynthesis in these cells. Incubations were typically carried out for 25 min with gentle shaking under a normal atmosphere. The incubation was processed as described, by either ethyl acetate extraction or initial extraction on C_{18} packing (16). Following additional purification by DEAE-LH20 and LH20 chromatography, the monocarboxylic acid fraction was analyzed by reverse-phase high-pressure liquid chromatography (RP-HPLC) on a Waters analytical C_{18} column with ultraviolet (UV) detection at 280 nm (Fig. 1). Four main UV-absorbing components were observed. Peak III was identified on the basis of UV spectroscopy and gas chromatography-mass spectrometry (GC-MS) as LTB_4, and peaks IIa and IIb were similarly identified as 5(S),12(R,S)-dihydroxy-6,8,10-*trans*-14-*cis*-eicosatetraenoic acids, formed by nonenzymatic breakdown of the epoxide intermediate LTA_4 (3). The polar compound, peak I, was further purified by reverse- or straight-phase (SP) HPLC and resolved into two separate UV-absorbing components, Ia and Ib. The UV spectra of components Ia and Ib were identical to each other and to a reference spectrum recorded

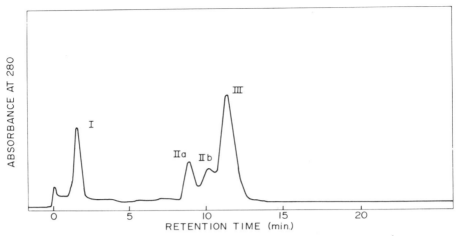

FIG. 1. RP-HPLC (Waters analytical C_{18}, 65% methanol, 35% water, 0.01% acetic acid, 2 ml/min) of the DEAE-LH20- and LH20-purified material obtained after stimulation of rat mononuclear cells with ionophore A23187. Peak I was further resolved into two separate conjugated triene compounds, Ia and Ib, by additional RP-HPLC.

on LTB_4, with $\lambda^{MeOH}_{max} = 270$ nm and shoulders at 261 and 281 nm. Each component was converted to the Me-Me$_3$Si derivative and analyzed by GC-MS. The mass spectra of components Ia and Ib were very similar and showed many of the same fragmentations observed in the spectrum of LTB_4 (Fig. 2). The molecular ion at m/z 582 established the parent compound as a trihydroxy acid with four double bonds, whereas the fragmentations at m/z 203 and 383, in conjunction with the UV data, established the positions of two of the hydroxy groups at C-5 and C-12, respectively; this was confirmed by the mass spectra of the hydrogenated compounds. The structures were provisionally assigned on the basis of GC retention time (12). Thus, the trihydroxy leukotriene of longer GC retention time, C-value 26.1, also more polar on RP-HPLC, Ia, is assigned as 5,12,20-trihydroxy-6,8,10,14-eicosatetraenoic acid. Peak Ib, giving a C-value of 25.8, was assigned as the analogous 5,12,19-trihydroxy compound.

Evidence for the origin of these metabolites from LTB_4, as opposed to the nonenzymatically formed 5,12-dihydroxy compounds, IIa and IIb, was provided by an experiment in which [^{3}H]LTB$_4$, prepared by incubation of cells with [^{3}H$_8$]arachidonic acid, was incubated with the rat mononuclear cells under conditions identical to those described above. Conversion of [^{3}H]radiolabel to the trihydroxy metabolites under these conditions was in excess of 50%.

Peaks Ia and Ib were tested for chemokinetic activity in a leukocyte migration assay from agar droplets using mouse peritoneal cells. Preliminary results suggest that 19- and 20-hydroxy LTB_4 are considerably less active than LTB_4 in inducing leukocyte migration. Additional experiments in our laboratory with human leukocytes and polymorphs have revealed that very similar, probably isomeric, ω-hy-

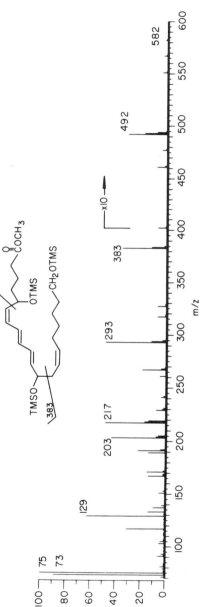

FIG. 2. Mass spectrum (70 eV) of peak la Me-Me₃Si showing fragmentation very similar to the Me-Me₃Si derivative of LTB₄ (m/e 203,383,293) (ref. 4) and indicating the presence of a third hydroxyl group.

droxylated 5,12-dihydroxy leukotrienes are also formed in these cells. The function of these interesting compounds remains open for further investigation.

FORMATION OF A NOVEL DIHYDROXY ACID (5,15-DHETE) BY DOUBLE LIPOXYGENATION OF ARACHIDONIC ACID

Identification of 5,15-DHETE in Preparations of Rat Mononuclear Cells

Further analysis of the dihydroxy acids obtained following incubation of rat mononuclear cells with ionophore A23187 and [^{3}H$_8$]arachidonic acid by RP-HPLC showed a major radioactive peak eluting about 4 ml prior to LTB$_4$, a separation too large to be accounted for by the normal separation of ^{3}H- and ^{1}H-labeled species under these conditions. The major radiolabeled peak was collected and further purified as the methyl ester by SP-HPLC, and the UV spectrum of the purified product recorded. The UV spectrum (Fig. 3) showed $\lambda^{MeOH}_{max} = 243$ nm, with a hypsochromic shoulder at 226 nm, compatible with the presence of a conjugated diene or other chromophore in the compound. The compound was converted to the

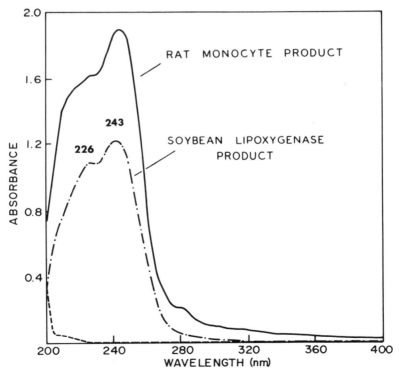

FIG. 3. UV spectrum recorded in methanol showing the double diene chromophore of the product isolated from rat mononuclear cells, and of that obtained from reaction of synthetic 5(R,S)-HETE with soybean lipoxygenase.

Me-Me$_3$Si derivative and analyzed by GC-MS. Substantial problems with thermal degradation on gas chromatographic analysis were encountered. The main peak chromatographed with a C-value of 23.6 (1% OV-1). The mass spectrum (Fig. 4) showed a molecular ion at m/z 494, establishing the parent compound as a dihydroxy acid, and gave intense ions at m/z 173 (Me$_3$SiO$^+$ = CH − (CH$_2$)$_4$CH$_3$) and m/z 203 (Me$_3$SiO$^+$ = CH − (CH$_2$)$_3$COOCH$_3$), indicating that the hydroxyl groups were at carbons 5 and 15. Although this structure was supported by mass spectra of several other derivatives, including those of the hydrogenated compound, because of the problems encou tered with thermal degradation and the rather unique UV spectrum further attempts were made to establish the structure of the cell product by preparing a reference compound with the proposed structure.

Reaction of Soybean Lipoxygenase with 5-HETE and 5-HPETE

The substrate specificity requirements of the soybean lipoxygenase had been previously shown to include a *cis,cis*-1,4-pentadiene unit (10). Thus such diverse structures as arachidonic acid and LTC$_4$ react readily with the soybean lipoxygenase at ω6 to form 15(S)-hydroperoxy derivatives. Accordingly, we found that synthetically prepared 5(R,S)-HETE also reacted with the soybean lipoxygenase as expected to give 5(R,S)-hydroxy-15(S)-hydroperoxy-6,13-*trans*-8,11-*cis*-eicosatetraenoic acid, which was fully characterized by ^{1}H- and ^{13}C-NMR and mass spectrometry. This product showed a UV spectrum identical to that recorded for the cell product and, after reduction of the C-15 hydroperoxy group, was found to be identical. We also found that synthetic 5(R,S)-HPETE was an efficient substrate for this enzyme and led to the formation of the analogous dihydroperoxyeicosatetraenoate. The direct conversion of arachidonic acid to 5(S),15(S)-dihydroperoxy-6,13-*trans*-8,11-*cis*-eicosatetraenoic acid by the soybean lipoxygenase was subsequently demonstrated (19).

Analysis of 5,15-DHETE After Incubation Under an ^{18}O$_2$ Atmosphere

It seemed reasonable to suppose that 5,15-DHETE isolated from the mononuclear cells originated from arachidonic acid by the successive action of C-5 and C-15 lipoxygenases. We investigated the biosynthesis of 5,15-DHETE by performing an incubation under an atmosphere of ^{18}O$_2$. An atom of ^{18}O was incorporated into the hydroxyl groups at C-5 and C-15. Moreover, it could be concluded that the oxygen atoms originated from different molecules of molecular oxygen because of the presence of high mass ions in the spectrum corresponding to mono- and bis- ^{18}O-labeled species. Although the stereochemistry of the hydroxyl groups was not determined, the double *cis-trans* diene geometry and the ^{18}O labeling data make it clear that 5,15-DHETE does originate from arachidonic acid by the action of C-5 and C-15 lipoxygenases.

The synthetic 5(R,S)-hydroxy and hydroperoxy-15(S)-hydroperoxy compounds were tested in preliminary fashion for bioactivity in leukocyte migration (mouse peritoneal cells in agar) and leukocyte enzyme release (β-glucuronidase and lyso-

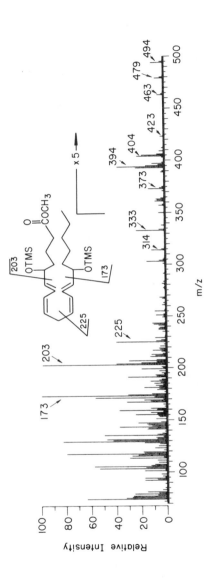

FIG. 4. Mass spectrum (70 eV) of 5,15-DHETE Me-Me₃Si isolated from rat monocytes.

zyme from guinea pig peritoneal polymorphs) assays. No biological activity could be observed. In view of the recent demonstration that 15(S)-HPETE is a potent inhibitor *in vitro* of the 12-lipoxygenase (20), it was of interest to test the 5,15-dihydroxy, hydroxy-hydroperoxy, and dihydroperoxy compounds for their ability to inhibit leukocyte HETE production. The hydroperoxy compounds had weak activity in selectively inhibiting human leukocyte 12-HETE formation, with an $IC_{50} \simeq 30$ μM. Additional studies are in progress to determine if the 5,15-mono or dihydroperoxy compounds might be substrates for other, as yet unidentified biotransformations.

FORMATION OF NOVEL LEUKOTRIENES FROM C-15 LIPOXYGENATION OF ARACHIDONIC ACID IN PORCINE LEUKOCYTES

When the dihydroxy acids obtained following incubation of porcine leukocytes with arachidonic acid were analyzed by RP-HPLC with UV detection at 280 nm, a chromatographic profile markedly different from that obtained after incubations with rat monocytes and human leukocytes was observed (Fig. 5) (17). Six prominent UV-absorbing peaks were observed, and each showed a conjugated triene chromophore. Thus it was clear that there were several additional leukotriene products present, in addition to the 5,12-dihydroxy compounds derived from LTA_4.

Each of the six peaks was converted to the methyl ester and further analyzed by SP-HPLC. The major RP-HPLC peak, peak 5, was found on the basis of UV and mass spectrometry and steric analysis to consist of a mixture of LTB_4 and larger amounts of 5(S),12(S)-dihydroxy-6,8,10,14-eicosatetraenoic acid, formed like 5,15-

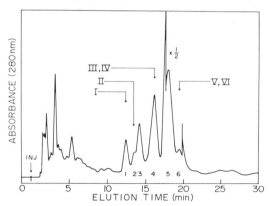

FIG. 5. RP-HPLC (Waters analytical C_{18}, 65% methanol, 35% water, 0.01% acetic acid, 2 ml/min) of the products obtained by ether extraction after incubation of porcine leukoctyes with arachidonic acid (100 μM) for 5 min. Six main UV-absorbing RP-HPLC peaks were observed (1 through 6). Compounds I through IV were 8,15-dihydroxy acids, and compounds V and VI were 14,15-dihydroxy acids.

DHETE by double lipoxygenation of arachidonic acid (5). Other 5,12-dihydroxy acids, in RP-HPLC peaks 3 and 4, appeared to correspond to the nonenzymatic hydrolysis products of LTA$_4$. Several other 5,12-dihydroxy compounds were also present.

Most noteworthy was the fact that RP-HPLC peaks 1 and 2 each consisted mainly of an 8,15-dihydroxy leukotriene, denoted as compounds I and II, whereas RP-HPLC peak 4 contained two such 8,15-dihydroxy leukotrienes, compounds III and IV. RP-HPLC peak 6 contained two 14,15-dihydroxy leukotrienes, compounds V and VI. Following incubation with [^{14}C]arachidonic acid, compounds I through VI were each found to be radiolabeled. UV, GC, and HPLC data for compounds I through VI are given in Fig. 6 and Table 1.

Compounds I and IV

The mass spectrum of compound I Me-Me$_3$Si showed a molecular ion at m/z 494 and prominent ions at m/z 353 [M-141; loss of ·CH$_2$ − CH = CH − (CH$_2$)$_3$COOCH$_3$] and m/z 173 (Me$_3$SiO$^+$ = CH − (CH$_2$)$_4$CH$_3$), establishing the structure of the parent compound as an 8,15-dihydroxy-5,9,11,13-eicosatetraenoic acid.

Following incubation of cells under an atmosphere of ^{18}O$_2$, mass-spectrometric analysis of compound I Me-Me$_3$Si revealed about 70% incorporation of ^{18}O into the hydroxyl group at C-15 but none into the hydroxyl group at C-8 (Fig. 7). A similar experiment, conducted in H$_2$^{18}O, confirmed the incorporation of ^{18}O from water at C-8. These findings suggested that compound I originated from 15(S)-HPETE by hydrolysis of an unstable epoxide intermediate in a fashion analogous to the formation of 5,12-dihydroxy acids from LTA$_4$.

Analysis of the stereochemistry of compound I by oxidative ozonolysis of the bis-menthoxycarbonyl derivative (Fig. 8) showed that it had the R configuration at

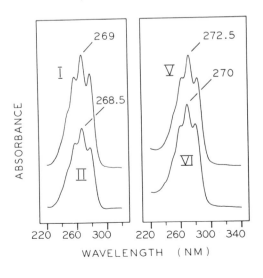

FIG. 6. UV spectra recorded in methanol of compounds I (identical to IV), II (identical to III), V, and VI.

TABLE 1. *Analytical data for compounds I–VI*

Compound	UV Spectrometric Data[a] Maximum; hypso-, bathochromic shoulder (nm)	Gas chromatographic data C-value (3% SP2100 or OV-1) Me-Me₃Si	Me-Me₃Si/cat.-H₂	SP-HPLC retention times[b] Me esters (min)
I, IV	269; 260, 280[c]	24.9	24.1(I-IV)	11.5, 15.0
II, III	268.5; 259, 279[c]	23.6		9.5, 12.3
V	272.5; 263, 283	23.9	23.8(V,VI)[d]	10.8
VI	270; 261, 281	24.9		11.4

[a]UV spectra recorded in methanol on a Beckman 25 spectrophotometer, calibrated with LTB₄.
[b]Waters μporasil, silicic acid, 10 μm particles, hexane:isopropanol, 100:3 (v/v), 2 ml/min, 100 psi. V-Me and VI-Me could be better resolved using hexane:isopropanol, 100:2 (v/v), 0.5 ml/min (separation factor, VI/V = 1.10).
[c]UV spectra for I and IV were identical; similarly for II and III. I and IV characteristically showed a more deeply defined bathochromic shoulder than did II and III.
[d]Synthetic *threo*-14,15-dihydroxyeicosanoate Me-Me₃ Si had a C-value of 23.6 under these conditions (3% OV-1, 215°C); *erythro* gave a C-value of 23.8.

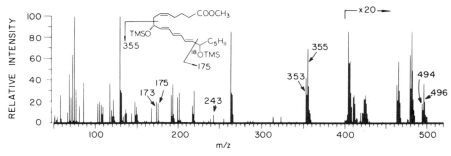

FIG. 7.   Mass spectrum of compound I Me-Me₃Si obtained following incubation of porcine leukocytes with ¹⁸O₂.

C-8 and the S configuration at C-15, and moreover that the $\Delta^5 cis$ double bond of arachidonic acid was retained in its original position.

UV, GC-MS, and ¹⁸O-labeling data for compound IV were identical to those for compound I. This fact, taken with the ready separation of these compounds by HPLC and their presence in the incubations in approximately equal amounts suggested that compounds I and IV might be diastereomers. This was confirmed by the steric analysis of compound IV, which showed that, like compound I, it had the S configuration at C-15, but that it was the epimer at C-8 (Fig. 8).

It is well known that acid-catalyzed opening of the epoxide ring of LTA₄ and subsequent reaction with water leads via a carbonium ion intermediate to the non-enzymatic formation of two 5,12-dihydroxy acids which are epimers at C-12 and which contain an all-*trans* conjugated triene (4). To test the surmise that compounds I and IV arose from the 14,15(S)-oxido analog of LTA₄ in similar fashion, we synthesized the presumptive epoxide intermediate, 14,15(S)-*trans*-oxido-5,8-*cis*-

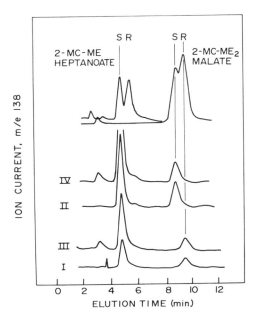

FIG. 8. Steric analysis of the four 8,15-dihydroxy acids isolated from porcine leukocytes, compounds I through IV, by oxidative ozonolysis of the bis-menthoxycarbonyl derivatives. The early eluting fragment indicates the configuration at C-15, and the later fragment at C-8.

10,12-*trans*-eicosatetraenoate using the biomimetic synthesis devised by Corey et al. with 15(S)-HPETE as a starting material (7,8). As expected, acid-water treatment of the epoxide intermediate led to the formation of four triene-containing compounds, 8(R),- and 8(S),15(S)-dihydroxy-5-*cis*-9,11,13-*trans*-eicosatetraenoates and *erythro*- and *threo*-14,15(S)-dihydroxy-5-*cis*-8,10,12-eicosatetraenoates. Compounds I and IV from porcine leukocytes were chromatographically and spectroscopically identical to the two synthetic 8,15-dihydroxy compounds obtained from the epoxide. These data indicated the formation of a 14,15(S)-oxido analog of LTA₄ in these cell preparations, arising from arachidonic acid via initial oxygenation by a C-15 lipoxygenase.

Compounds II and III

Compounds II and III were also 8,15-dihydroxy-5,9,11,13-eicosatetraenoates, but surprisingly each showed incorporation of ^{18}O from molecular oxygen into both hydroxyl groups (Fig. 9). Compounds II and III were also epimeric at C-8 (Fig. 8), and both had the S configuration at C-15.

It is known that soybean lipoxygenase can react with arachidonic acid to give, after reduction, an 8(S),15(S)-dihydroxy-5,11-*cis*-9,13-*trans*-eicosatetraenoate by sequential 15(S)- and 8(S)-lipoxygenation (1,19). We prepared this reference compound and found it to be chromatographically and spectroscopically indistinguishable from compound II, which was also 8(S),15(S). This suggests but does not prove that the geometry of the conjugated triene in compounds II and III may be *trans-cis-trans*, which would be the geometry expected in the case of formation of

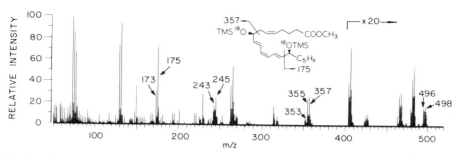

FIG. 9.   Mass spectrum of compound II Me-Me₃Si after incubation of porcine leukoctyes with ¹⁸O₂.

these compounds by either double lipoxygenation or by auto-oxidation of 15(S)-HPETE or HETE. In view of the lack of stereochemical purity in compounds II and III, and the finding that a high concentration of ETYA (100 μM) was required to inhibit formation of compounds II and III when 15(S)-HPETE was added to the cells, we consider that enzymatic double lipoxygenation is an unlikely mechanism for their formation. It is more probable that compounds II and III originate from 15(S)-HPETE by a previously undescribed mechanism of oxygenation that is not stereoselective. The possibilities include formation of compounds II and III via co-oxygenation or from the 14,15-epoxide intermediate by reaction with an oxygen metabolite. This possibility is further discussed in connection with compounds V and VI.

Compounds V and VI

The mass spectra of compounds V and VI were very similar, and indicated that they were 14,15-dihydroxy-5,8,10,12-eicosatetraenoic acids, assuming the Δ^5 *cis* double bond in arachidonic acid is retained unchanged, which seems likely. The clear differences in UV and GC retention time indicated that compounds V and VI were probably geometric isomers within the conjugated triene.

¹⁸O-labeling experiments revealed the very surprising finding that compounds V and VI incorporated an atom of ¹⁸O from molecular oxygen into the hydroxyl groups at C-14 and C-15. A partial mass spectrum (obtained by rapid repetitive scanning of selected fragments) of compound V following incubation of porcine leukocytes with ¹⁶O₂-labeled 15(S)-HPETE under an atmosphere of ¹⁸O₂ is shown in Fig. 10 and demonstrates the incorporation of ¹⁸O into the C-14 hydroxyl group. This finding made it clear that compounds V and VI do not arise from the 14,15-epoxide intermediate by reaction with water, as do, for example, the 5,6-dihydroxy acids formed from LTA₄ (3). Additionally, compounds V and VI differed from the synthetic 14,15(S)-dihydroxy-5-*cis*-8,10,12 eicosatetraenoates prepared from acid-water treatment of the postulated epoxide intermediate by UV, HPLC, and GC comparisons. Nonetheless, in view of the structural similarity of compounds V and VI to the synthetic 14,15-dihydroxy acids and to the 14,15-epoxide intermediate,

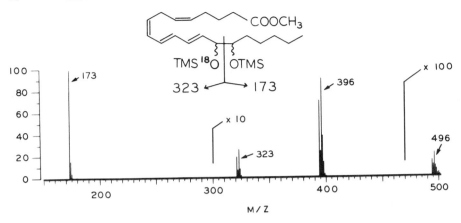

FIG. 10. Partial mass spectrum of compound V Me-Me$_3$Si obtained after incubation of porcine leukocytes with ^{16}O$_2$-labeled 15(S)-HPETE under an atmosphere of ^{18}O$_2$. The spectrum thus shows incorporation of ^{18}O from molecular oxygen at C-14. The spectrum was obtained by scanning over selected ion ranges.

it seemed very possible that compounds V and VI might still in fact originate from the latter compound.

Mechanistic Considerations Concerning Compounds V and VI

The biosynthesis of compounds V and VI in porcine leukocytes was further investigated by analyzing the inhibition of compounds I through VI by different concentrations of ETYA using RP-HPLC with UV detection, following addition of either arachidonic acid or 15(S)-HPETE to the cells. In the first instance, following incubation with arachidonic acid the formation of compounds I through VI was virtually eliminated by 1 μM ETYA, consistent with their formation via a 15(S)-lipoxygenase. In the latter case, following incubation with 15(S)-HPETE, compounds I, IV, V, and VI (measured as V + VI) were each inhibited virtually in concert over a dose range of 1 to 100 μM ETYA, with an IC$_{50}$ = 5 μM, consistent with their formation by the same biosynthetic process and with inhibition of the 14,15-epoxide synthetase enzyme by ETYA. The latter is not unreasonable because ETYA has been previously shown to inhibit LTA$_4$ synthetase in another species (2).

A final critical point in the understanding of the formation of compounds V and VI was obtained by determining the relative configuration of these two compounds. *Threo* and *erythro* 14,15-dihydroxy eicosanoates were synthesized in a stereochemically unambiguous manner via 14,15-*cis*-oxido-5,8,11-eicosatrienoic acid using the selective internal epoxidation reaction of arachidonic acid described by Corey et al. (6). Following catalytic hydrogenation, compounds V and VI were found to have the *erythro* configuration, in distinct contrast to the synthetic 14,15-dihydroxy acids obtained from acid-water treatment of the epoxide which, as expected, were *threo* and *erythro* (Fig. 11).

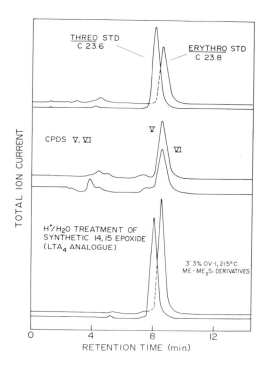

FIG. 11. Determination of the relative configuration in the two 14,15-dihydroxy leukotrienes, compounds V and VI. Both compounds have the *erythro* configuration.

This important finding would be compatible with the formation of compounds V and VI from the proposed 14,15-*trans* epoxide intermediate, by nucleophilic attack at the allylic C-14 position with inversion of configuration. The *erythro* configuration in compounds V and VI and the apparent absence of any diastereomeric *threo* compounds in these incubations precludes their formation by either radical or carbonium ion mechanisms, as either would result in racemization at C-14 and lead to the formation of diastereomers. Taken with the $^{18}O_2$-labeling and ETYA inhibition experiments, the stereochemical data thus lead us to propose that compounds V and VI originate from the epoxide intermediate 14,15(S)-*trans*-oxido-5,8,10,12-eicosatetraenoic acid by reaction with a form of activated molecular oxygen acting as a nucleophile (Fig. 12). Superoxide anion, which is a good nucleophile (11), and which is formed in appreciable quantities in leukocytes following stimulation by arachidonic and other fatty acids (14), therefore appears to be a potential candidate to react with the epoxide to give compounds V and VI. The reason for and nature of the geometric isomerism of compounds V and VI is not known. It is possible that compounds II and III also arise by this mechanism, although the evidence for this is less compelling.

It should be pointed out that the stereochemical data for compounds I, II, III, and IV argues strongly for their nonenzymatic formation—from the epoxide in the case of compounds I and IV. In the case of compounds V and VI, the stereochemical

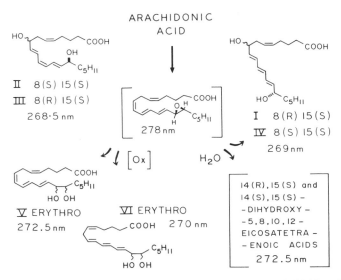

FIG. 12. Hypothetical scheme showing the formation of compounds I, IV, V, and VI from an epoxide intermediate, the 14,15(S)-*trans*-oxido analog of LTA$_4$. The exact manner of formation of compounds II and III is not known. The geometry of the conjugated triene is not assigned in II, III, V, and VI; the absolute configuration of the two *erythro* diols, compounds V and VI, was not assigned.

data do not preclude the possibility that their formation could be enzymatic; however, this possibility will require additional investigation.

CONCLUSIONS

This report describes several new lipoxygenase products of arachidonic acid metabolism. Of considerable potential import is the finding of a new pathway of leukotriene biosynthesis, proceeding via 15(S)-oxygenation of arachidonic acid. It should be pointed out that in addition to the leukotrienes originating from 5(S)-HPETE, other families of leukotrienes arising from 12(S)- and 15(S)-lipoxygenation have always been mechanistically possible. Previously, for example, compounds tentatively identified as 14,15- and 8,15-dihydroxy trienes were isolated following incubation of dihomo-γ-linolenic acid with human platelets (9), and more recently the 12- and 15-lipoxygenase derived analogs of LTA$_4$ and LTC$_4$ have been chemically synthesized (8). In addition, some 8,15- and 14,15-dihydroxy leukotrienes were recently identified in preparations of human leukocytes, although the stereochemistry and origins of the hydroxyl oxygens were not determined (13,15).

The pathway of leukotriene biosynthesis described here as originating from 15(S)-HPETE appears to differ from the "classical" LTA$_4$ pathway in several important ways, at least in porcine leukocytes. Although two 8,15-dihydroxy acids which are analogous to the 5,12-dihydroxy acids formed nonenzymatically from LTA$_4$ were identified, no compound analogous to LTB$_4$ could be detected. Moreover, it is clear that the 14,15-dihydroxy acids which were isolated, compounds V and VI, originate from arachidonic acid and 15(S)-HPETE via the postulated 14,15(S)-epoxide intermediate, not by reaction with water but rather by means of a novel and hitherto unprecedented mechanism of oxygenation.

In order to investigate the possibility that a 14,15-analog of LTC_4 might be formed in these cells, we synthesized this compound. No evidence for its formation in porcine leukocytes could be obtained. It still seems possible, of course, that the 14,15-analog of LTC_4 might be synthesized in cell systems other than the one examined here. However, these findings, taken together, suggest that it may be the unstable 14,15(S)-epoxide intermediate which is the biologically important compound.

Finally, it should be pointed out that additional studies in our laboratory have confirmed that the formation of 8,15- and 14,15-dihydroxy leukotrienes is not limited to porcine leukocytes. Mouse peritoneal cells and human leukocytes, isolated from a patient with mastocytosis, also synthesize these compounds. Additional studies are under way to define the biological activities of these novel compounds and to identify their role, if any, in human disease.

ACKNOWLEDGMENTS

We wish to thank Mrs. Christiana Ingram, Mr. Arthur Porter, and Mr. John Lawson for superior technical assistance, and Dr. Douglass Taber for valuable advice on chemical matters. We thank Dr. Thomas Harris for use of the JEOL FX 90Q FT NMR, purchased with funds from a USPHS Biomedical Research Support Grant and the Dreyfus Foundation. R.L.M. is supported by the Vivian Allen Fund of Vanderbilt Medical School. J.A.O. is the Joe and Morris Werthan Professor of Investigative Medicine. This work was supported by grant GM 15431.

REFERENCES

1. Bild, G. S., Ramadoss, C. S., Lim, S., and Axelrod, B. (1977): Double dioxygenation of arachidonic acid by soybean lipoxygenase-1. *Biochem. Biophys. Res. Commun.*, 74:949–954.
2. Bokoch, G., and Reed, P. W. (1981): Evidence for inhibition of leukotriene A_4 synthesis by 5,8,11,14-eicosatetraynoic acid in guinea pig polymorphonuclear leukoctyes. *J. Biol. Chem.*, 256:4156–4159.
3. Borgeat, P., and Samuelsson, B. (1979): Arachidonic acid metabolism in polymorphonuclear leukocytes: unstable intermediate in formation of dihydroxy acids. *Proc. Natl. Acad. Sci. USA*, 76:3213–3217.
4. Borgeat, P., and Samuelsson, B. (1979): Metabolism of arachidonic acid in polymorphonuclear leukoctyes. *J. Biol. Chem.*, 254:7865–7869.
5. Borgeat, P., Picard, S., Vallerand, P., and Sirois, P. (1981): Transformation of arachidonic acid in leukocytes. Isolation and structural analysis of a novel dihydroxy derivative. *Prostaglandins Med.*, 6:557–570.
6. Corey, E. J., Niwa, H., and Falck, J. R. (1979): Selective epoxidation of eicosa-*cis*-5,8,11,14-tetraenoic (arachidonic) acid and eicosa-*cis*-8,11,14-trienoic acid. *J. Am. Chem. Soc.*, 101:1586–1587.
7. Corey, E. J., Barton, A. E., and Clark, D. A. (1980): Synthesis of the slow reacting substance of anaphylaxis leukotriene C-1 from arachidonic acid. *J. Am. Chem. Soc.*, 102:4278–4279.
8. Corey, E. J., Marfat, A., and Goto, G. (1980): Simple synthesis of the 11,12-oxido and 14,15-oxido analogues of leukotriene A and the corresponding conjugates with glutathione and cysteinylglycine, analogues of leukotrienes C and D. *J. Am. Chem. Soc.*, 102:6607–6608.

9. Falardeau, P., Hamberg, M., and Samuelsson, B. (1976): Metabolism of 8,11,14-eicosatrienoic acid in human platelets. *Biochem. Biophys. Acta*, 441:193–200.
10. Hamberg, M., and Samuelsson, B. (1967): On the specificity of the oxygenation of unsaturated fatty acids catalyzed by soybean lipoxidase. *J. Biol. Chem.*, 242:5329–5335.
11. Hamilton, G. A. (1974): Chemical models and mechanisms for oxygenases. In: *Molecular Mechanisms of Oxygen Activation*, edited by O. Hayaishi, pp. 410–415. Academic Press, New York.
12. Israelsson, U., Hamberg, M., and Samuelsson, B. (1969): Biosynthesis of 19-hydroxy-prostaglandin A. *Eur. J. Biochem.*, 11:390–394.
13. Jubiz, W., Rådmark, O., Lindgren, J. Å., Malmsten, C., and Samuelsson, B. (1981): Novel leukotrienes: products formed by initial oxygenation of arachidonic acid at C-15. *Biochem. Biophys. Res. Commun.*, 99:976–986.
14. Kakinuma, K., and Minakami, S. (1978): Effects of fatty acids on superoxide radical generation in leukocytes. *Biochim. Biophys. Acta*, 538:50–59.
15. Lundberg, U., Rådmark, O., Malmsten C., and Samuelsson, B. (1981): Transformation of 15-hydroperoxy-5,9,11,13-eicosatetraenoic acid into novel leukotrienes. *FEBS Lett.*, 126:127–132.
16. Maas, R. L., Brash, A. R., and Oates, J. A. (1981): Novel leukotrienes and lipoxygenase product from rat mononuclear cells. In: *SRS-A and Leukotrienes*, edited by P. J. Piper, pp. 151–159. J. Wiley, New York.
17. Maas, R. L., Brash, A. R., and Oates, J. A. (1981): A second pathway of leukotriene biosynthesis in porcine leukocytes. *Proc. Natl. Acad. Sci. USA*, 78:5523–5527.
18. Samuelsson, B., Borgeat, P., Hammarström, S., and Murphy, R. C. (1980): Leukotrienes: a new group of biologically active compounds. In: *Advances in Prostaglandin and Thromboxane Research*, Vol. 6, edited by B. Samuelsson, P. W. Ramwell and R. Paoletti, pp. 1–18. Raven Press, New York.
19. Van Os, C. P. A., Rijke-Schilder, G. P. M., Van Halbeek, H., Verhagen, J., and Vligenthart, J. F. G. (1981): Double dioxygenation of arachidonic acid by soybean lipoxygenase-1 kinetics and regio-stereo specificities of the reaction steps. *Biochim. Biophys. Acta*, 663:177–193.
20. Vanderhoek, J. Y., Bryant, R. W., and Bailey, J. M. (1980): 15-Hydroxy-5,8,11,13-eicosatetraenoic acid: A potent and selective inhibitor of platelet lipoxygenase. *J. Biol. Chem.*, 255:5996–5998.

Leukotrienes and Other Lipoxygenase Products,
edited by B. Samuelsson and R. Paoletti.
Raven Press, New York © 1982.

Double Dioxygenation of Arachidonic Acid in Leukocytes by Lipoxygenases

P. Borgeat, B. Fruteau de Laclos, S. Picard, P. Vallerand, and *P. Sirois

*Departement d'Endocrinologie Moléculaire, Le Centre Hospitalier de l'Université Laval, Ste-Foy, Quebec, Canada G1V 4G2; and *Unité de Recherche Pulmonaire, Centre Hospitalier Universitaire, Sherbrooke, Quebec, Canada J1H 5N4*

The recent discovery of leukotrienes (LTs), a new family of bioactive metabolites of arachidonic acid formed in a lipoxygenase-type reaction (3), has greatly stimulated interest in these transformations of the fatty acid. In this chapter we report some other lipoxygenase-catalyzed transformations of arachidonic acid in leukocytes and discuss their possible biological significance.

MATERIALS AND METHODS

Hydroxy acids were prepared as follows: the 12S-hydroxy-5,8,10,14(Z,Z,E,Z)-eicosatetraenoic acid (12-HETE) was obtained by incubating human platelets with arachidonic acid (7). The 5S-hydroxy-6,8,11,14(E,Z,Z,Z)-eicosatetraenoic acid (5-HETE) (4) and the 5S,12R-dihydroxy-6,8,10,14(Z,E,E,Z)-eicosatetraenoic acid (leukotriene B$_4$, LTB$_4$) (2) were generated by incubating swine peripheral blood leukocytes with arachidonic acid and the ionophore A23187. The compounds were purified to homogeneity by successive silica gel and reverse-phase high-pressure liquid chromatography (RP-HPLC). (Detailed procedures for the preparation of these compounds will be reported separately.)

Human peripheral blood leukocytes were obtained from normal donors by leukapheresis; after ammonium chloride treatment for red cell lysis, the suspension was centrifuged at 250 × g for 15 min and the pellet resuspended in Dulbecco's phosphate-buffered saline (PBS) for incubation at a concentration of 100 × 10^6 cells/ml. The cell suspension contained lymphocytes, monocytes, polymorphonuclear leukocytes, and platelets (platelet/leukocyte ratio ≃ 1).

Swine peripheral blood leukocytes were prepared as described before (5) except that the concentration of the final cell suspension was 50 × 10^6 cells/ml.

Ether extractions and preparation of samples by column chromatography for HPLC analysis were also performed as described (5) but on a smaller scale, using 1 g silicic acid and 30 ml each solvent.

Measurements of myotropic activity of hydroxy acids on guinea pig lung parenchymal strips were performed as described previously (11).

RESULTS

Isolation of 5S,12S-DHETE from Human Leukocytes

A suspension of human leukocytes (225 ml) was incubated with arachidonic acid and the ionophore A23187. After extraction and fractionation of the metabolites by column chromatography, a sample (1%) of the fraction containing the monohydroxy- and dihydroxy derivatives of arachidonic acid was esterified and analyzed by HPLC (Fig. 1). In addition to compounds previously identified [i.e., 12-HETE, HHT (12S-hydroxy-5(Z),8,10(E)-heptadecatrienoic acid), 5-HETE, and LTB$_4$], a novel metabolite was detected. Analysis of the compound by ultraviolet (UV) photometry, HPLC, and gas chromatography-mass spectrometry (GC-MS) indicated that the compound was identical to the 5S,12S-dihydroxy-6(E),8(Z),10(E),14(Z)-eicosatetraenoic acid (5S,12S-DHETE) recently isolated in swine leukocytes (5).[1]

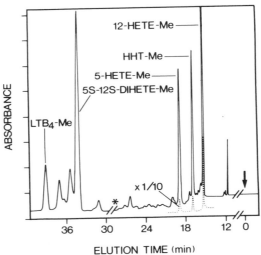

FIG. 1. Silica gel HPLC chromatogram (Radial-Pak B column, 8 × 100 mm, 10 μm particles (Waters Scientific, Milford, MA) of the products (methyl esters, Me) contained in the diethyl ether/methanol, 95:5 (v/v) fraction of the silicic acid fractionation of an ether extract. Human peripheral blood leukocytes were incubated (4 min) with arachidonic acid (165 μM) and the ionophore A23187 (20 μM) before adding 1.5 volumes of methanol. The sample was dissolved in diethyl ether for injection. The compounds were eluted (3 ml/min, 20°C) using a linear gradient of 99.9% hexane (0.1% acetic acid) to 92.9:7, hexane/isopropanol, v/v (0.1% acetic acid) in 45 min. The elution was monitored by UV photometry at 235 nm (0 to 28 min) and 280 nm (28 to 40 min). The lower trace *(broken line)* was recorded at 10× higher attenuation. Asterisk (*) indicates the change of wavelength. Arrow indicates the injection. The peaks between 5S-12S-DHETE and LTB$_4$ are the Δ⁶-*trans*-LTB$_4$ epimers (1).

---

[1]The complete stereochemistry of this compound was recently confirmed, and the data will be reported separately.

Figure 1 clearly indicates that 5S,12S-DHETE is a major metabolite of arachidonic acid in this cell preparation; the total amount of the various DHETEs formed in this experiment (22×10^9 cells) was 790, 210, 160, and 205 μg, respectively, for 5S,12S-DHETE, Δ^6-*trans*-LTB$_4$, Δ^6-*trans*-12-epi-LTB$_4$, and LTB$_4$.

Detection and Isolation of 5,15-DHETE

Other incubations were performed using swine peripheral blood leukocytes. The careful HPLC analysis (detection at 235 nm) of the fraction containing the DHETEs revealed the presence of another unknown metabolite (data not shown). The compound was eluted from silica gel columns between 5-HETE and 5S,12S-DHETE, and showed a UV spectrum with a λ_{max} at 243 nm. GC-MS analysis of the compound and of its hydrogenated derivative clearly indicated that the metabolite was a 5,15-dihydroxy-6,8,11,13-eicosatetraenoic acid (5,15-DHETE). The details of the isolation procedures and structural analysis will be reported separately. The yield of the 5,15-DHETE was 20 to 50 times lower than that of LTB$_4$.

Metabolism of 12-HETE and 15-HETE by Leukocytes

The structures of 5S,12S-DHETE and 5,15-DHETE suggested that the compounds could be the products of the successive reactions of arachidonic acid with two lipoxygenases. In order to validate this hypothesis, swine leukocytes were incubated with HETEs in the presence of the ionophore A23187 to activate the C-5 lipoxygenase, and of 5,8,11,14-eicosatetraynoic acid (ETYA) to inhibit cyclooxygenase and C-12 lipoxygenase (7) [ETYA does not inhibit C-5 lipoxygenase (4,13)]. Figure 2A shows the effect of the ionophore alone on ETYA-treated cells: Peaks corresponding to 5-HETE, LTB$_4$, and Δ^6-*trans*-LTB$_4$ (1) (Δ^6-*trans*-12-epi-LTB$_4$ cochromatographed with PGB$_2$) were detected. Addition of arachidonic acid caused increased formation of these compounds (Fig. 2B). Incubation of leukocytes with 15-HETE led to the formation of 5,15-DHETE (Fig. 2C). Similarly, the addition of 12-HETE to leukocytes caused the formation of substantial amounts of 5S,12S-DHETE (Fig. 2D). Concentrations of 12-HETE as low as 1 μg/ml (3 μM) produced detectable formation of the 5S,12S-DHETE. The identity of the compounds was ascertained by GC-MS analysis. These experiments confirmed that 5S,12S-DHETE and the 5,15-DHETE could be formed by double dioxygenation of arachidonic acid. In other experiments (not shown), we demonstrated that 5S,12S-DHETE incorporates two atoms of ^{18}O from molecular oxygen at C-5 and C-12, and that ETYA completely abolished the formation of 5S,12S-DHETE from arachidonic acid.

Experiments similar to those shown in Fig. 2 further indicated that the synthesis of DHETEs from HETEs in leukocytes was accompanied by the progressive inhibition of the formation of 5-HETE and LTB$_4$. This inhibitory effect of the HETEs is obvious in Figs. 2A, 2C, and 2D; at concentrations of 30 and 45 μM, 15-HETE and 12-HETE, respectively, completely inhibited the formation of the two metabolites of the C-5 lipoxygenase.

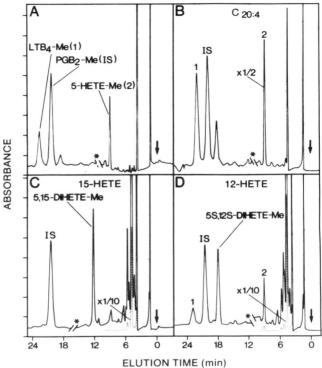

FIG. 2. Silica gel HPLC chromatogram (Radial-Pack B column, 5 × 100 mm, 10 μm particles) of the methyl esters of the products contained in the diethyl ether/methanol, 95:5 (v/v) fraction of the silicic acid fractionation of an ether extract. Swine peripheral blood leukoctyes (2 ml, 50 × 10⁶ cells/ml) were preincubated 5 min at 37°C with ETYA (10 μM) before adding the (**A**) ionophore A23187 (20 μM), (**B**) ionophore plus arachidonic acid (33 μM), (**C**) ionophore plus 15-HETE (30 μM), and (**D**) ionophore plus 12-HETE (30 μM). The reactions were stopped with methanol after 10 min and 2 μg prostaglandin B₂ (PGB₂) was added as internal standard (IS) before extraction. For HPLC analysis, the samples were dissolved in diethyl ether and injected. The compounds were eluted (1.5 ml/min, 20°C) using a gradient of isopropanol/hexane, 1:99 (v/v) to 3:97 (0 to 6 min) and to 4.5:95.5 (6 to 6.5 min). The first peak on the right side of PGB₂ in **A** and **B** is Δ⁶-*trans*-LTB₄; Δ⁶-*trans*-12-epi-LTB₄ cochromatographed with PGB₂. The elution was monitored at 235 nm (~ 0 to 12 min) and 280 nm (~ 12 to 24 min). The attenuation of the signal was two times higher at 235 nm than at 280 nm, except in **B** (four times higher, as indicated). The lower trace *(broken line)* was recorded at 10 times higher attenuation, as indicated. Asterisk (*) indicates the change of wavelength. Arrows indicate injections.

DISCUSSION

We reported here the occurrence of two novel metabolites of arachidonic acid in leukocytes, 5S,12S-DHETE and 5,15-DHETE. Figure 3 shows the structures of these compounds and a possible mechanism of formation involving two reactions with lipoxygenases as demonstrated in these studies. In other experiments, we have observed that platelet C-12 lipoxygenase can form 5S,12S-DHETE from 5-HETE, leaving some uncertainty about the order of reaction of arachidonic acid with the

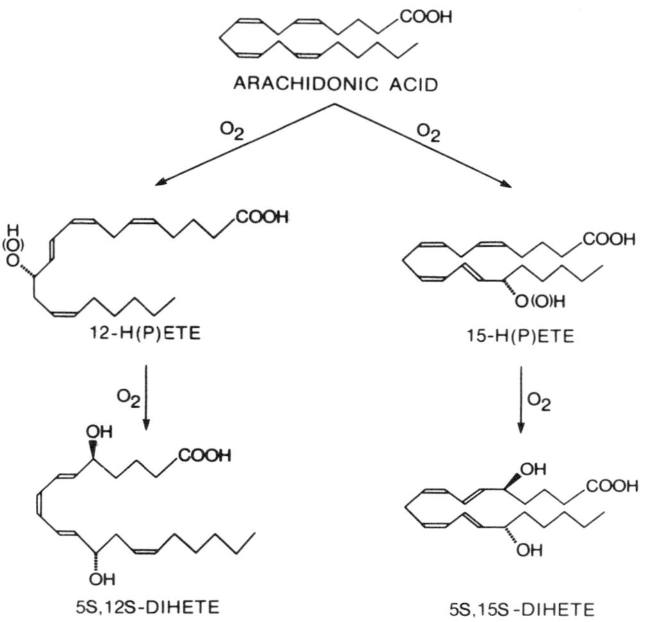

FIG. 3. Proposed mechanism of formation of 5S,12S-DHETE and 5S,15S-DHETE. The tentative stereochemistry of 5S,15S-DHETE was based on biogenetic considerations (see *Discussion*).

C-5 and C-12 lipoxygenases in the formation of the 5S,12S-DHETE. It is clearly established, however, that the compound is formed by double dioxygenation of arachidonic acid; this is also the most likely mechanism of formation for 5,15-DHETE, although the [18]O incorporation experiment was not performed in this case. It is recognized that leukocytes contain a C-5 lipoxygenase (4), but the presence of C-12 and C-15 lipoxygenases in these cells has never been clearly demonstrated. Some 15-HETE (with poor stereochemical purity) was isolated from human leukocytes (2), but it was not established whether the compound was a product of the cyclo-oxygenase or of a lipoxygenase. Concerning the C-12 lipoxygenase, some experiments strongly support the presence of the enzyme in swine leukocytes (not shown); indeed, the amount of 12-HETE formed by these cells largely exceeds (10- to 100-fold) the amount expected from the contaminating platelets.

The other point of interest of the present studies is the biological significance of these new transformations of arachidonic acid by lipoxygenases. Recently it was shown that LTB$_4$ (a stereoisomer of the 5S,12S-DHETE), is a potent contractile agent of the guinea pig lung parenchymal strip (11,12). LTB$_4$ was also found to have several proinflammatory properties (6,8–10). Measurements of the effects of 5S,12S-DHETE and 5,15-DHETE on lung parenchymal strips indicated no myotropic activity (data not shown). In addition, 5S,12S-DHETE did not alter calcium homeostasis in rabbit peritoneal neutrophils and did not affect lysosomal enzyme

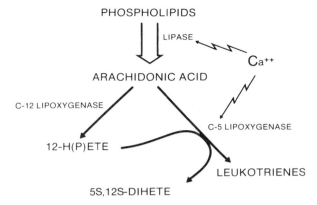

FIG. 4. Interaction between C-5 and C-12 lipoxygenases: a hypothetical mechanism for the control of leukotriene synthesis.

release, in contrast to LTB_4.[2] These preliminary results indicate that the profiles of the biological properties of the novel metabolites and of LTB_4 show some differences.

The observations that 15-HETE and 12-HETE are efficiently transformed into dihydroxy derivatives by leukocytes and concomitantly inhibit the synthesis of 5-HETE and LTB_4 (Fig. 2) are suggestive of a mechanism of control of leukotriene synthesis. Figure 4 illustrates this hypothesis where 12-HETE competes with arachidonic acid as substrate of the C-5 lipoxygenase, thus limiting the production of leukotrienes. It was reported recently that 15-HETE is a potent inhibitor of LTB_4 synthesis (13), and the authors raised the possibility of a modulation of leukotriene synthesis by this hydroxy acid. We believe that 12-HETE might be a better candidate in this respect, as under our experimental conditions it is produced in much higher yield (100-fold) than 15-HETE in swine peripheral blood leukocytes. Further studies concerning this hypothetical mechanism of control of leukotriene synthesis are in progress.

ACKNOWLEDGMENTS

This project was supported by a grant of the National Cancer Institute of Canada to P.B.

REFERENCES

1. Borgeat, P., and Samuelsson, B. (1979): Metabolism of arachidonic acid in polymorphonuclear leukocytes: structural analysis of novel hydroxylated compounds. *J. Biol. Chem.*, 254:7865–7869.
2. Borgeat, P., and Samuelsson, B. (1979): Arachidonic acid metabolism in polymorphonuclear leukocytes: effect of ionophore A 23187. *Proc. Natl. Acad. Sci. USA*, 76:2148–2152.
3. Borgeat, P., and Sirois, P. (1981): Leukotrienes: a major step in the understanding of immediate hypersensitivity reactions. *J. Med. Chem.*, 24:121–126.

[2]Personal communication with Drs. Sha'afi and Naccache, Connecticut Health Center.

4. Borgeat, P., Hamberg, M., and Samuelsson, B. (1976): Transformation of arachidonic acid and homo-γ-linoleic acid by rabbit polymorphonuclear leukocytes: monohydroxy acids from novel lipoxygenase. *J. Biol. Chem.*, 251:7816–7820.
5. Borgeat, P., Picard, S., Vallerand, P., and Sirois, P. (1981): Transformation of arachidonic acid in leukocytes: isolation and structural analysis of a novel dihydroxy derivative. *Prostaglandins Med.* 6:557–570.
6. Ford-Hutchinson, A. W., Bray, M. A., Doig, M. V., Shipley, M. E., and Smith, M. J. H. (1980): Leukotriene B, a potent chemokinetic and aggregating substance released from polymorphonuclear leukocytes. *Nature*, 286:264–265.
7. Hamberg, M., and Samuelsson, B. (1974): Prostaglandin endoperoxides: novel transformation of arachidonic acid in human platelets. *Proc. Natl. Acad. Sci. USA.*, 71:3400–3404.
8. König, W., Kroegel, C., Kunau, H. W., and Borgeat, P. (1981): Comparison of the eosinophil chemotactic factor (ECF) with endogenous hydroxyeicosatetraenoic acids of leukoctyes. *Monogr. Allergy (in press).*
9. Naccache, P. H., Borgeat, P., Goetzl, E. J., and Sha'afi, R. I. (1981): Mono- and di-hydroxy eicosatetraenoic acids alter calcium homeostasis in rabbit neutrophils. *J. Clin. Invest.* 67:1584–1587.
10. Palmer, R. M. J., Stepney, R. J., Higgs, G. A., and Eakins, K. E. (1980): Chemokinetic activity of arachidonic acid lipoxygenase products on leukocytes of different species. *Prostaglandins*, 20:411–418.
11. Sirois, P., Borgeat, P., Jeanson, A., Roy, S., and Girard, G. (1980): The action of leukotriene B₄ (LTB₄) on the lung. *Prostaglandins Med.*, 5:429–444.
12. Sirois, P., Roy, S., Borgeat, P., Picard, S., and Corey, E. J. (1981): Structural requirement for the action of leukotriene B₄ on the guinea-pig lung: importance of double bond geometry in the 6,8,10-triene unit. *Biochem. Biophys. Res. Commun.*, 99:385–390.
13. Vanderhoek, J. Y., Bryant, R. W., and Bailey, J. M. (1980): Inhibition of leukotriene biosynthesis by the leukocyte product 15-hydroxy-5,8,11,13-eicosatetraenoic acid. *J. Biol. Chem.*, 255:10064–10066.

Leukotrienes and Other Lipoxygenase Products,
edited by B. Samuelsson and R. Paoletti.
Raven Press, New York © 1982.

Formation of Novel Biologically Active Leukotrienes by ω-Oxidation in Human Leukocyte Preparations

Jan Åke Lindgren, Göran Hansson, Hans-Erik Claesson,
and Bengt Samuelsson

*Department of Physiological Chemistry, Karolinska Institute,
S-104 01 Stockholm, Sweden*

Human polymorphonuclear leukocytes (PMNLs) convert arachidonic acid into the biologically active leukotrienes (LT): LTA_4, LTB_4, and LTC_4 (2,9,13). The epoxide intermediate LTA_4 can be hydrolyzed either enzymatically to LTB_4 or nonenzymatically to isomeric 5,12- and 5,6-dihydroxy acids (3,4,7). These compounds are easily separated by reverse-phase high-performance liquid chromatography (RP-HPLC). However, we now report that human PMNL preparations produce an additional 5,12-dihydroxy acid, cochromatographing with LTB_4 in RP-HPLC (12). This dihydroxy acid, 5S,12S-dihydroxy-6-*trans*,8-*cis*,10-*trans*,14-*cis*-eicosatetraenoic acid (5S,12S-DHETE) is not formed via an epoxide intermediate but by a double oxygenation of arachidonic acid.

LTB_4 and 5S,12S-DHETE are further metabolized by ω-oxidation in preparations of human leukocytes (10,12). In the present chapter we describe the formation of 5S,12R,20-trihydroxy-6-*cis*,8,10-*trans*,14-*cis*-eicosatetraenoic acid (20-OH-LTB_4), 5S,12R-dihydroxy-6-*cis*,8,10-*trans*,14-*cis*-eicosatetraen-1,20-dioic acid (20-COOH-LTB_4), and 5S,12S,20-trihydroxy-6-*trans*,8-*cis*,10-*trans*,14-*cis*-eicosatetraenoic acid (5S,12S,20-THETE) in this system. The ω-oxidized metabolites were identified after stimulation of the leukocytes with either ionophore A23187 plus arachidonic acid or serum-treated zymosan. The biological activity of LTB_4, 20-OH-LTB_4, and 20-COOH-LTB_4 on guinea pig lung strips is reported.

MATERIALS AND METHODS

Preparations of human leukocyte suspensions (60 × 10⁶ cells/ml phosphate-buffered saline containing 0.87 mM $CaCl_2$, pH 7.4) and serum-treated zymosan were carried out as described (6). The incubation procedure, purification of the products, analytical methods (12), and bioassay using guinea pig subpleural parenchymal strips (10) were described recently.

54 ω-*OXIDATION OF LEUKOTRIENES*

RESULTS AND DISCUSSION

Incubation of human leukocytes (37°C, 10 min) with ionophore A23187 (5 μM) and arachidonic acid (150 μM) led to leukotriene formation. After purification of the products using ether extraction and silicic acid chromatography, the samples were subjected to RP-HPLC. The chromatogram contained two major peaks eluted after 12 and 40 min, respectively. The material corresponding to these peaks was collected separately and converted to methyl esters.

Straight-phase (SP) HPLC of the less-polar material resulted in the appearance of two major peaks (compounds A and B; Fig. 1). Ultraviolet (UV) spectrometry of the compounds showed absorption maxima at 258, 268, and 278 nm (compound A) and at 260, 270, and 281 nm (compound B), respectively, indicating that the compounds contained conjugated triene structures. Gas chromatography-mass spectrometry (GC-MS) of the trimethylsilyl ether methyl ester of compound A showed a peak with a C-value of 22.4 (1% SE-30). The mass spectrum (12) had prominent ions at m/e 494 (M), 479 (M-15), 463 (M-31), 404 (M-90), 393 [M-101, loss of $\cdot CH_2 - (CH_2)_2 - COOCH_3$], 383 [M-111, loss of $\cdot CH_2 - CH = CH - (CH_2)_4 - CH_3$], 354 [probably $\cdot CH = CH - (CH = CH)_2 - C(OSiMe_3) - (CH_2)_3 - C(OCH_3) = O^+ SiMe_3$ from a rearrangement], 293 [M-(111 + 90)], 279, 213 [$Me_3SiO^+ = CH - CH_2 - CH =$

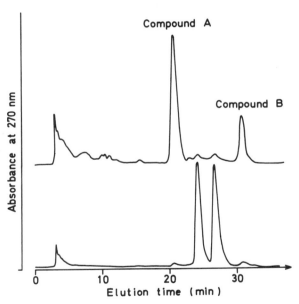

FIG. 1. The upper curve represents an SP-HPLC chromatogram of the methyl esters of the less-polar products isolated from human leukocyte suspensions after incubation with ionophore A23187 and arachidonic acid. The lower curve represents an SP-HPLC chromatogram of the methyl esters of the two epimers (at C-12) of 5,12-dihydroxy-6-*trans*,8-*trans*,10-*trans*,14-*cis*-eicosatetraenoic acid.

$CH-(CH_2)_4-CH_3$], 203 [base peak; $Me_3SiO^+ = CH-(CH_2)_3-COOCH_3$ and M-$(111+90+90)$], 171, and 129. The mass spectrum of the Me_3Si derivative of the methyl ester of hydrogenated compound A (C-value 24.0) was in agreement with the mass spectrum of the same derivative of hydrogenated LTB_4 (3), indicating that compound A was a new isomer of 5,12-dihydroxy-6,8,10,14-eicosatetraenoic acid. UV absorbance and GC-MS data confirmed that compound B was LTB_4. For steric analysis of the alcohol groups, oxidative ozonolysis of the (−)-menthoxycarbonyl derivative of compound A was performed, followed by diazomethane treatment of the fragments and gas chromatographic analysis. The products cochromatographed with the reference compounds (−)-menthoxycarbonyl dimethyl S-malate and (−)-menthoxycarbonyl dimethyl 2-S-hydroxy-adipate, demonstrating that the structure of compound A was 5S,12S-dihydroxy-6,8,10,14-eicosatetraenoic acid (5S,12S-DHETE).

The chromatogram obtained after SP-HPLC of the methyl esters of the polar products contained three main peaks (compounds C, D and E, Fig. 2). The UV spectrum of compound C was almost identical to that of 5S,12S-DHETE, whereas the spectra of compounds D and E were in agreement with the spectrum of LTB_4. GC-MS of the Me_3Si derivatives of the methyl esters of compounds C, D, and E (C-values 25.4, 26.7, and 26.5, respectively) was performed. The trimethylsilyl ether methyl ester of compound C showed a mass spectrum similar to that of the corresponding derivative of 5S,12S-DHETE (containing ions at m/e 129, 171, 203, 292, 354, and 383), although several ions were shifted 88 units to higher m/e values ($213 \rightarrow 301$, $393 \rightarrow 481$, $404 \rightarrow 492$, $463 \rightarrow 551$, $479 \rightarrow 567$, $494 \rightarrow 582$). This indicated that compound C was a metabolite of 5S,12S-DHETE containing an additional hydroxyl group at a position beyond C-13. Evidence for an ω1-hydroxylation was obtained from the equivalent chain length (C-value 25.4) which was increased with 3.0 C (8,11) as compared to the C-value of compound A (C-value 22.4). Thus the indicated structure was 5S,12S,20-trihydroxy-6,8,10,14-eicosatetraenoic acid. The proposed structure was further indicated by the mass spectrum of the Me_3Si derivative of the methyl ester of hydrogenated compound C. The mass spectra of the trimethylsilyl ether methyl esters of compounds D and E (10) were similarly related to the corresponding derivative of LTB_4, containing ions at m/e 129, 191, 203, 217, 229, 267, 293, and 383. In the spectrum of the derivative of compound D, other ions were shifted 88 units to higher m/e values ($404 \rightarrow 492$, $463 \rightarrow 551$, $479 \rightarrow 567$, $494 \rightarrow 582$), whereas the same ions were shifted 44 units to higher m/e values in the mass spectrum of the corresponding derivative of compound E ($404 \rightarrow 448$, $463 \rightarrow 507$, $479 \rightarrow 523$, $494 \rightarrow 538$). The spectra together with the increases of equivalent chain length (increases with 3.1 and 2.9 C, respectively, as compared to the corresponding derivative of LTB_4) suggested that compounds D and E were ω1-oxidized metabolites of LTB_4 containing a hydroxyl (compound D) or a carboxyl (compound E) group at the C-20 position. Thus the structures are 5S,12R,20-trihydroxy-6,8,10,14-eicosatetraenoic acid (20-OH-LTB_4) and 5S,12R-dihydroxy-6,8,10,14-eicosatetraen-1,20-dioic acid (20-COOH-LTB_4). The proposed structure of compound D was further supported by the mass spectrum

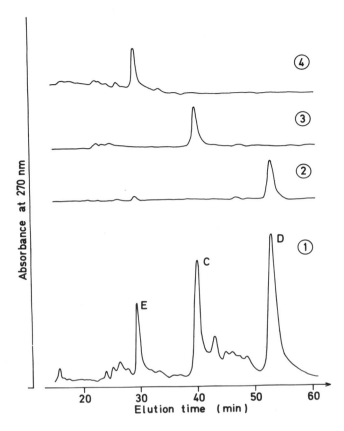

FIG. 2. Curve 1 represents an SP-HPLC chromatogram of the methyl esters of the polar products isolated from human leukocyte suspensions after incubation with ionophore A23187 and arachidonic acid. Curves 2, 3, and 4 represent corresponding chromatograms obtained after incubation of human leukocyte preparations with LTB₄, 5S,12S-DHETE, and 20-OH-LTB₄, respectively.

of the trimethylsilyl ether methyl ester of the hydrogenated compound (C value 27.3), which was in agreement with the mass spectrum of the same derivative of hydrogenated 5S,12S,20-THETE (12). Oxidative ozonolysis of the (-)-menthoxy-carbonyl derivatives of compounds C and D supported the proposed positions of the four double bonds and confirmed the configuration of the alcohol groups at C-5 and C-12.

In order to study the pathway of formation of the novel trihydroxy acids, human leukocyte suspensions were incubated with LTB₄ or 5S,12S-DHETE (10 μM, 37°C, 10 min). After initial purification, RP-HPLC revealed polar peaks which were converted to methyl esters and subjected to SP-HPLC (Fig. 2). LTB₄ gave rise to one major peak, corresponding to compound D, whereas 5S,12S-DHETE was converted mainly to a compound with a retention time identical to that of compound

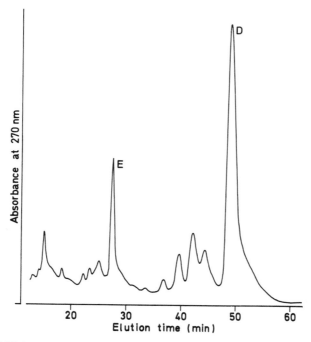

FIG. 3. SP-HPLC chromatogram of the methyl esters of the polar products isolated from human leukocyte suspensions after incubation with serum-treated zymosan.

C. GC-MS analysis of the material confirmed the identity with 20-OH-LTB$_4$ and 5S,12S,20-THETE, respectively. In other experiments, human leukocyte suspensions were incubated with compound D (10 μM, 37°C, 30 min). After ether extraction and silicic acid chromatography the products were subjected to anion-exchange chromatography using diethyl-aminohydroxypropyl Sephadex LH-20 (DEAP-LH-20) (1). The dicarboxylic fraction from this column was extracted at pH 3 with ethyl acetate, converted to methyl esters, and subjected to SP-HPLC. One major peak appeared with the same retention time as compound E (Fig. 2). This material and compound E were identical, as judged by GC-MS analysis.

To investigate the mechanism of formation of 5S,12S-DHETE, leukocyte preparations were incubated with ionophore A23187 and arachidonic acid under an atmosphere of $^{18}O_2$ (12). The mass spectrum of the trimethylsilyl ether methyl ester of hydrogenated compound A formed in these experiments contained several ions, which were shifted 4 units to higher m/e values (389 → 393, 401 → 405, 471 → 475, 487 → 491), whereas other ions were shifted 2 units to higher m/e values (203 → 205, 215 → 217, 299 → 301, 311 → 313). This showed that 5S,12S-DHETE carried two atoms of ^{18}O, located at C-5 and C-12. Thus the hydroxyl groups at C-5 and C-12 in 5S,12S-DHETE were both derived from molecular oxygen, showing that the compound is formed by a double oxygenation of arachidonic acid. The mass spectra of the Me$_3$Si derivatives of the methyl esters of hydrogenated compounds

C and D formed in $^{18}O_2$-labeling experiments indicated that these compounds were metabolites of 5S,12S-DHETE and LTB_4, respectively, containing additional atoms of ^{18}O at the C-20 position (10,12).

We recently showed that LTB_4 is formed by human leukocyte suspensions after addition of serum-treated zymosan (STZ) particles (6). Therefore we incubated leukocyte suspensions with STZ (6.7 mg/ml, 37°C, 10 min) to investigate if the ω-oxidized metabolites were formed under these conditions. The RP-HPLC chromatogram of the extracted and purified products was similar to that obtained after incubation with ionophore A23187 and arachidonic acid, although the peaks were smaller. The material corresponding to a polar peak (retention time 12 min) was collected and converted to methyl esters prior to SP-HPLC. As shown in Fig. 3, STZ induced the formation of 20-OH-LTB_4 (compound D) and 20-COOH-LTB_4 (compound E) (5). The structures were confirmed by GC-MS of the trimethylsilyl ether methyl esters of the compounds.

The biological activity of LTB_4, 20-OH-LTB_4, and 20-COOH-LTB_4 was compared with that of LTC_4 and histamine on the guinea pig lung strip (10). According to dose-response relations assessed noncumulatively LTB_4, 20-OH-LTB_4, and 20-COOH-LTB_4 were at least 10 times more active than histamine, although 50 to 100 times less potent than LTC_4 (Fig. 4). Even 5S,12S-DHETE and 5S,12S,20-THETE were active, although to a lesser extent (Dahlén et al., *unpublished observation*).

In summary, a novel leukotriene-like dihydroxy acid, 5S,12S-DHETE, was formed in leukocyte suspensions together with ω-oxidized metabolites of this dihydroxy acid and LTB_4 (Fig. 5). Thus the polar compounds 5S,12S,20-THETE, 20-OH-LTB_4, and 20-COOH-LTB_4 were identified and found to be biologically active. Preliminary results indicate the conversion of 5S,12S,20-THETE to a 1,20-dioic acid as well as the formation of 19-hydroxylated metabolites of 5S,12S-DHETE

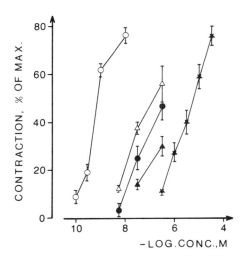

FIG. 4. Noncumulative dose-response relations for LTB_4 (●), 20-OH-LTB_4 (Δ), 20-COOH-LTB_4 (▲), LTC_4 (○), and histamine (★) in guinea pig lung parenchymal strips. Indicated drug concentrations refer to the final concentration in bath fluid surrounding the tissue. Log dose-response relations for leukotrienes were established from mean values of four or five observations in separate tissues at each concentration. Data are presented as the mean ± SEM, in percent of maximal concentration.

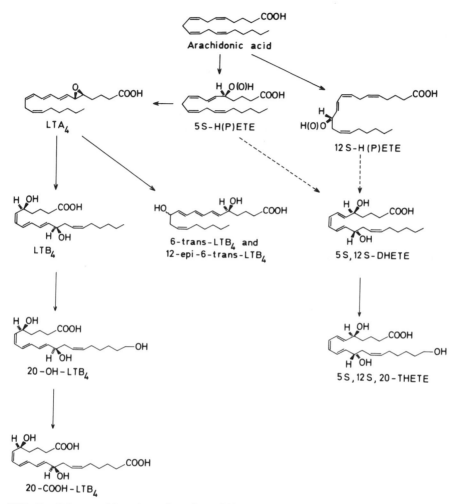

FIG. 5. Scheme of transformation of arachidonic acid by preparations of human PMNLs. Dashed lines indicate proposed reactions.

and LTB_4. The importance of ω-oxidation in leukotriene metabolism and function must be further investigated.

ACKNOWLEDGMENTS

The skillful technical assistance of Ms. Margareta Hovgard and Ms. Siv Andell is greatly appreciated. This study was supported by a grant from the Swedish Medical Research Council (project 03X-217), the Swedish National Association against Rheumatism, The King Gustaf the Vth 80-Years Fund, and the World Health Organization.

REFERENCES

1. Almé, B., and Hansson, G. (1978): Analysis of metabolic profiles of prostaglandins in urine using a lipophilic anion exchanger. *Prostaglandins*, 15:199–217.
2. Borgeat, P., and Samuelsson, B. (1979): Arachidonic acid metabolism in polymorphonuclear leukocytes: effects of the ionophore A23187. *Proc. Natl. Acad. Sci. USA*, 76:2148–2152.
3. Borgeat, P., and Samuelsson, B. (1979): Transformation of arachidonic acid by rabbit polymorphonuclear leukocytes: formation of a novel dihydroxy-eicosatetraenoic acid. *J. Biol. Chem.*, 254:2643–2646.
4. Borgeat, P., and Samuelsson, B. (1979): Metabolism of arachidonic acid in polymorphonuclear leukocytes: structural analysis of novel hydroxylated compounds. *J. Biol. Chem.*, 254:7865–7869.
5. Claesson, H.-E., Hansson, G., Lindgren, J. Å., and Samuelsson, B. (1981): To be published.
6. Claesson, H.-E., Lundberg, U., and Malmsten, C. (1981): Serum-coated zymosan stimulates the synthesis of leukotriene B$_4$ in human polymorphonuclear leukocytes: inhibition by cyclic AMP. *Biochem. Biophys. Res. Commun.*, 99:1230–1237.
7. Corey, E. J., Marfat, A., Goto, G., and Brion, F. (1981): Leukotriene B. Total synthesis and assignment of stereochemistry. *J. Am. Chem. Soc.*, 102:7984–7985.
8. Gréen, K. (1971): Metabolism of prostaglandin E$_2$ in the rat. *Biochemistry*, 10:1072–1086.
9. Hansson, G., and Rådmark, O. (1980): Leukotriene C$_4$: isolation from human polymorphonuclear leukocytes. *FEBS Lett.*, 122:87–90.
10. Hansson, G., Lindgren, J. Å., Dahlén, S.-E., Hedqvist, P., and Samuelsson, B. (1981): Identification and biological activity of novel ω-oxidized metabolites of leukotriene B$_4$ from human leukocytes. *FEBS Lett.*, 130:107–112.
11. Israelsson, U., Hamberg, M., and Samuelsson, B. (1969): Biosynthesis of 19-hydroxy-prostaglandin A$_1$. *Eur. J. Biochem.*, 11:390–394.
12. Lindgren, J. Å., Hansson, G., and Samuelsson, B. (1981): Formation of novel hydroxylated eicosatetraenoic acids in preparations of human polymorphonuclear leukocytes. *FEBS Lett.*, 128:329–335.
13. Rådmark, O., Malmsten, C., Samuelsson, B., Goto, G., Marfat, A., and Corey, E. J. (1980): Leukotriene A: isolation from human polymorphonuclear leukocytes. *J. Biol. Chem.*, 255:11828–11831.

Leukotrienes and Other Lipoxygenase Products,
edited by B. Samuelsson and R. Paoletti.
Raven Press, New York © 1982.

New Group of Leukotrienes Formed by Initial Oxygenation at C-15

O. Rådmark, U. Lundberg, W. Jubiz,[1] C. Malmsten,
and B. Samuelsson

*Department of Physiological Chemistry, Karolinska Institute,
S-104 01 Stockholm, Sweden*

The leukotrienes are biologically active compounds of particular interest in immediate hypersensitivity reactions and inflammation (15). These derivatives of arachidonic acid and other polyunsaturated fatty acids are formed by initial oxygenation at C-5. The 5-hydroperoxyeicosatetraenoic acid (5-HPETE) obtained from arachidonic acid is converted to leukotriene A_4 (LTA_4) (4,14) by dehydration. This unstable allylic epoxide (LTA_4) is either hydrolyzed, forming leukotriene B_4 (LTB_4) (2), or conjugated with glutathione, to form leukotriene C_4 (LTC_4) (9,12).

Thus human leukocytes synthesize various leukotrienes via the 5-hydroperoxy acid intermediate. However, these cells also produce 15-hydroperoxyeicosatetraenoic acid (15-HPETE) (3). In view of the symmetrical location of the two diene structures next to the methylene group (C-10) subject to hydrogen abstraction in leukotriene synthesis, it seemed conceivable that a 15-hydroperoxy derivative might serve as substrate as well (Fig. 1).

In this chapter we describe the formation of a new group of leukotrienes formed by initial oxygenation at C-15 of arachidonic acid. Preliminary reports of this work have appeared elsewhere (10,11).

EXPERIMENTAL PROCEDURES

Human leukocytes, either intact cells or the 10,000 × *g* supernatant from the homogenate, were incubated with 15-HPETE (100 μM) for 30 min at 37°C. Preparation of cells and 15-HPETE, extractions, and purifications were as described (10,11).

To prepare the 10,000 × *g* supernatant, the cells were suspended in tris-HCl buffer (10 mM, pH 7.4, 100 × 10^6 cells/ml). After 30 min in this buffer on ice, the cells were disrupted using a Potter-Elvehjem homogenizer and subsequently

---

[1]Present address: Veterans Administration Medical Center, University of Utah, College of Medicine, Salt Lake City, Utah

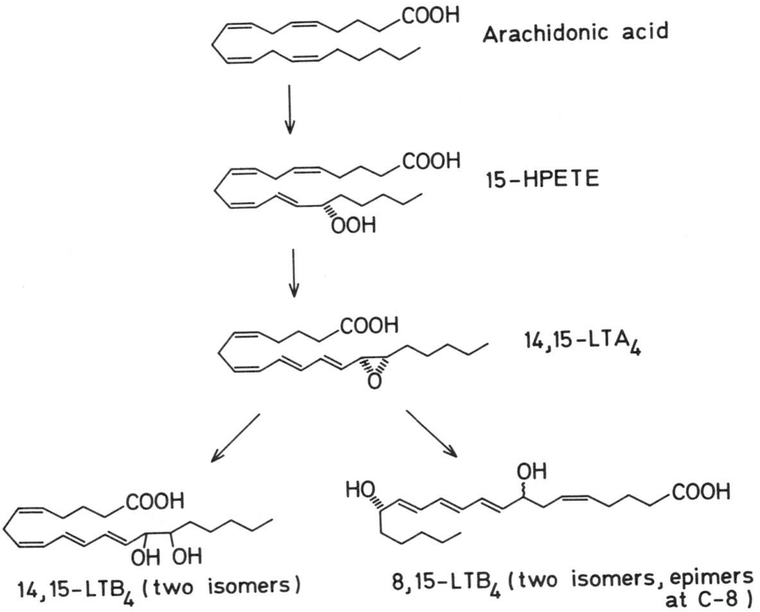

FIG. 1. Biosynthetic pathways of leukotrienes via 15-HPETE. Regarding the detailed structures of 14,15-LTA$_4$ and 14,15-LTB$_4$, see text.

centrifuged at $10,000 \times g$ for 15 min. The incubations were done in the presence of nordihydroguaiaretic acid (NDGA), 10^{-5}M.

The procedures for trapping the intermediate and the experiments with $^{18}O_2$ gas were as described (4); 100 ml (100×10^{-6}/ml) of cells was used in both instances. For trapping the incubation time was 25 sec, and for $^{18}O_2$ labeling 30 min. Analytical methods used were reported previously (4,8,10,11).

RESULTS AND DISCUSSION

When the $10,000 \times g$ supernatant from a human leukocyte homogenate was incubated with 15-HPETE, a number of compounds exhibiting a leukotriene-type ultraviolet (UV) spectrum were produced (Fig. 2). The estimated yield of these compounds (summation of peak areas, comparison with internal standard) ranged between 20 and 50 μg; i.e., about 1% of the added substrate was converted to compounds A, B, C, and D. The same compounds were also obtained in incubations of intact leukocytes with 15-HPETE. In intact cells, compounds A and B were the major products, whereas compounds C and D dominated the supernatant incubations.

In control incubations—i.e., supernatant incubated with substrate solvent (ethanol), buffer incubated with 15-HPETE, and supernatant incubated with 15-hydroxyeicosatetraenoic acid (15-HETE)—none of these compounds were obtained.

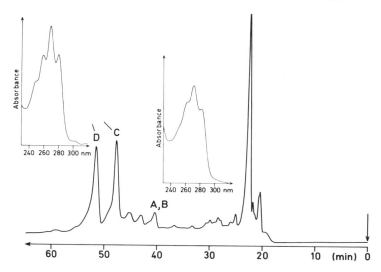

FIG. 2. SP-HPLC chromatogram obtained from incubation of 10,000 *g* human leukocyte supernatant with 15-HPETE. Mobile phase: hexane/isopropanol/acetic acid, 92:8:0.01 (v/v/v), 2 ml/min. UV detection, 280 nm. **Insets:** UV spectra (in methanol) of the denoted fractions. Prior to SP-HPLC, the incubate was extracted with ether and the extract purified by column chromatography. The ethyl acetate fraction (see text) was evaporated and the residue treated with diazomethane before injection on the HPLC column.

Structure of Compound A and B

The UV spectrum of the high-pressure liquid chromatography (HPLC) fraction denoted A,B showed maxima at 282, 272, and 262 nm, indicating a conjugated triene structure. Further purification on reverse-phase (RP) HPLC did not separate the compounds; however, when analyzed by gas chromatography-mass spectrometry (GC-MS), two compounds with C-values, respectively, of 23.8 and 24.7 (SE-30) that exhibited practically identical mass spectra were obtained. In Fig. 3 (upper panel) the mass spectrum of the compound with a C-value 23.8 is shown.

When the same material was hydrogenated prior to GC-MS analysis, only one compound with a mass spectrum was obtained (Fig. 3, lower panel). The C-value of the hydrogenated compound was 23.6 (SE-30). For detailed discussion of the mass spectra see refs. 10 and 11. The mass spectra show that compounds A and B are two isomers of 14,15-diOH-eicosatetraenoic acid, with the four double bonds located between C-1 and C-14. Based on biogenetic considerations (see below) and the UV spectrum, the exact positions of the double bonds are assumed to be at Δ^5, Δ^8, Δ^{10}, and Δ^{12}. Compounds A and B thus are two isomers of 14,15-diOH-5,8,10,12-eicosatetraenoic acid (14,15-leukotriene B_4, 14,15-LTB$_4$).

Structure of Compounds C and D

Further purification of compounds C and D was achieved on RP-HPLC, where compound C migrated before compound D. The UV spectra of the two compounds

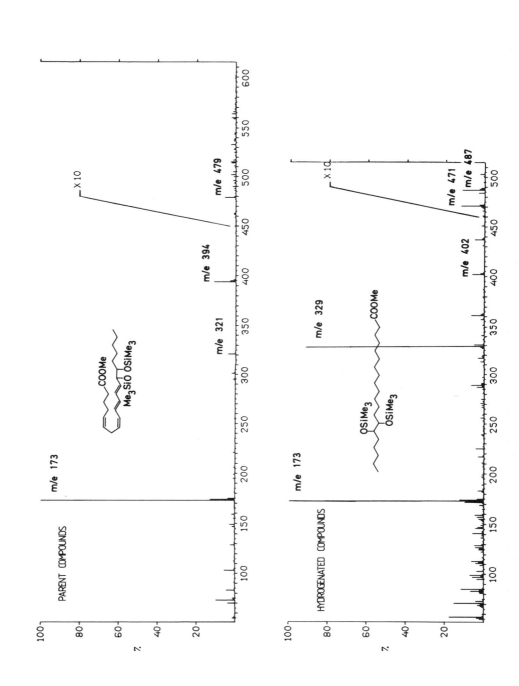

were practically identical, with maxima at 280, 269, and 259 nm, indicating conjugated trienes in the structures.

The C-values of the parent compounds (24.8, SE-30) and of the hydrogenated derivatives (24.0, SE-30) were also practically identical, as were the mass spectra. The mass spectra of compound C are shown in Fig. 4; for detailed discussion, see refs. 10 and 11. It can be concluded from the mass spectra that compounds C and D are two isomers of 8,15-dihydroxyeicosatetraenoic acid, with one double bond between C-1 and C-7, and three double bonds between C-8 and C-14.

Oxidative ozonolysis of the menthoxy carbonyl derivatives of compounds C and D was performed to elucidate the stereochemistry at C-8 of the structures. After derivatization of the hydroxyl groups with an optically active reagent (menthylchloroformate), the compounds were degraded by oxidative ozonolysis. The resulting diastereoisomeric products were identified by gas chromatographic comparison with known standards (methyl esters).

A fragment was obtained from compound C that cochromatographed with the menthoxycarbonyl derivative of D-malic acid on GC (OV-210). Hence compound C had the R-configuration at C-8. The corresponding fragment from compound D cochromatographed with L-malic acid and accordingly had the S-configuration at C-8. Concerning the stereochemistry at C-15 of compounds C and D, it is assumed that the S-configuration of the original 15-HPETE is retained.

Information was also obtained regarding positions of the double bonds. The formation of malic acid fragments from compounds C and D fits with the concept of double bonds being positioned at Δ^5 and Δ^9 of the parent structures.

The infrared spectra of compounds C and D, respectively (in CS_2), showed absorption bands *inter alia* at 996 cm^{-1}, indicating that the three conjugated double bonds in the structures all have the *trans* configuration (6). The remaining double bond that is located at Δ^5 is assumed to retain its original *cis* configuration.

It is thus clear from the data that compounds C and D are two isomers, distinguished by different retention times in straight-phase (SP)- and RP-HPLC. The structures of compounds C and D are 8(R),15(S)-dihydroxy-(Z,E,E,E)-5,9,11,13-eicosatetraenoic acid and 8(S), 15(S)-dihydroxy-(Z,E,E,E)-5,9,11,13-eicosateraenoic acid, respectively; they are thus epimers at C-8.

Trapping of Proposed Intermediate

Biosynthesis of the compounds described above from arachidonic acid and 15-HPETE is presumed to involve formation of the allylic epoxide 14,15-oxido-5,8,10,12-eicosatetraenoic acid (14,15-LTA$_4$) (Fig. 1). To obtain evidence for the existence of this structure in these transformations, cells were incubated for a short time with

---

FIG. 3. **Upper panel**: Mass spectrum of compound A. **Lower panel**: Mass spectrum of compounds A and B after catalytic hydrogenation. Methyl esters, trimethyl silyl ethers.

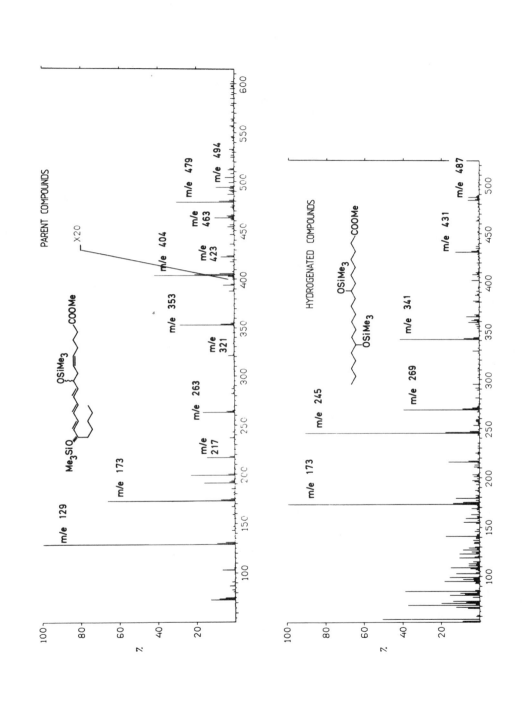

PARENT COMPOUNDS

m/e 129

m/e 173

m/e 217

m/e 263

m/e 321

m/e 353

m/e 404

m/e 423

m/e 463

m/e 479

m/e 494

X20

HYDROGENATED COMPOUNDS

m/e 173

m/e 245

m/e 269

m/e 341

m/e 431

m/e 487

15-HPETE and the incubation was stopped by adding acidic methanol, as previously described (4).

Following extraction and column chromatography, a pair of novel compounds were detected in the SP-HPLC chromatogram. These compounds had shorter retention times than those obtained for compounds C and D, and they exhibited UV spectra that were practically identical to those of compounds C and D (see above). When subjected to GC-MS, both compounds had a C-value of 24.5 (OV-210) and gave practically identical mass spectra.

The mass spectrum of the less polar of the two compounds is shown in Fig. 5. Ions are seen at (m/e): 436 (M^+), 421 (M^+-15), 405 (M^+-31), 404 (M^+-32, loss of methanol), 365 [M^+-71, loss of $\cdot(CH_2)_4CH_3$], 346 (M^+-90, loss of trimethylsilanol), 338, 333 [M^+-(71 + 32)], 314 [M^+-(90 + 32)], 295 [M^+-141, loss of $\cdot CH_2$ $(CH)_2(CH_2)_3COOMe$], 263 [M^+-(141 + 32), alternatively loss of $Me_3SiO\cdot CH$ $(CH_2)_4CH_3$], 251 [$Me_3Si\overset{+}{O}CHCH_2(CH)_2(CH_2)_3COOCH_3$], 224 [$M^+$-(141 + 71)], 205 [$M^+$-(141 + 90)], 199, 173 [$Me_3Si\overset{+}{O}CH(CH_2)_4CH_3$], 159 (probably $Me_3Si\overset{+}{O} =$ $CH - CH = CH - OCH_3$, from a rearrangement), 133, 129, and 71 (base peak).

The mass spectrum was compatible with a C_{20} tetraunsaturated fatty acid containing one hydroxy group and one methoxy group. The hydroxy group was located at C-15 because of the ions at m/e 365, 363, 173, and 71; and the methoxy group was located at C-8 because of the ions at m/e 295 and 251.

The ions at m/e 365, 263, 173, and 71 also indicated that the molecule is saturated between C-15 and C-20, whereas the region C-1 to C-14 contains four double bonds. The ions at m/e 295 and 251 indicated that three of these double bonds are located between C-8 and C-14. These data indicated that the two derivatives were isomeric 8-methoxy-15-hydroxy-5,9,11,13-eicosatetraenoic acids, epimeric at C-8. The formation of these two compounds in the trapping experiments is analogous to the reaction of the 5,6-epoxide and provides strong evidence for the intermediary formation of 14,15-LTA$_4$ in these incubations.

In consideration of the stereochemistry of 14,15-LTA$_4$ (Fig. 1), it is assumed that the S-configuration is retained at C-15, that the epoxide has the *trans* configuration, and that the double bonds are 5-*cis*,8-*cis*,10-*trans*, and 12-*trans* in analogy with the structure of 5,6-LTA$_4$ (13).

$^{18}O_2$-Labeling Experiments

Intact cells were incubated with arachidonic acid under $^{18}O_2$ atmosphere. When analyzed by GC-MS, the ions in the mass spectra of compounds A and B (14,15-LTB$_4$) were shifted (Fig. 6). Thus prominent ions were seen at (m/e) 483 (479 + 4), 396 (394 + 2), 323 (321 + 2), and 175 (173 + 2). Less intensive (10%) ions at m/e 394, 321, and 175 were also seen. This clearly shows incorporation of molecular oxygens at C-14 and C-15 in the 14,15-LTB$_4$.

FIG. 4. **Upper panel**: Mass spectrum of compound C. **Lower panel**: Mass spectrum of compound C after catalytic hydrogenation. Methyl esters, trimethyl silyl ethers.

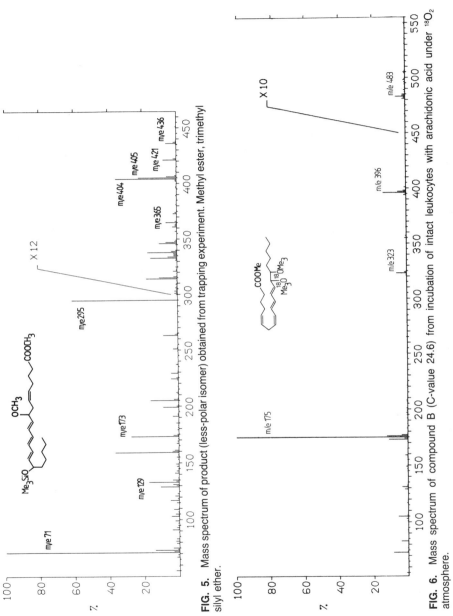

FIG. 5. Mass spectrum of product (less-polar isomer) obtained from trapping experiment. Methyl ester, trimethyl silyl ether.

FIG. 6. Mass spectrum of compound B (C-value 24.6) from incubation of intact leukocytes with arachidonic acid under $^{18}O_2$ atmosphere.

In another study intact cells were incubated with 15-HPETE under $^{18}O_2$ atmosphere. The mass spectra of 14,15-LTB$_4$ then were as follows (m/e): 481 (479 + 2), 396 (394 + 2), 323 (321 + 2), and 173. This indicates incorporation of molecular oxygen at C-14.

The expected mode of formation of 14,15-LTB$_4$ would be hydrolysis of the epoxide 14,15-LTA$_4$, with incorporation of molecular oxygen at only one of the carbons C-14 or C-15. These results indicate, however, that at least 90% of the 14,15-LTB$_4$ formed under these circumstances is formed by another mechanism. A plausible alternative could be the reaction of 14,15-LTA$_4$ with an active oxygen species. Leukocytes have the capacity to produce several such species (1). This presumed insertion of active oxygen seems to be specific to C-14 of 14,15-LTA$_4$, as no ion of m/e 175 was detected in the product of the 15-HPETE incubation (see above). It is thus conceivable that 14,15-LTB$_4$ is formed in two ways. A minor part is due to hydrolysis of 14,15-LTA$_4$, whereas the major part stems from reaction of 14,15-LTA$_4$ with an active oxygen species. This concept fits with the observation that 14,15-LTB$_4$ is a major product in incubations of intact cells, whereas it is a minor product in supernatant incubations. Presumably the active oxygen is produced to a much lesser degree in the supernatant preparation.

It is not known if 14,15-LTA$_4$ is formed by the same enzyme as 5,6-LTA$_4$. However, 14,15-LTA$_4$ does not seem to undergo enzymatic hydrolysis to any appreciable extent, as a major third isomer of 8,15-LTB$_4$ is not found. In some incubations, however, small amounts of a third compound with a mass spectrum practically identical to those of compounds C and D have been detected. It differs, though, in that it exhibits a shorter C-value (23.4, OV-101).

It is also of interest that the calcium ionophore A23187 has different effects on the two leukotriene pathways. Leukotriene biosynthesis via 5-HPETE as well as biosynthesis of 5-HETE in human leukocytes is markedly increased by the ionophore (3), whereas leukotriene biosynthesis via 15-HPETE and the production of 15-HETE seem to be unaffected.

Further studies regarding this new group of leukotrienes are in progress.

ACKNOWLEDGMENTS

We wish to thank Ms. Ulla Andersson and Ms. Inger Tollman-Hansson for excellent help in these studies. This work was supported by a grant from the Swedish Medical Research Council (project 03X-217) and by a grant from Torsten and Ragnar Söderbergs Stiftelser.

REFERENCES

1. Badwey, J. A., and Karnovsky, M. L. (1980): Active oxygen species and the functions of phagocytic leukocytes. *Annu. Rev. Biochem.*, 49:695–726.
2. Borgeat, P., and Samuelsson, B. (1979): Transformation of arachidonic acid by rabbit polymorphonuclear leukocytes. *J. Biol. Chem.*, 254:2643–2646.
3. Borgeat, P., and Samuelsson, B. (1979): Arachidonic acid metabolism in polymorphonuclear leukocytes: effects of ionophore A23187. *Proc. Natl. Acad. Sci. USA*, 76:2148–2152.

4. Borgeat, P., and Samuelsson, B. (1979): Arachidonic acid metabolism in polymorphonuclear leukocytes: unstable intermediate in formation of dihydroxy acids. *Proc. Natl. Acad. Sci. USA*, 76:3212–3217.
5. Corey, E. J., Marfat, A., and Goto, G. (1980): Simple synthesis of the 11,12-oxido and 14,15-oxido analogues of leukotriene A and the corresponding conjugates with glutathione and cysteinylglycine, analogues of leukotrienes C and D. *J. Am. Chem. Soc.*, 102:6607–6608.
6. Crombie, L., and Jacklin, A. G. (1957): Lipids. VI. Total synthesis of α- and β-elaeostearic and pucinic (trichosanic) acid. *J. Chem. Soc.*, II:1632–1646.
7. Hamberg, M., and Samuelsson, B. (1967): On the specificity of the oxygenation of unsaturated fatty acids catalyzed by soybean lipoxydase. *J. Biol. Chem.*, 242:5329–5335.
8. Hamberg, M. (1971): Steric analysis of hydroperoxides formed by lipoxygenase oxygenation of linoleic acid. *Anal. Biochem.*, 43:515–536.
9. Hammarström, S., Murphy, R. C., Samuelsson, B., Clark, D. A., Mioskowski, C., and Corey, E. J. (1979): Structure of leukotriene C: identification of the amino acid part. *Biochem. Biophys. Res. Commun.*, 9:1266–1272.
10. Jubiz, W., Rådmark, O., Lindgren, J. Å., Malmsten, C., and Samuelsson, B. (1981): Novel leukotrienes: products formed by initial oxygenation of arachidonic acid at C-15. *Biochem. Biophys. Res. Commun.*, 99:976–986.
11. Lundberg, U., Rådmark, O., Malmsten, C., and Samuelsson, B. (1981): Transformation of 15-hydroperoxy-5,9,11,13-eicosatetraenoic acid into novel leukotrienes. *FEBS Lett.*, 126:127–132.
12. Murphy, R. C., Hammarström, S., and Samuelsson, B. (1979): Leukotriene C: a slow reacting substance from murine mastocytoma cells. *Proc. Natl. Acad. Sci. USA*, 76:4275–4279.
13. Rådmark, O., Malmsten, C., Samuelsson, B., Clark, D. A., Goto, G., Marfat, A., and Corey, E. J. (1980): Leukotriene A: stereochemistry and enzymatic conversion to leukotriene B. *Biochem. Biophys. Res. Commun.*, 92:954–961.
14. Rådmark, O., Malmsten, C., Samuelsson, B., Goto, G., Marfat, A., and Corey, E. J. (1980): Leukotriene A: isolation from human polymorphonuclear leukocytes. *J. Biol. Chem.*, 255:11828–11831.
15. Samuelsson, B., Hammarström, S., Murphy, R. C., and Borgeat, P. (1980): Leukotrienes and slow reacting substance of anaphylaxis (SRS-A). *Allergy*, 35:375–381.

Leukotrienes and Other Lipoxygenase Products,
edited by B. Samuelsson and R. Paoletti.
Raven Press, New York © 1982.

Isolation of an Enzyme System in Rat Lung Cytosol Which Converts Arachidonic Acid into 8,11,12-Trihydroxyeicosatrienoic Acid

C. R. Pace-Asciak, *K. Mizuno, and *S. Yamamoto

*Research Institute, Hospital for Sick Children, Toronto, Canada M5G 1X8;
and *Department of Biochemistry, Tokushima University School of Medicine,
Tokushima, Japan*

In 1976 Falardeau et al. described the isolation of two novel open-chain trihydroxy derivatives of 8,11,14-eicosatrienoic acid from human platelets, i.e., 8,11,12- and 8,9,12-trihydroxyeicosatrienoic acids (2). Subsequently Jones et al. and Bryant and Bailey reported similar results using arachidonic acid as substrate (1,3). Little information is available on the characteristics of this oxidative pathway, i.e., if it is enzymatic, whether the transformation is carried out by particulate or soluble enzymes, and if this transformation is restricted to the platelet. We recently showed that a similar enzyme system is present in rat lung (5). In the current chapter we describe the detection and partial purification of an enzyme system from rat lung cytosol which catalyzes the conversion of arachidonic acid into only one of the open triols, i.e., 8,11,12-trihydroxyeicosatrienoic acid.

MATERIALS AND METHODS

Adult Wistar rats (250 to 300 g) used in this study were purchased from a local supplier in Tokushima. In order to reduce platelet contamination from the lungs, the rats were perfused *in vivo* by injecting 25 ml 50 mM KH_2PO_4-NaOH (pH 7.4) through the right ventricle. The lungs were rapidly dissected and then inflated several times by injecting ice-cold buffer through the trachea to remove surfactant and alveolar macrophages.

Enzyme Preparation

Lungs were homogenized in 2 volumes of buffer, and the homogenate was centrifuged sequentially at 600 × g for 15 min, 8,000 × g for 30 min, and finally 100,000 × g for 60 min. The high-speed supernatant fraction was fractionated by adding ammonium sulfate. All activity was retained in the 30 to 50% fraction. This fraction was taken up in a small volume of buffer and chromatographed on Sephadex G-200 (1 × 22 cm) or Sepharose 6B (1 × 12 cm) pre-equilibrated with buffer at 4°C. The active fractions eluting from these columns were combined and chromatographed on DEAE-cellulose. The column was eluted with phosphate buffer of increasing ionic strength: 20, 100, and 500 mM.

Enzyme Assay

The standard assay consisted of 60 μl enzyme preparation, 40 μl (66%) 50 mM KH_2PO_4-NaOH (pH 7.4), and 2 μl 1-^{14}C-arachidonic acid (Amersham, specific activity, 50 mCi/mmole, 2 nmoles in ethanol). The mixture was incubated 40 min at 37°C in air in a shaking water bath. Incubation was terminated with the addition of 300 μl ethyl acetate and 30 μl 0.2 M citric acid. The organic phase was dried with anhydrous sodium sulfate (40 mg), and after centrifugation the ethyl acetate phase was spotted directly on silica gel G thin-layer plates. Plates were developed with ethyl acetate/acetic acid (99:1, v/v), and radioactive products were localized by scanning the plates on a Packard radiochromatogram scanner. Quantitation of products was achieved by scraping zones of silica gel into scintillation vials and adding 1 ml methanol-water (1:1) and 10 ml Brays scintillation cocktail. Radioactivity in each vial was measured in a Beckman scintillation counter.

Identification of 8,11,12-Trihydroxyeicosatrienoic Acid

The zone on thin-layer chromatography (TLC) migrating as products I, II, and III (Fig. 1) were scraped, eluted with methanol, and derivatized into methyl esters by ethereal diazomethane and subsequently into the t-butyldimethylsilyl ethers (TBDMS) by reaction of the methyl esters with t-butyldimethylchlorosilane in imidazole as described previously (4). The samples were analyzed first by gas chromatography with on-line radioactivity detection (Packard model 891) and subsequently by gas chromatography-mass spectrometry (JEOL-D300) using a column of 1.5% SE30 on Chromosorb W-HP.

RESULTS

Arachidonic acid is converted by a rat lung high-speed supernatant fraction into several products of varying polarity; of these we identified 12-hydroxyeicosatrienoic acid (product I) and two products (II and III) having identical mass spectra (Fig. 1). Products II and III are isomers of 8,11,12-trihydroxyeicosatrienoic acid, probably isomeric at carbon 8 (5).

The enzyme system that converts arachidonic acid into the open-chain triols is present in the high-speed supernatant fraction, as shown in the TLC profile presented in Fig. 2A. Enzyme activity is sedimented by 30 to 50% saturation with ammonium sulfate (Fig. 2B). It is absent in the 0 to 30% fraction (not shown). Enzyme activity could be partially retained on gel filtration, Sephadex G-200 (Fig. 2C) and Sepharose 6B (Fig. 2D), but in either case it could not be resolved from 12-lipoxygenase.

FIG. 1 Properties of the epimeric 8,11,12-triols formed by rat lung. **A:** TLC—free acids. **B:** Gas chromatography—MeTBDMS derivative. **C:** Mass spectrometry—MeTBDMS derivative.

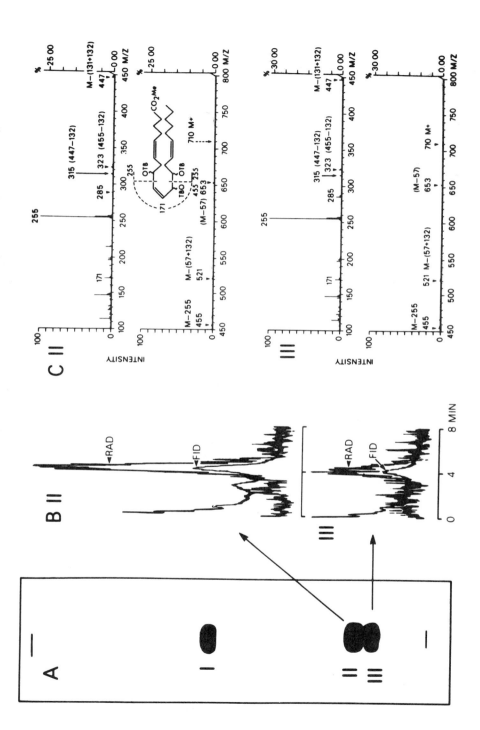

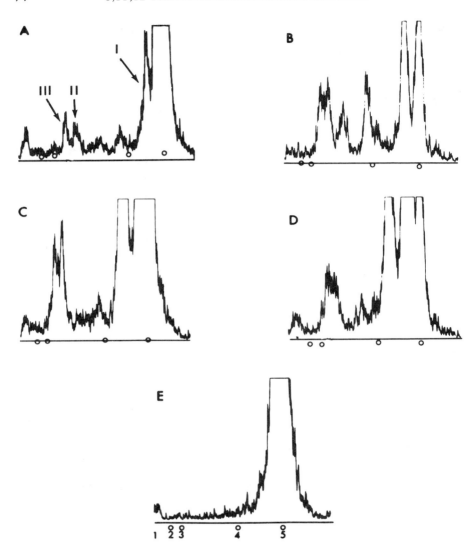

FIG. 2. Assay by radio-TLC of 8,11,12-triol formation (II,III) by (**A**) rat lung 100,000 × *g* supernatant; (**B**) 30 to 50% ammonium sulfate precipitate; (**C**) fraction 8 to 12 ml from Sephadex G-200; (**D**) fraction 8 to 11 ml from Sepharose 6B; and (**E**) fraction eluted with 20 mM buffer from DEAE-cellulose. I = 12-HETE. II and III = 8,11,12-trihydroxyeicosatrienoic acid. 1 = origin; 2 = PGF$_{2\alpha}$; 3 = PGE$_2$; 4 = PGA$_2$; 5 = AA.

Chromatography on DEAE cellulose was unsuccessful, leading to complete loss of activity (Fig. 2E). Enzyme activity was also lost from the 30 to 50% ammonium sulfate fraction during heating for 10 min at 50°C.

Using the 30 to 50% ammonium sulfate fraction as enzyme source, we found that maximal conversion of arachidonic acid into the open-chain triols and 12-HETE occurred at 20 min at 37°C (Fig. 3A). Under our assay conditions (2 nmoles substrate, 40 min, 37°C), maximal activity is reached with 60 μl (20 mg protein) enzyme fraction (Fig. 3B).

DISCUSSION

We showed that rat lung contains a novel enzyme system which transforms arachidonic acid into 8,11,12-trihydroxyeicosatrienoic acid. Although a similar enzyme activity was previously shown to occur in platelets from several species (1–3), that tissue transforms arachidonic acid into *two* positionally isomeric triols, i.e., 8,9,12- and 8,11,12-triols. Our preparation differs in that it is a soluble enzyme which converts arachidonic acid into only one product (i.e., 8,11,12-triol) present as two isomers probably epimeric at carbon 8. These results suggest that the 8,9,12- and 8,11,12-triols are formed by two separate enzyme systems with only the latter pathway present in rat lung cytosol.

SUMMARY

An enzyme system in rat lung is described which transforms arachidonic acid into 8,11,12-trihydroxyeicosatrienoic acid. The enzyme system is present in the high-speed supernatant fraction of a rat lung homogenate and can be precipitated with ammonium sulfate (30 to 50% saturation) and concentrated. Preliminary ex-

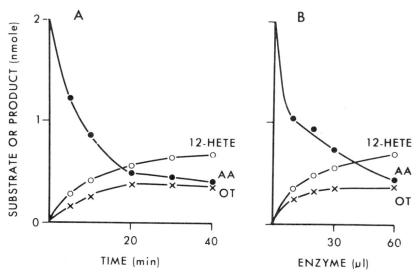

FIG. 3. Formation of 12-HETE and the mixture of epimeric 8,11,12-triols (OT) by a 30 to 50% ammonium sulfate fraction as a function of time of incubation (**A**) and enzyme concentration (**B**). Points represent the mean of duplicate analyses.

periments indicate that the enzyme activity can be eluted from Sephadex G-200 and Sepharose 6B gel chromatography, whereas all activity is lost on DEAE cellulose chromatography. Enzyme activity is heat-labile, being lost after 10 min at 50°C.

ACKNOWLEDGMENT

This study was supported by grants to C. R. Pace-Asciak from the Medical Research Council of Canada (MT-4181) and the Josiah Macy Jr. Foundation and to S. Yamamoto from the Ministry of Education, Science and Culture of Japan. We wish to acknowledge the assistance of Mrs. Yoshido, Tokushima University School of Pharmacology, in recording the mass spectra. This study was carried out while C. R. Pace-Asciak was on sabbatical leave at Tokushima University.

REFERENCES

1. Bryant, R. W., and Bailey, J. M. (1979): Isolation of a new lipoxygenase metabolite of arachidonic acid, 8,11,12-trihydroxy-5,9,14-eicosatrienoic acid from human platelets. *Prostaglandins*, 17:9–18.
2. Falardeau, P., Hamberg, M., and Samuelsson, B. (1976): Metabolism of 8,11,14-eicosatrienoic acid in human platelets. *Biochim. Biophys. Acta*, 411:193–200.
3. Jones, R. L., Kerry, P. J., Poyser, N. L., Walker, I. C., and Wilson, N. H. (1978): The identification of trihydroxyeicosatrienoic acids as products from the incubation of arachidonic acid with washed blood platelets. *Prostaglandins*, 16:583–589.
4. Pace-Asciak, C. R., and Edwards, N. S. (1980): Tetranor thromboxane B₂ is the major urinary metabolite formed after IV infusion of thromboxane B₂ in the rat. *Biochem. Biophys. Res. Commun.*, 17:81–86.
5. Pace-Asciak, C. R., Mizuno, K., and Yamamoto, S. (1981): Formation of 8,11,12-trihydroxyeicosatrienoic acid by a rat lung high speed supernatant fraction. *Biochim. Biophys. Acta*, 665:352–354.

Leukotrienes and Other Lipoxygenase Products,
edited by B. Samuelsson and R. Paoletti.
Raven Press, New York © 1982.

Purification and Properties of Arachidonate-15-Lipoxygenase from Rabbit Peritoneal Polymorphonuclear Leukocytes

Shuh Narumiya,[1] John A. Salmon, Roderick J. Flower, and John R. Vane

Department of Prostaglandin Research, The Wellcome Research Laboratories, Langley Court, Beckenham, Kent BR3 3BS, England

There are several specific lipoxygenases present in mammalian tissues which can convert arachidonic acid to a variety of hydroperoxy derivatives (hydroperoxy eicosatetraenoic acids; HPETEs), and these may be further metabolized to the corresponding hydroxy acids (HETEs) or in some cases to the leukotrienes. One of these products, 15-HPETE, is a potential bioregulator of arachidonic acid metabolism since it inhibits prostacyclin synthetase (6,9), diglyceride lipase (8), and arachidonate-12-lipoxygenase (10). Furthermore, it has been suggested that 15-HPETE is the primary intermediate in the biosynthesis of a novel group of leukotrienes (4,5). However, 15-HPETE itself has not been isolated from mammalian tissues, although its generation has been implicated by the detection of 15-HETE. We now wish to report the purification of the enzyme which generates 15-HPETE, arachidonate-15-lipoxygenase, from rabbit peritoneal leukocytes.

METHODS

Rabbit peritoneal polymorphonuclear leukocytes were collected according to the method of Hirsch (3). After washing once with saline, the cells were suspended in 50 mM potassium phosphate (pH 7.0) containing 1 mM EDTA and disrupted at 4°C by sonication (12 times for 30 sec). The homogenate was centrifuged at $10,000 \times g$ for 20 min to remove nuclei, cell debris, and granules, and the supernatant was then centrifuged at $105,000 \times g$ for 60 min. The arachidonate-15-lipoxygenase activity was found in the $105,000 \times g$ supernatant (the cytoplasm fraction) and purified as described in detail elsewhere (7). Briefly, the enzyme was precipitated from the cytosol with acetone (40%, final concentration) at $-10°C$. The precipitated enzyme was applied to a CM52 column equilibrated with 10 mM potassium phosphate, pH 6.0, and eluted with a linear gradient from 10 to 200 mM potassium phosphate, pH 6.0. The enzyme was further purified by gel filtration on a column

---

[1]Present address: Department of Medical Chemistry, Kyoto University, Faculty of Medicine, Kyoto 606, Japan

of Sephadex G-150 with 50 mM potassium phosphate, pH 7.0. The active fraction from the Sephadex column was then applied on a hydroxyapatite column and eluted with a linear gradient from 50 to 500 mM potassium phosphate, pH 7.0.

The enzyme assay was carried out by measuring the conversion of 1-[^{14}C]arachidonic acid (30 nmole, 15,000 cpm/nmole) to 15-HPETE and its decomposition products in 50 mM potassium phosphate, pH 7, containing 5 mM CaCl$_2$ (total volume 0.1 ml). The incubation was carried out at 30°C for 2 min and was terminated by adding 1 ml methanol. The mixture was acidified with 2 M citric acid to pH 3 and extracted with diethyl ether. The extracted products were separated by silica gel thin-layer chromatography (TLC) using diethyl ether/*n*-hexane/acetic acid (60:40:1) as developing solvent. After detection of the radioactivity by autoradiography, the silica gel zones corresponding to each radioactive spot were scraped off and the radioactivity determined by conventional liquid scintillation counting.

RESULTS AND DISCUSSION

Incubations of arachidonic acid with 105,000 × *g* supernatant or purified enzyme produced multiple products which were separated by TLC (Fig. 1). Large-scale incubations were performed to identify these products; the compounds designated I through IV and VI in Fig. 1 were isolated by normal-phase high-pressure liquid chromatography (HPLC) (Fig. 2). Compound V was purified by reverse-phase (RP) HPLC. Products VI and II had retention volumes identical to those of 15-HPETE and 15-HETE, respectively. Ultraviolet (UV) spectrophotometry and gas-liquid chromatography-mass spectrometry (GLC-MS) data for several derivatives of compounds VI and II confirmed that these were 15-HPETE and 15-HETE, respectively. The other unknown compounds were similarly subjected to GLC-MS and other analyses. The details of these investigations are presented elsewhere (7); briefly, the compounds were identified as follows: compound I was 15-keto-5,9,11,13-eicosatetraenoic acid; compounds III and IV were diastereomers of 13-hydroxy-14,15-epoxy-5,8,11-eicosatrienoic acid; and compound V was 11,14,15-trihydroxy-5,8,12-eicosatrienoic acid. Thus incubation of arachidonic acid with the purified enzyme produced 15-HPETE and compounds I through V, each of which had an oxygen atom at carbon 15. In order to establish if all these products were derived from a common enzymic reaction, 1-[^{14}C]15-HPETE (prepared by incubating 1-[^{14}C]arachidonic acid with soybean lipoxygenase) was incubated with (a) purified enzyme, (b) enzyme heated at 90°C for 5 min, and (c) buffer alone. The conversion of the hydroperoxy acid to products I through V was observed in all three incubations, suggesting that 15-HPETE decomposed nonenzymically to compounds I through V. Similar decomposition of other hydroperoxy fatty acids has been reported (1,2). A scheme depicting the conversion of arachidonic acid by the purified enzyme to 15-HPETE and its secondary decomposition is shown in Fig. 3.

Using the successive chromatographic procedures described in *Methods*, the enzyme was purified about 250-fold with a recovery of 7% (Table 1). The purified

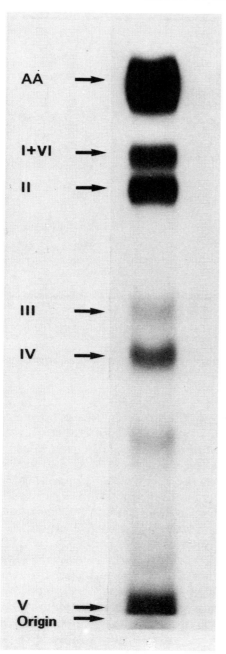

FIG. 1. TLC separation of the products obtained after incubation of 1-[¹⁴C]arachidonic acid with purified arachidonate-15-lipoxygenase. The radioactive zones were detected by autoradiography. The structures of the six major products (I through VI) were elucidated during the course of the study and are shown in Fig. 3.

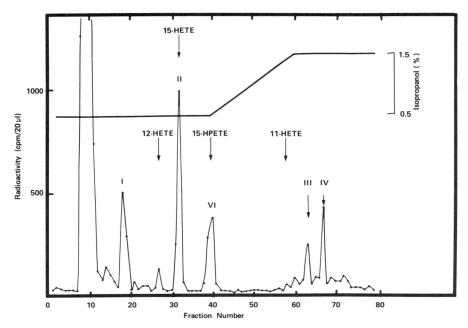

FIG. 2. HPLC of the products (I through IV and VI) obtained after incubation of 1-[^{14}C]arachidonic acid with purified enzyme. The separation was performed on a Zorbax Sil column (Dupont) using a gradient from *n*-hexane/isopropanol/acetic acid (994:5:1, v/v/v) to *n*-hexane/isopropanol/acetic acid (984:15:1, v/v/v) as eluting solvents (3 ml/min).

enzyme lost 30% of the original activity when stored at pH 7 and 4°C for 24 hr, and lost 60% of its activity when kept at -40°C for 1 month. The time course of the reaction was linear for less than 2 min at 30°C. The K_m value of the purified enzyme for arachidonic acid was estimated to be 28 μM, and the reaction had a pH optimum of approximately pH 6.5. The molecular weight of the enzyme was determined to be 61,000 by gel filtration on a Sephadex G-150 column calibrated with authentic markers. Although the 105,000 × g supernatant from rabbit peritoneal leukocytes exhibited a dependence on divalent cations for maximum conversion of arachidonic acid (7), this was not shared by the purified enzyme. The enzyme activity was inhibited by the sulfydryl blocking agent PCMB (IC$_{50}$ 0.5 mM). The enzyme was also inhibited by 5,8,11,14-eicosatetraenoic acid (ETYA; IC$_{50}$ = 0.5 μg/ml) and 3-amino-1-[*m*-(trifluoromethyl)-phenyl]-2-pyrazoline (BW755C; IC$_{50}$ = 20 μg/ml) but not by indomethacin (up to 200 μg/ml) or *o*-phenanthroline (up to 5 mM). Hematin, L-tryptophan (cofactors for the cyclooxygenase reaction), reduced glutathione (GSH), and cysteine did not effect the enzyme reaction.

Recently Samuelsson and his colleagues reported the formation of novel leukotrienes, 8,15-LTB$_4$ and 14,15-LTB$_4$, from 15-HPETE in human leukocytes (4,5). Therefore the enzyme studied during the present investigation may control the initial

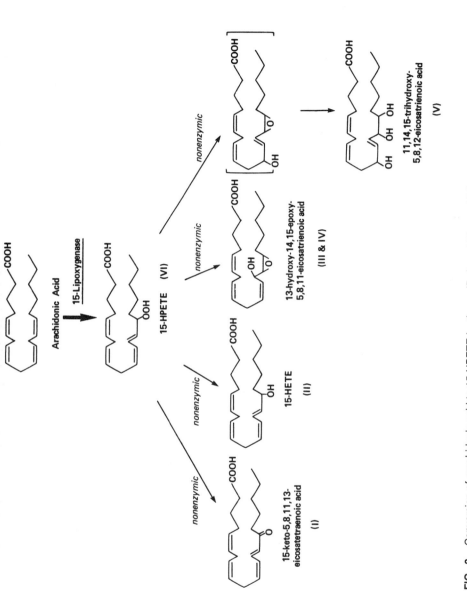

FIG. 3. Conversion of arachidonic acid to 15-HPETE by the purified arachidonate-15-lipoxygenase and the subsequent decomposition of this hydroperoxy acid.

TABLE 1. *Purification of arachidonic acid-15-lipoxygenase from rabbit leukoctyes*

Step	Protein (mg)	Total activity (munits)	Specific activity (munits/mg)	Yield (%)	Purification (-fold)
1. Crude enzyme	1,104	828	0.75	100	1
2. Acetone	504	786	1.56	95	2
3. CM-52	96	254	2.64	31	3.5
4. Sephadex G-150	7	140	20.0	17	27
5. Hydroxyapatite	0.3	56	187	7	250

step in the formation of these leukotrienes from arachidonic acid. Neither 8,15-LTB_4 nor 14,15-LTB_4 were detected after incubating arachidonic acid with purified arachidonate-15-lipoxygenase, indicating that the epoxidation to 14,15-LTA_4 and the subsequent hydrolysis to the LTB_4 analogs is controlled by a different enzyme(s).

ACKNOWLEDGMENT

The authors wish to thank Drs. F. H. Cottee and B. C. Weatherley for the GLC-MS analyses. This work was performed during S. N.'s postdoctoral study at Wellcome Research Laboratories.

REFERENCES

1. Gardner, H. W., Kleinman, R., and Weisleder, D. (1974): Homolytic decomposition of linoleic acid hydroperoxide: Identification of fatty acid products. *Lipids*, 9:696–706.
2. Hamberg, M. (1975): Decomposition of unsaturated acid hydroperoxides by hemoglobin: structures of major products of 13L-hydroperoxy-9,11 octadecadienoic acid. *Lipids*, 10:87–92.
3. Hirsch, J. G. (1956): Phagocytin: A bacterial substance from polymorphonuclear leucocytes. *J. Exp. Med.*, 103:589–611.
4. Jubiz, W., Rådmark, O., Lindgren, J. A., Malmsten, O., and Samuelsson, B. (1981): Novel leukotrienes: products formed by initial oxygenation of arachidonic acid at C-15. *Biochem. Biophys. Res. Commun.*, 99:976–986.
5. Lundberg, U., Rådmark, O., Malmsten, C., and Samuelsson, B. (1981): Transformation of 15-hydroperoxy-5,9,11,13-eicosatetraenoic acid into novel leukotrienes. *FEBS Lett.*, 126:127–132.
6. Moncada, S., Gryglewski, R. J., Bunting, S., and Vane, J. R. (1976): A lipid peroxide inhibits the enzyme in blood vessel microsomes that generates from prostaglandin endoperoxides substance (Prostaglandin X) which prevents platelet aggregation. *Prostaglandins*, 12:715–737.
7. Narumiya, S., Salmon, J. A., Cottee, F. H., Weatherley, B. C., and Flower, R. J. (1981): Arachidonic acid 15-lipoxygenase from rabbit peritoneal polymorphonuclear leukocytes: Partial purification and properties. *J. Biol. Chem.* , 256:9583-9592.
8. Rittenhouse-Simmons, S. (1980): Indomethacin-induced accumulation of diglyceride in activated human platelets: the role of diglyceride lipase. *J. Biol. Chem.*, 255:2259–2262.
9. Salmon, J. A., Smith, D. R., Flower, R. J., Moncada, S., and Vane, J. R. (1978): Further studies on the enzymatic conversion of prostaglandin endoperoxides into prostacyclin by porcine aorta microsomes. *Biochim. Biophys. Acta*, 523:250–262.
10. Vanderhoek, J. Y., Bryant, R. W., and Bailey, J. M. (1980): Inhibition of leukotriene biosynthesis by the leukocyte product 15-hydroxy-5,8,11,13-eicosatetraenoic acid. *J. Biol. Chem.*, 255:5996–5998.

Leukotrienes and Other Lipoxygenase Products,
edited by B. Samuelsson and R. Paoletti.
Raven Press, New York © 1982.

Metabolism of Leukotriene C_3

Sven Hammarström

Department of Physiological Chemistry, Karolinska Institute,
S-104 01 Stockholm, Sweden

Structural work on slow-reacting substance of anaphylaxis (SRS-A) showed that this presumed mediator of allergic and anaphylactic reactions (3) is a novel derivative of arachidonic acid, 5(S)-hydroxy-6(R)-S-glutathionyl-7,9-*trans*-11,14-*cis*-eicosatetraenoic acid (6,15,16,19). This compound has been designated leukotriene C_4 (22). An isomer, 11-*trans* leukotriene C_4, was also identified as a constituent of slow-reacting substance (SRS) from mastocytoma cells (5). The transformation of other polyunsaturated fatty acids to leukotrienes (Fig. 1) was recently described (10–12). Eicosatrienoic acid (n-9) is converted to leukotriene C_3, which differs from leukotriene C_4 by the absence of a double bond at Δ^{14} (11). Leukotriene C_5, having a double bond in addition to those at $\Delta^{7,9,11,\ \&14}$ is formed from eicosapentaenoic acid (n-3) (10). Eicosatrienoic acid (n-6), (derived from linoleic acid), was converted to a positional isomer of leukotriene C_3 with the oxygen and sulfur substituents at C-8 and C-9, respectively (12). These leukotrienes have biological activities comparable to that of leukotriene C_4 on guinea pig ileum (10,11) and lung strips (8).

CATABOLISM OF LEUKOTRIENE C_4 BY KIDNEY γ-GLUTAMYL TRANSPEPTIDASE AND DIPEPTIDASE

The structural work on SRS was performed on material isolated from murine mastocytoma cells (19). Additional investigations on SRS from basophilic leukemia

TABLE 1. *Amino acid composition of leukotriene C_4 and its products after treatment with GGTP or GGTP plus CG:ase*

		Product after	
Amino acid[a]	Leukotriene C_4	GGTP[b]	GGTP + CG:ase[b]
Glutamic acid	0.93	< 0.01	< 0.01
Glycine	1.04	0.92	0.06
½ Cystine	0.35	0.41	0.36

[a]Moles/mole leukotriene; other residues < 0.01.
[b]GGTP = γ-glutamyl transpeptidase; dipeptidase.

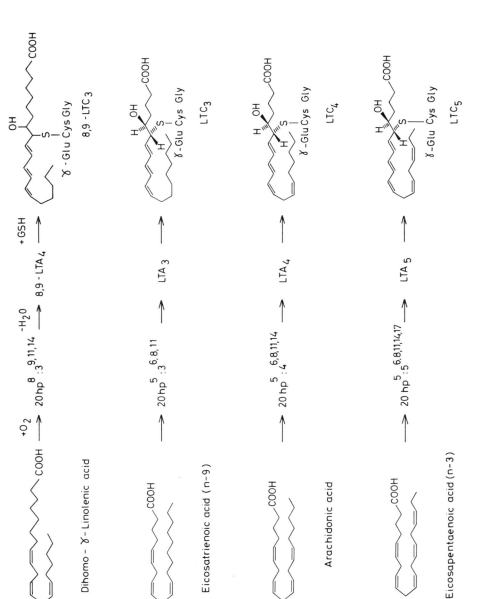

FIG. 1. Transformation of polyunsaturated fatty acids to leukotrienes.

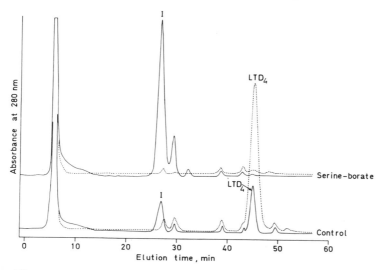

FIG. 2. Effect of L-serine borate on leukotriene biosynthesis by basophilic leukemia (RBL) cells. (From ref. 20.)

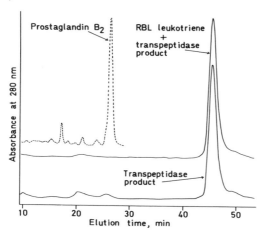

FIG. 3. RP-HPLC of product obtained from leukotriene C_4 by treatment with γ-glutamyl trans-peptidase and the same product plus leukotriene from RBL cells. (From ref. 21.)

(RBL) cells revealed that this compound was less polar than leukotriene C_4. The analyses indicated that the fatty acid part of the molecule was identical to that of leukotriene C_4 and that the peptide substituent differed by the lack of glutamic acid (21) (Table 1). Addition of an inhibitor of γ-glutamyl transpeptidase to RBL cells prevented biosynthesis of the new product [leukotriene D_4 (22)] and led to increased formation of leukotriene C_4 (20) (Fig. 2). This indicated that leukotriene D_4 was formed from leukotriene C_4 by γ-glutamyl transpeptidase-catalyzed degradation.

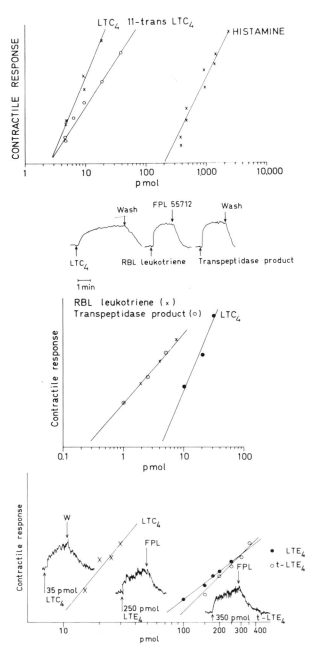

FIG. 4. Bioassay on isolated guinea pig ileum of leukotrienes C_4, 11-*trans* C_4, and histamine **(top)**; leukotriene C_4 (●), transpeptidase product from leukotriene C_4 (○), leukotriene from RBL cells (x) (LTD₄) **(middle)**; and leukotrienes C_4 (X), E_4 (●), and 11-*trans* E_4 (○) **(bottom)**.

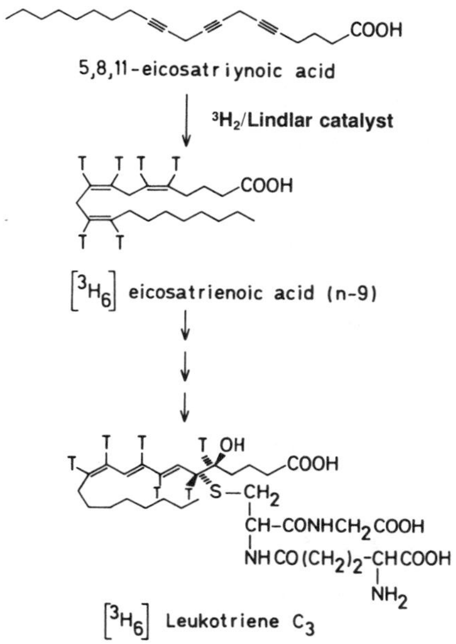

5,8,11-eicosatriynoic acid

$^3H_2$/Lindlar catalyst

$[^3H_6]$ eicosatrienoic acid (n-9)

$[^3H_6]$ Leukotriene C_3

FIG. 5. Synthesis of tritium-labeled leukotriene C_3.

TABLE 2. *Autoradiographic distribution of radioactivity from $[^3H_6]$ leukotriene C_3 in mouse tissues*

Organ or secretion	Relative degree of blackening[a]					
	5 min	20 min	1 hr	4 hr	24 hr	4 days
Blood	> 64	64	> 32	16	4	0
Lung	> 64	64	> 128	16	4	—
Liver	> 256	> 256	256	128	16	4
Bile	> 512	> 512	> 512	> 512	—	—
Kidney	> 128	> 128	> 64	32	> 8	—
Pleural fluid	128	> 64	> 32	> 16	—	—
Spleen white pulp	32	16	16	8	4	—
"Spots" in blood sinus	> 64	> 32	32	16	16	—
Salivary glands	128	64	> 16	—	—	—
Myocardium	32	32	16	—	—	—

From ref. 1.

[a] The relative degree of blackening in different organs was compared with the blackening of an ^3H isotope "staircase" exposed under similar conditions (strongest level = 2^9 (512), next level = 2^8 (256), 2^7 (128) etc.).

Leukotriene C_4 was also converted by partially purified γ-glutamyl transpeptidase to leukotriene D_4 as judged by chromatographic (Fig. 3) and bioassay comparisons (Fig. 4, middle panel). A dipeptidase, partially purified from porcine kidney, metabolized leukotriene D_4 further by removal of glycine to 5-hydroxy-6-S-cysteinyl-7,9-*trans*-11,14-*cis*-eicosatetraenoic acid, leukotriene E_4 (4). The structural deter-

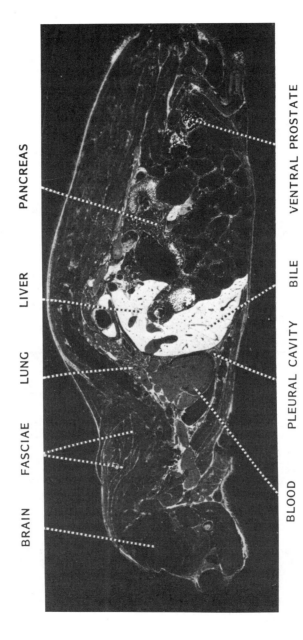

BRAIN   FASCIAE   LUNG   LIVER   PANCREAS

BLOOD   PLEURAL CAVITY   BILE   VENTRAL PROSTATE

FIG. 6. Whole-body autoradiogram of a male mouse 5 min after intravenous injection of 66 μCi [$^3H_6$]leukotriene C_3. (From ref. 1.)

mination was based on amino acid analyses (Table 1), gas-liquid chromatography-mass spectrometry of the product obtained after Raney nickel desulfurization, conversion by soybean lipoxygenase, and the detection of a free amino group. In addition to transforming leukotriene C_4 to leukotriene D_4, γ-glutamyl transpeptidase also converted leukotrienes C_3, 11-*trans* C_4, and C_5 to D-type leukotrienes. These products were further transformed to corresponding leukotrienes E by treatment with dipeptidase (4).

Leukotrienes C_4, D_4, and E_4 induce contractions of slow onset and long duration in isolated guinea pig ileum (Fig. 4). The contractions were reversed by the SRS antagonist FPL 55712 (2). The potency of leukotriene D_4 was 3 to 10 times greater than that of leukotriene C_4, whereas leukotriene E_4 had only 1 to 3% of the activity of leukotriene D_4. The 11-*trans* isomers of leukotrienes C_4 and E_4 were somewhat less potent than leukotrienes C_4 and E_4, respectively.

SYNTHESIS OF ³H-LABELED LEUKOTRIENE C₃

[5,6,8,9,11,12-$^3H_6$]Leukotriene C_3 was synthesized as outlined in Fig. 5. The label was inexpensively introduced from $^3H_2$ gas by reduction of 5,8,11-eicosatrienoic acid to the corresponding *cis* ethylenic acid in the presence of Lindlar catalyst (18). $^3H_6$-Eicosatrienoic acid (n-9) of close to the theoretical specific activity was obtained in good yield (10% of added $^3H_2$ gas). The labeled acid was biosynthetically converted to leuketriene C_3 by mastocytoma cells stimulated with calcium ionophore A23187. The yield was 2%, and the specific activity of the product was 50 Ci/mmole as judged by UV spectroscopy and liquid scintillation spectrometry. Co-injections (into a high-performance liquid chromatograph) with previously characterized unlabeled leukotriene C_3 or after γ-glutamyl transpeptidase treatment with leukotriene D_3 demonstrated identity. Furthermore, a labeled product identical with 5-hydroxyarachidic acid was obtained by treatment with Raney nickel as judged by gas-liquid radiochromatography of the methyl ester, trimethylsilylether derivative (C-value 21.7 on 3% OV-17).

DISTRIBUTION AND EXCRETION OF ³H-LABELED LEUKOTRIENE C₃

Distribution studies by whole-body autoradiography were performed in collaboration with Appelgren (1). Following intravenous injection of [$^3H_6$]leukotriene C_3 into mice, the radioactivity was rapidly eliminated from the blood circulation. Figures 6 and 7 show autoradiograms obtained from animals sacrificed by rapid freezing 5 and 20 min after the injections. Most of the radioactivity was taken up by the liver and excreted into the bile. This was evident from a remarkably high concentration of biliary radioactivity 5 min after administration (Fig. 6, Table 2). A high uptake of radioactivity also occurred in the cortical parts of the kidney (Figs. 7A and B, Table 2). The data on excretion routes for leukotriene C_3 (metabolites) in the mouse were confirmed by radioactivity measurements on feces and urine

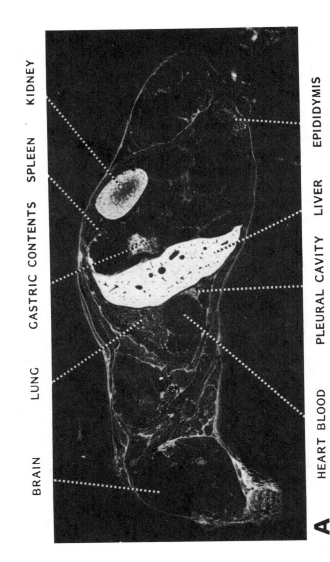

BRAIN   LUNG   GASTRIC CONTENTS   SPLEEN   KIDNEY

A   HEART BLOOD   PLEURAL CAVITY   LIVER   EPIDIDYMIS

FIG. 7A: Whole-body autoradiogram of a male mouse 20 min after intraveneous injection of ^{3}H-leukotriene C_3.

THYMUS   HEART BLOOD   MYOCARDIUM   PLEURAL CAVITY

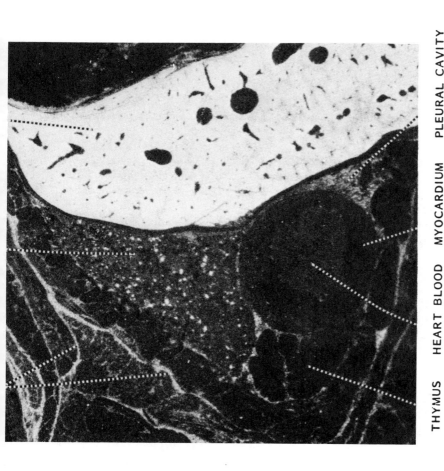

FIG. 7B: Detail of **A** showing the localization of radioactivity in the lung and pleural cavity. (From ref. 1.)

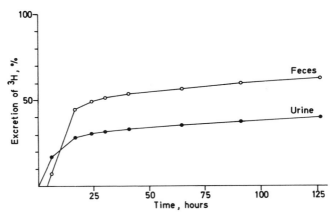

FIG. 8. Excretion of tritium in feces and urine after subcutaneous administration of 20 μCi [³H₆]leukotriene C_3 to a male guinea pig. (From ref. 13.)

after subcutaneous administration of [³H₆]leukotriene C_3 to guinea pigs (Fig. 8). Quantitative excretion of the injected amount of ³H had been achieved 5 days after administration, with approximately 60% of the radioactivity appearing in feces and 40% in urine.

The concentration of radioactivity in the lungs was comparable to that in blood of mice killed 5 and 20 min after injection of the ³H-labeled leukotriene (Figs. 6 and 7). However, the radioactivity in the lung appeared in scattered strong spots (Fig. 7). This pattern was less pronounced after intramuscular administration.

Radioactivity was also taken up by connective tissue in various organs, e.g., muscle, fascia, salivary glands, pancreas, accessory sex glands (Fig. 7). A similar uptake by connective tissue has been observed for prostaglandin E_1, an oxygenated derivative of eicosatrienoic acid (n-6) (17).

IN VITRO METABOLISM OF LEUKOTRIENE C_3

[³H₆]Leukotriene C_3 was incubated with the 950 × g supernatant fraction from guinea pig lung, liver, and kidney homogenates (13). Efficient conversion to a less-polar metabolite (CI) with a relative elution time of 1.44 compared to that of leukotriene C_3 was observed after incubations with lung (Fig. 9). CI was identified as leukotriene D_3 by cochromatography on reverse-phase high-performance liquid chromatography (RP-HPLC) and by its conversion to leukotriene E_3 after incubation with the renal dipeptidase. The failure of liver and kidney to metabolize leukotriene C_3 was probably due to competition by endogenous glutathione for the γ-glutamyl

FIG. 9. HPLC recordings of products from ³H-leukotriene C_3 after incubation with 950 × g supernatant fractions from guinea pig lung, liver, and kidney homogenates. (From ref. 13.)

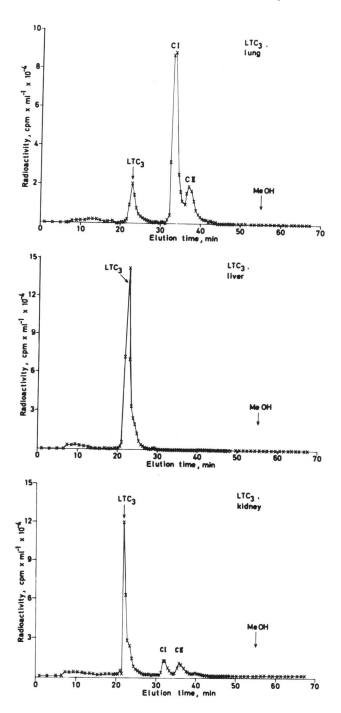

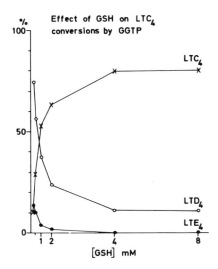

FIG. 10. Effect of glutathione on the conversion of leukotriene C_4 to leukotriene D_4 by γ-glutamyl transpeptidase. (From ref. 13.)

transpeptidase in these tissues. Figure 10 shows the effect of varying concentrations of glutathione on the conversion of leukotriene C_4 to leukotriene D_4 by γ-glutamyl transpeptidase. Tissue levels of 6 to 8 mM and 1 to 4 mM glutathione have been reported for liver and kidney, respectively, in the rat, whereas lung levels are considerably lower (9).

IN VITRO METABOLISM OF LEUKOTRIENE D₃

[$^3H_6$]Leukotriene D_3 was prepared by treating ^3H-labeled leukotriene C_3 with γ-glutamyl transpeptidase. The product was isolated in 80% yield by RP-HPLC. Incubations of 950 × *g* supernatants of lung, liver, and kidney homogenates (aliquots of the same preparations that were used for the experiments with ^3H-leukotriene C_3) gave the results shown in Fig. 11. Lung yielded a relatively small conversion to a product (DII) with the chromatographic properties of leukotriene E_3. Greater conversion to DII was observed with liver and in particular with kidney homogenates. In addition, conversion to a product (DI) eluting prior to leukotriene D_3 was observed as well as conversion to more polar metabolites (a substantial amount of polar metabolites eluting close to the solvent front did not adsorb to XAD-8 and are therefore not seen in Fig. 11). DI was identified as leukotriene C_3 by cochromatography with unlabeled reference compound and by conversion of DI to leukotriene D_3 using γ-glutamyl transpeptidase. The transformation of leukotriene D_3 to leukotriene C_3 by liver especially but also by lung homogenates was apparently catalyzed by γ-glutamyl transpeptidase. This enzyme reversibly transfers a γ-glu-

FIG. 11. HPLC recordings of products from ^3H-leukotriene D_3 after incubation with 950 × *g* supernatant fractions from guinea pig lung, liver, and kidney homogenates. (From ref. 13.)

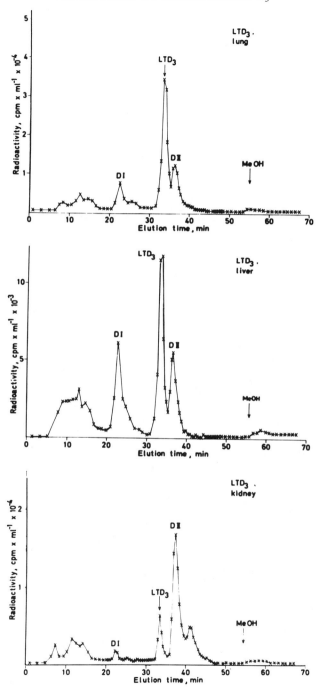

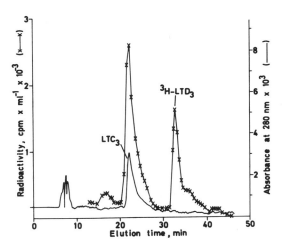

FIG. 12. Conversion of [³H₆]leukotriene D₃ by γ-glutamyl transpeptidase in the presence of glutathione (8 mM). Unlabeled leukotriene C₃ was added as internal reference after the incubation. (From ref. 13.)

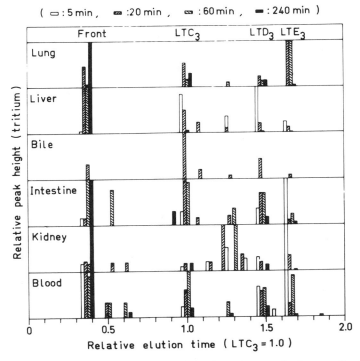

FIG. 13. HPLC analyses of radioactive products from tissues of mice sacrificed by rapid freezing 5, 20, 60, and 240 min after intravenous injections of 60 μCi [³H₆]leukotriene C₃ per animal. (From ref. 1.)

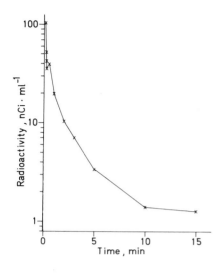

FIG. 14. Concentrations of tritium in blood from the aorta of a male monkey following injection of 5 μCi [$^3H_6$]leukotriene C_3 into the right atrium. (From ref. 14.)

tamyl group from glutathione to various amino acid or dipeptide acceptors (23). Further support was obtained by incubating [$^3H_6$]leukotriene C_3 with partially purified γ-glutamyl transpeptidase in the presence of 8 mM glutathione. About 60% conversion to leukotriene C_3 was observed (Fig. 12). The experiments with leukotrienes C_3 and D_3 demonstrated a substantial difference in the metabolism of these compounds in liver and kidney. The inertness to metabolism of leukotriene C_3 compared to that of leukotriene D_3 suggests that the γ-glutamyl group of leukotriene C_3 protects it from metabolism, and that the removal of this group by γ-glutamyl transpeptidase will permit further degradations to occur.

IN VIVO METABOLISM OF LEUKOTRIENE C_3

In connection with the distribution experiments in mice, described above, the chromatographic properties of the radioactive metabolites present in various parts of the animals were determined as functions of time after administration (1) (Fig. 13). At 5, 20, and 60 min after injection, appreciable amounts of the radioactivity present in lung, liver, bile, intestine, and heart blood had the same elution times as leukotrienes C_3, D_3, and/or E_3. Additional metabolites eluting between leukotrienes C_3 and D_3 were observed in the kidney. After 4 hr more-polar metabolites, appearing close to the solvent front in this system, were seen.

The presence of unaltered leukotriene C_3 in liver 5 min after injection and in liver, bile, and intestine 20 min after injection was established by cochromatography with unlabeled leukotriene C_3 and after conversion with γ-glutamyl transpeptidase with unlabeled leukotriene D_3. In addition, radioactive material from lung after 20 min and from blood and intestine after 60 min cochromatographed with leukotriene C_3. These results indicate that leukotriene C_3 can be taken up by the liver and excreted into the bile unchanged. The lack of metabolism of leukotriene C_3 by the

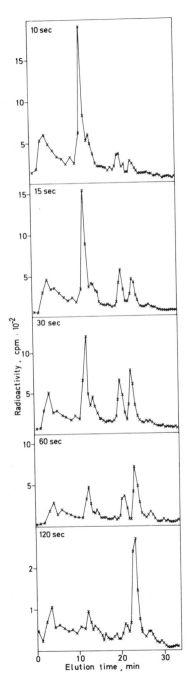

FIG. 15. HPLC analyses of radioactive products in blood from the aorta of a male monkey receiving 5 μCi [³H₆]leukotriene C_3 into the right atrium at various times after the injection. (From ref. 14.)

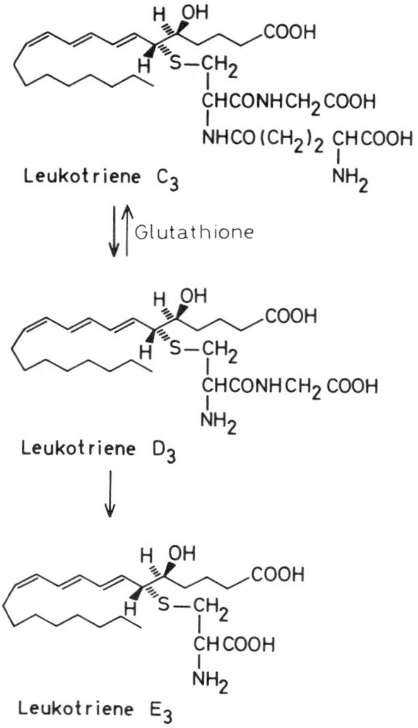

FIG. 16. Metabolic transformations of leukotriene C_3.

liver agrees with the *in vitro* results described above. Presumably, metabolism will occur in the gut by intestinal γ-glutamyl transpeptidase (7) as well as by dipeptidases present in the intestinal contents.

In another set of experiments, the metabolism of [$^3H_6$]leukotriene C_3 was investigated in the primate *Macaca irus* (14). The animal was anesthetized, and catheters were introduced under X-ray monitoring into the right and left atria, the pulmonary artery, and the aorta. [$^3H_6$]Leukotriene C_3 (5 μCi) was injected into the right atrium, and blood samples were withdrawn from the aorta. Figure 14 shows the decrease of radioactivity in aortic blood as a function of time after injection. The circulation time was 20 sec and an increase in aortic blood radioactivity was observed between 20 and 30 sec. The rapid decline of radioactivity (from 100 to 1 μCi/ml in 15 min) suggested that leukotriene C_3 was efficiently eliminated from the circulation in the monkey. Chromatographic analyses of blood samples indicated that leukotriene C_3 was metabolized to two less-polar products (Fig. 15). About 40% conversion was observed 15 sec after injection. At this time the tritium labeled material had not recirculated. Thus it is likely that the conversions occurred in the lung since leukotriene C_3 was injected into the *right* atrium and blood was collected from the aorta. This is also in agreement with the *in vitro* experiments since the elution behaviors of the metabolites corresponded to those of leukotrienes D_3 and E_3, respectively. Further conversion to the latter metabolite occurred, and 2 min after injection this product was the predominating radioactive component in blood. The results of these investigations demonstrate that leukotriene C_3 is rapidly degraded by successive elimination of glutamic acid and glycine (Fig. 16). The former reaction occurs readily in lungs of guinea pigs and monkeys and is essential for the further catabolism of leukotriene C_3. It also leads to increased biological activity. The enzymatic removal of glycine, which occurred predominantly in kidney, leads to a considerable reduction of biological activity in the guinea pig ileum assay. Several unidentified metabolites, including some very polar ones, were formed under *in vitro* and *in vivo* conditions. The nature of these products is currently under investigation.

ACKNOWLEDGMENTS

Supported by grants from the Swedish Medical Research Council (03X-5914), the Swedish Cancer Society (1503–02X), and Magnus Bergwalls Stiftelse.

REFERENCES

1. Appelgren, L. E., and Hammarström, S. (1981): Distribution and metabolism of ³H-labeled leukotriene C_3 in mice. *J. Biol. Chem. (in press)*.
2. Augstein, J., Farmer, J. B., Lee, T. B., Sheard, P., and Tattersall, M. L. (1973): Selective inhibitor of slow reacting substance of anaphylaxis. *Nature New Biol.*, 245:215–217.
3. Austen, K. F. (1977): Biological implications of the structural and functional characteristics of the chemical mediators of immediate-type hypersensitivity. *Harvey Lect.*, 73:93–161.
4. Bernström, K., and Hammarström, S. (1981): Metabolism of leukotriene D by porcine kidney. *J. Biol. Chem.*, 256:9579–9582.
5. Clark, D. A., Goto, G., Marfat, A., Corey, E. J., Hammarström, S., and Samuelsson, B. (1980): 11-*trans* leukotriene C: A naturally occurring slow reacting substance. *Biochem. Biophys. Res. Commun.*, 94:1133–1139.
6. Corey, E. J., Clark, D. A., Goto, G., Marfat, A., Mioskowski, C., Samuelsson, B., and Hammarström, S. (1980): Stereospecific total synthesis of a slow reacting substance of anaphylaxis, leukotriene C-1. *J. Amer. Chem. Soc.*, 102:1436–1439.
7. Cornell, J. S., and Meister, A. (1976): Glutathione and γ-glutamyl cycle enzymes in crypt and villus tip cells of rat jejunal mucosa. *Proc. Natl. Acad. Sci. USA*, 73:420–422.
8. Dahlén, S. E., Hedqvist, P., and Hammarström, S. (1981): Contractile activity of various cysteine-containing leukotrienes in guinea pig lung. Submitted for publication.
9. Davidson, D. E., and Hird, F. J. R. (1964): The estimation of glutathione in rat tissues. A comparison of a new spectrophotometric method with the glyoxalase method. *Biochem. J.* (London), 93:232–236.
10. Hammarström, S. (1980): Leukotriene C_5: a slow reacting substance derived from eicosapentaenoic acid. *J. Biol. Chem.*, 255:7093–7094.
11. Hammarström, S. (1981): Conversion of 5,8,11-eicosatrienoic acid to leukotrienes C_3 and D_3. *J. Biol. Chem.*, 256:2275–2279.
12. Hammarström, S. (1981): Conversion of dihomo-γ-linolenic acid to an isomer of leukotriene C_3, oxygenated at C-8. *J. Biol. Chem.*, 256:7712–7714.
13. Hammarström, S. (1981): Metabolism of leukotriene C_3 in the guinea pig. Identification of metabolites formed by lung, liver and kidney. *J. Biol. Chem.*, 256:9573–9578.
14. Hammarström, S., Bernström, K., Örning, L., Dahlén, S. E., Hedqvist, P., Smedegård, G., and Revenis, M. (1981): Rapid *in vivo* metabolism of leukotriene C_3 in the monkey, *Macaca irus*. *Biochem. Biophys. Res. Commun.*, 102:808–809.
15. Hammarström, S., Murphy, R. C., Samuelsson, B., Clark, D. A., Mioskowski, C., and Corey, E. J. (1979): Structure of leukotriene C: identification of the amino acid part. *Biophys. Res. Commun.*, 91:1266–1272.

16. Hammarström, S., Samuelsson, B., Clark, D. A., Goto, G., Marfat, A., Mioskowski, C., and Corey, E. J. (1980): Stereochemistry of leukotriene C-1. *Biochem. Biophys. Res. Commun.*, 92:946–953.

17. Hansson, E., and Samuelsson, B. (1965): Autoradiographic distribution studies of ³H-labeled prostaglandin E_1 in mice. *Biochim. Biophys. Acta*, 106:379–385.

18. Lindlar, H. (1952): Ein neuer Katalysator für selektive Hydrierungen. *Helv. Chim. Acta*, 35:446–450.

19. Murphy, R. C., Hammarström, S., and Samuelsson, B. (1979): Leukotriene C: a slow reacting substance from murine mastocytoma cells. *Proc. Natl. Acad. Sci. USA*, 76:4275–4279.

20. Örning, L., and Hammarström, S. (1980): Inhibition of leukotriene C and leukotriene D biosynthesis. *J. Biol. Chem.*, 255:8023–8026.

21. Örning, L., Hammarström, S., and Samuelsson, B. (1980): Leukotriene D: a slow reacting substance from rat basophilic leukemia cells. *Proc. Natl. Acad. Sci. USA*, 77:2014–2017.

22. Samuelsson, B., and Hammarström, S. (1980): Nomenclature for leukotrienes. *Prostaglandins*, 19:645–648.

23. Tate, S. S., and Meister, A. (1974): Interaction of γ-glutamyl transpeptidase with amino acids, peptides, and derivatives and analogs of glutathione. *J. Biol. Chem.*, 249:7593–7602.

Leukotrienes and Other Lipoxygenase Products,
edited by B. Samuelsson and R. Paoletti.
Raven Press, New York © 1982.

Formation of Leukotrienes C and D and Pharmacologic Modulation of Their Synthesis

Michael K. Bach, John R. Brashler, Douglas R. Morton,
*Linda K. Steel, *Michael A. Kaliner, and **Tony E. Hugli

*Hypersensitivity Diseases Research and Experimental Chemistry, The Upjohn Company, Kalamazoo, Michigan 49001; *Allergic Diseases Section, National Institutes for Allergy and Infectious Diseases, Bethesda, Maryland 20205; and **The Scripps Clinic and Research Foundation, La Jolla, California 92037*

The finding that rat mononuclear cells can be induced to produce leukotrienes upon challenge with the calcium ionophore A23187 (1,2,26) has become one of the bases of the current flurry of activity in leukotriene research. More recent findings with murine macrophages (5,6,35) and rat alveolar macrophages (33), particularly the observation that pulmonary mononuclear cells from disaggregated human lung may be necessary for immunoglobulin E (IgE)-dependent leukotriene synthesis in that system (22), have added considerably to the relevance of the mononuclear cells' capability to the pathophysiology of disease. Nevertheless, our efforts to find a physiologically meaningful method for inducing leukotriene production in peritoneal mononuclear cells have continued to be unsuccessful. Although these efforts constitute negative results, we are presenting them here in the hope that, at some point, an explanation might be found for these failures.

The elucidation of the structures of the leukotrienes (12,13,20–24,28,30) and the demonstration of their metabolic interconversions (3,27,31) raise interesting questions regarding the enzymology of these steps. Specifically, the last step in leukotriene C formation, the formation of a thiolether linkage between glutathione and leukotriene A (LTA), is catalyzed by a glutathione S-transferase (EC 2.5.1.18). This step in leukotriene synthesis is relatively susceptible to pharmacologic modulation. In fact, several reports have already appeared concerning the effectiveness of glutathione depletion in inhibiting leukotriene formation (29,36). Given the ubiquitous nature of glutathione S-transferases and their role in detoxification reactions, one naturally must wonder what effects such maneuvers might have on the well-being of the recipient animal beyond those effects on leukotriene formation.

Glutathione S-transferases have received considerable attention in the last few years (for a review see ref. 15). Specifically, rat and human liver contain at least six forms of the enzyme, and their total concentration accounts for a significant part of total liver protein. The different forms of the enzyme, all very similar in

M_r (about 45,000), differ somewhat in their specificity for substrates of various general types (aryl halides, aryl epoxides, alkyl halides, etc.). Although the belief was once held that such substrate specificities reflected specific subtypes of the enzyme, this is no longer valid. Despite the differences in specificity, all subtypes of the enzyme in human liver cross react with antibody produced to any one of the forms. Interestingly, the same is not true for the rat, in which antibody to type A cross reacts with type C but the antibodies to type B (ligandin) and to type E appear to be specific.

We were interested in attempting to establish which form(s) of glutathione S-transferase might be present in leukotriene-synthesizing cells. In this chapter we present evidence from selective inhibition by specific antibody that the major portion of the enzyme in rat basophilic leukemia (RBL) cells is of type E. We also further document the conversion of arachidonic acid and LTA to LTC + LTD by cell-free preparations from RBL cells and rat liver and show that inhibitors of glutathione S-transferase influence these reactions.

MATERIALS AND METHODS

Procedures for the partial purification of rat peritoneal mononuclear cells and their challenge with ionophore A23187 in the presence of cysteine in order to generate leukotrienes are described elsewhere (2). RBL cells were cultured in basal medium (Eagle) with 7.5% newborn, heat-inactivated calf serum, penicillin, streptomycin, gentamycin, and glutamine. Cells were harvested 72 hr after feeding and washed twice in a Tyrode's salt solution (without calcium or magnesium) which was 10 mM with respect to HEPES buffer, pH 7.0, and 14 μM with respect to indomethacin. Cell suspensions (5 × 10^7 cells/ml) were placed in a nitrogen bomb and equilibrated at 800 pounds/square inch for 15 min at 0°C. Cells were then broken by nitrogen cavitation upon extrusion from the bomb through a fine needle valve (40). The preparation was routinely centrifuged (1,600 × g for 20 min), and the supernatant fraction was used for generation of leukotrienes. In certain experiments this preparation was further fractionated into high-speed supernatant and pellet fractions by centrifugation (100,000 × g for 60 min). The incubation conditions for leukotriene generation consisted of 1 mM glutathione and 1 mM calcium chloride. Reaction was started (at 37°C) by adding 5 μl ethanolic arachidonic acid to a final concentration of 0.1 mM. Where LTA served as the substrate, it was added as 2 or 5 μl of a 0.1 mM ethanolic solution of the lithium salt. The reaction was terminated by adding sufficient absolute methanol to yield an 80% methanolic solution. After flocculation of the protein precipitate and centrifugation, the supernatant solutions were concentrated to dryness in a Savant Instrument Co. Speed Vac Concentrator.

The colorimetric assays for glutathione S-transferase activity, using 1,2-dichloro-4-nitrobenzene (DCNB) or 1,2-epoxy-3-(p-nitrophenoxy)-propane (ENPP) were carried out on a Varian model 219 spectrophotometer equipped with a thermostated sample chamber (30°C) and a five-sample turret arrangement. The incubation conditions have been previously described (10,11). Protein concentration was estimated by the Lowry procedure using bovine serum albumin as a standard.

Leukotriene A was prepared by published procedures (8) as were human anaphylatoxins C3a (14) and C5a (9). The prostaglandin-generating factor preparations were the crude extracts of anaphylactically challenged human lung, and the Amicon YM5 retentate and UMO5 ultrafiltrates were prepared from these, as described by Steel and Kaliner (37). All other reagents were obtained commercially.

RESULTS

Attempts to Induce Leukotriene Formation

We previously reported that, in studies carried out in collaboration with Drs. Michel Joseph and Andre Capron, we were unable to induce rat peritoneal mononuclear cells to produce leukotrienes when these cells were sensitized with high concentrations of rat myeloma IgE and then challenged with anti IgE. Under the same conditions lysosomal enzyme release from these cells was readily demonstrable (4). Control incubations, using heat-stable IgG_{1a} type homocytotropic antibody were also ineffective. Two other potential means of inducing leukotriene formation by these cells were investigated. One of these was the use of the anaphylatoxins C3a and C5a from human plasma. Neither of these substances, at concentrations up to 1 μg/ml (C5a) and 10 μg/ml (C3a) caused production of leukotrienes or thromboxane B_2 (TXB_2) by the rat mononuclear cells. Neither did they cause leukotriene formation in RBL cells. This is in contrast to the findings with guinea pig tissue (34,39), where leukotriene formation could be shown when the tissue was exposed to considerably smaller amounts of the anaphylatoxins (7). Perhaps this failure is due to the well-known insensitivity of rat tissue to heterologous anaphylatoxins. However, it was not possible to test higher anaphylatoxin concentrations, nor was rat anaphylatoxin available for testing.

The prostaglandin-generating factors of anaphylaxis were of particular interest to us in that these substances, which are produced as a consequence of anaphylactic challenge, might represent the postulated signal that transmits the message from mast cells to mononuclear cells causing them to participate in the anaphylactic response and to produce the leukotrienes. In contrast to the results with the anaphylatoxins, the high-molecular-weight fraction of this preparation (YM 5 retentate, $M_r > 5,000$) was active in inducing TXB_2 formation in the monocytes. Thromboxane levels of approximately 3 ng/ml were obtained, compared to the approximately 30 ng/ml routinely formed when the cells are challenged with the calcium ionophore. However, neither the crude preparation nor the YM 5 retentate or the UM 05 filtrate caused any leukotriene formation. The apparent dissociation between the ability of the UM 05 ultrafiltrate to cause prostaglandin formation (37) and the ability of the retentate and not the ultrafiltrate to cause thromboxane formation is of some interest in that this clearly points to the existence of multiple factors in the preparation, each with its own selective ability to induce the formation of different arachidonate-derived products. Elucidation of the mechanisms of action of these materials should prove to be of considerable interest.

Characterization of the Cell-Free Leukotriene-Generating System from RBL Cells and Rat Liver

Dose-response plots for the formation of LTC + LTD from arachidonic acid or from LTA by RBL cell homogenates, and from LTA by liver homogenates, are shown in Fig. 1. In contrast to the reports from Jakschik and Lee (16) and Steinhof et al. (38), we found that at least one of the enzymes in the RBL homogenate which was required for the conversion of arachidonic acid to leukotrienes was attached to particles which were sedimented upon centrifugation at 10,000 × g for 20 min. However, as this activity was retained in the supernatants after centrifugation at 1,600 × g for 20 min, this procedure was adopted for routine use. There was only marginal reconstitution of the ability to convert arachidonate to LTC + LTD when the 100,000 × g pellet fraction and the 100,000 × g soluble fraction were combined. Although the 100,000 × g soluble fraction retained all the activity in the colorimetric assay for glutathione S-transferase, most of the capacity to convert LTA to LTC + LTD was found in the 100,000 × g pellet fraction.

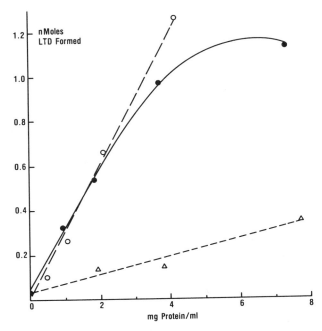

FIG. 1. LTC + LTD formation as a function of enzyme concentration under standard conditions. (●) Liver enzyme, 75 μM LTA methyl ester as substrate. (○) RBL cell enzyme, 100 μM arachidonate as substrate. (Δ) RBL cell enzyme, 20 μM LTA_4 lithium salt as substrate. Results expressed as LTD_4.

The effect of several variables on the rate of LTA or arachidonate conversion to LTC + LTD was examined next. Figure 2 is a composite summary of these results. In each case, results with the standard incubation mixture (i.e., 1 mM calcium, 1 mM glutathione, 0.1 mM arachidonate) are taken as 100% relative activity. As can be seen, each of the components of the incubation was essential and was present at optimal concentration. The situation when LTA served as the precursor for LTC + LTD was more complex because LTA has an exceedingly short half-life in aqueous solution. Although there was very extensive conversion of LTA even when 20 to 50 μM was added, it was often difficult to demonstrate dependence on enzyme concentration or on LTA concentration even though the addition of enzyme was absolutely essential. This also resulted in poor results in experiments where inhibition of this reaction was tested. The LTA-dependent reaction did not require calcium and was optimal at 1 mM glutathione (data not shown). The course of LTC + LTD synthesis from LTA (Fig. 3) was nearly complete after a 3-min incubation, whereas the reaction starting with arachidonate peaked only at 10 min. Addition of further aliquots of LTA or glutathione failed to cause any further formation of LTC + LTD.

The efficiency of conversion of either arachidonic acid or LTA to LTC + LTD was actually quite striking. Even though the reaction continued to be limited by the enzyme concentration (Fig. 1), the highest enzyme concentration which was used routinely resulted in conversion of as much as 2 to 5% of the applied LTA or arachidonate to leukotrienes (expressed as LTC). These results extend the findings of Rådmark et al. (32) who previously demonstrated conversion of LTA to LTC in murine mastocytoma and human polymorphonuclear leukocyte cell suspensions.

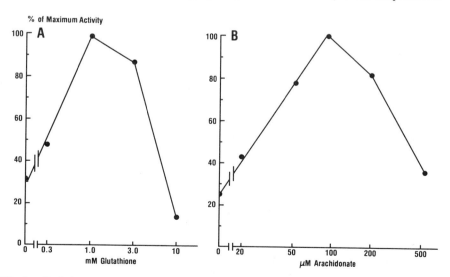

FIG. 2. Optimization of substrate mixture for LTC + LTD formation from arachidonate. **A:** Arachidonate 16 μM. **B:** Glutathione 1 mM. Calcium concentration (1 mM) was constant. 100% represents 71 pmoles LTD in **A** and 45 pmoles in **B**.

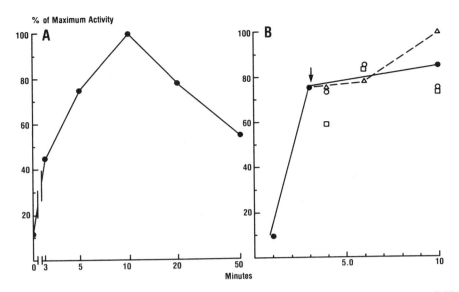

FIG. 3. Time course of leukotriene formation by RBL cell enzyme. **A:** From arachidonate (100 μM). **B:** From LTA$_4$ lithium salt. Initial LTA concentration was 3.3 μM. Where indicated, a second aliquot of LTA (Δ), glutathione 1 mM final concentration ($\circ$), or both ($\square$) were added. Maximum leukotriene production in **A** was 0.4 nmoles and in **B** 0.13 nmoles.

Characterization of Glutathione S-transferase Activity of RBL Cells

As already pointed out, liver contains a series of glutathione S-transferases which can be separated from one another on the basis of molecular charge and which differ to a certain extent in their substrate specificity. Form B (ligandin) is by far the most prevalent form in liver; this is followed, in order of decreasing abundance, by forms C, A, AA, E, and D. DCNB is a good substrate for forms A and C, and a relatively poor substrate for forms B and AA; it is not attacked at all by form E of the enzyme. On the other hand, ENPP is an excellent substrate for form E and a poor substrate for form A. The other forms of the enzyme are incapable of derivatizing this substrate molecule (15). Thus we hoped that the relative ability of RBL cell extracts to derivatize these two chromogenic substrates might give an indication of the relative abundance of the E form of the enzyme in these cells. The enzyme dose-response plots for RBL cell extracts and for a crude liver preparation are shown in Fig. 4. It can be seen that the enzyme from RBL cells formed 1.2 as much product from ENPP than from DCNB, and the ratio for the enzymes from rat liver was 1.4; hence there was no difference in the abundance of these respective specificities between the preparations. However, as shown in Fig. 5, antibody to type E enzyme effectively interfered with the colorimetric assay using DCNB with enzyme from RBL cells. This antibody was also the only one which interfered with the ability of the RBL cell extract to convert arachidonic acid to LTC + LTD. Unfortunately, not enough of the precious antibody remained to

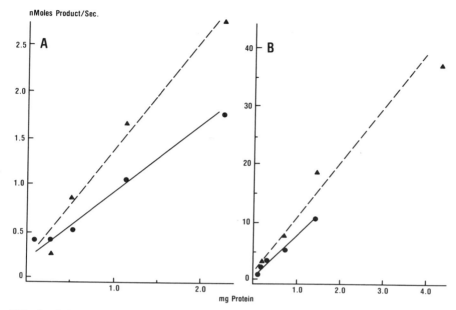

FIG. 4. Substrate specificity of glutathione S-transferase of liver compared with that of RBL cells. **A:** Enzyme from RBL cells. **B:** Enzyme from liver. *Solid line* = DCNB pH 8.0. *Dotted line* = ENPP pH 6.5. Glutathione concentration was 1.0 mM for DCNB, 5.0 mM for ENPP.

permit us to determine if it is the E form in liver homogenates which is responsible for the conversion of LTA to LTC + LTD. It should again be stressed that, according to the literature, the E form of the enzyme is not able to utilize DCNB (15), so that the results with enzyme from RBL cells form a bit of a paradox.

Studies with Inhibitors of Glutathione S-Transferase

Several categories of glutathione S-transferase inhibitors are known, including bromosulfophthalein (17), *p*-aminohippuric acid, and probenecid (19). Other compounds bind to the enzyme but are not themselves substrates; among these are various steroid sulfates (25) and bilirubin (18). Accordingly, we evaluated the effects of a series of such inhibitors on the colorimetrically assayed glutathione S-transferase reaction using DCNB or ENPP and on the conversion of arachidonic acid to LTC + LTD in RBL cells. We then compared the results with the activities of these compounds on the total glutathione S-transferase activity of liver homogenates. The results, summarized in Fig. 6, are not easily interpreted. Bilirubin was exceptionally active in inhibiting the formation of LTC + LTD and was strikingly less active when ENPP was the substrate. On the other hand, the steroid sulfates were inactive when they were tested on the leukotriene-generating system but were reasonably active in inhibiting all the colorimetric reactions. The other inhibitors affected all the reactions, giving more-or-less parallel dose-response plots, although

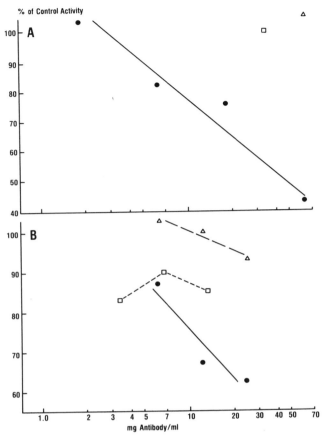

FIG. 5. Effect of preincubation (10 min) with antibodies to glutathione S-transferase on activity of RBL cell extracts. **A:** Colorimetric assay with DCNB. **B:** LTC + LTD formation from arachidonate. (Δ) Anti-B enzyme. (□) Anti-C (and A) enzyme. (●) Anti-E enzyme.

the order of activity for the different tests differed from one inhibitor to the next. Clearly, to make sense out of these observations it will be necessary to purify the glutathione S-transferases from RBL cells, as the possibility that these cells contain additional forms of the enzyme has not been ruled out.

CONCLUSIONS

Despite considerable effort, we still do not know what physiologically meaningful inducer will initiate leukotriene synthesis in rat peritoneal monocytes. It is certainly possible that homologous anaphylatoxins might be active, but this remains to be demonstrated. In a more general sense, the role of mononuclear cells as the source of anaphylactically produced leukotrienes, if this should turn out to be the case, requires the existence of a "signal" which would transmit the message to produce

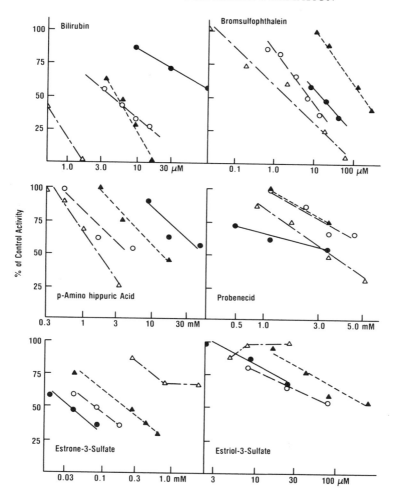

FIG. 6. Effect of various inhibitors on RBL cell extract and rat liver glutathione S-transferase activity and on LTC + LTD formation. (●) Rat liver using ENPP as substrate. (○) Rat liver enzyme using DCNB as substrate. (▲) RBL cell extract, DCNB substrate. (△) RBL cell extract, LTC + LTD formation using 100 μM arachidonate as substrate.

leukotrienes from the mast cells or basophils to the mononuclear cells. The prostaglandin-generating factors of anaphylaxis, which may be derived from mast cells, could serve this function provided a factor can be found which selectively activates the leukotriene pathway. The very concept that these factors can exercise highly selective control (e.g., the differential induction of prostaglandin versus thromboxane formation) offers a fascinating new wrinkle in our understanding of the homeostatic controls that play a role in modulating the many pathways of arachidonate metabolism in these cells.

Our studies with cell-free leukotriene formation and with glutathione S-transferase may have raised more questions than they answered. The discrepancy between the soluble nature of the DCNB derivatizing enzyme and the apparently particulate nature of the LTC-generating enzyme is intriguing: The relatively wide spread in ID_{50} values observed with the various inhibitors offers the possibility that selective inhibitors might be found which inhibit LTC + LTD formation without unduly compromising the detoxification pathways in the liver. Our results are clearly preliminary, and considerable work remains to be done. The next few years should prove exciting and fruitful as our quest to understand the mechanisms of leukotriene production continues.

SUMMARY

Our attempts to find a physiologically relevant means for inducing leukotriene synthesis in rat peritoneal mononuclear cells have thus far been unsuccessful. In addition to the IgE-anti-IgE challenge which we previously reported, we have now tried human C3a and C5a as well as crude and semipurified fractions of the prostaglandin-generating factor of anaphylaxis. In each case, it was possible to show that these substances activated the cells even though no leukotrienes were formed.

A cell-free system in which LTC + LTD formation can be studied was developed as a modification of published methods. Arachidonic acid and LTA_4 served as precursors in this system in the presence of added glutathione. Calcium was required for LTC and LTD synthesis from arachidonic acid but was not required for the glutathione S-transferase terminal step in the synthesis. Using inhibitor profiles, substrate specificity for chromogenic substrates, and inactivation by selective antibodies, we tried to identify the subtype of the glutathione S-transferase in RBL cells. Although antibody to type E of the enzyme was most effective in neutralizing the enzyme activity, neither the substrate specificity nor the inhibition profiles agreed with the conclusion that the E-type enzyme was the major form in these cells. The effect of known inhibitors of glutathione S-transferase on the conversion of arachidonate to LTC and LTD was examined. Bilirubin, an inhibitor which binds to the enzyme and is not a substrate, was much more active in inhibiting LTC + LTD formation than were steroid sulfates, which were markedly less active in inhibiting this reaction. The inhibitory activities of the other compounds were similar on all substrates tested.

ACKNOWLEDGMENTS

We are indebted to Dr. William B. Jakoby for a generous gift of antibodies to the subforms of rat glutathione S-transferase, and to Dr. Francis A. Fitzpatrick for performing the radioimmunoassays for thromboxane B_2. The RBL cells were originally obtained from Dr. T. Ishizaka, The Johns Hopkins School of Medicine.

REFERENCES

1. Bach, M. K., and Brashler, J. R. (1974): In vivo and in vitro production of a slow reacting substance in the rat upon treatment with calcium ionophores. *J. Immunol.*, 113:2040–2044.

2. Bach, M. K., and Brashler, J. R. (1978): Ionophore A23187 induced production of slow reacting substance of anaphylaxis (SRS-A) by rat peritoneal cells in vitro: evidence for production by mononuclear cells. *J. Immunol.*, 120:998–1005.

3. Bach, M. K., Brashler, J. R., Johnson, M. A., and Drazen, J. M. (1980): Two rat mononuclear cell-derived slow reacting substances: kinetic evidence that the peripheral airway-selective spasmogen is derived from a nonselectively acting precursor. *Immunopharmacology*, 2:361–373.

4. Bach, M. K., Brashler, J. R., Johnson, M. A., and Drazen, J. M. (1981): Slow reacting substances from rat leukocytes: selective action on the airways, and progress toward elucidation of their structure. In: *Biochemistry of the Acute Allergic Reactions*, edited by E. L. Becker, A. S. Simon, and K. F. Austen, pp. 37–48. Alan R. Liss, New York (*in press*).

5. Benveniste, J., Hadji, L., Jouvin, E., Mercia-Huerta, J. M., Pirotzky, E., and Roubin, R. (1981): Release of slow-reacting substances (SRS) from various cell types and organs. *Fed. Proc.*, 40:1022 (abstr. 4487).

6. Bretz, U., Dewald, B., Payne, T., and Schnyder, J. (1980): Phagocytosis stimulates the release of slow reacting substance (SRS) in cultured macrophages. *Br. J. Pharmacol.*, 71:631–634.

7. Chenoweth, D. E., Lane, T. A., Rowe, J. G., and Hugli, T. E. (1980): Quantitative comparisons of neutrophil chemotaxis in four animal species. *Clin. Immunol. Immunopathol.*, 15:525–535.

8. Corey, E. J., Clark, D. A., Goto, G., Marfat, A., Mioskowski, C., Samuelsson, B., and Hammarström, S. (1980): Stereospecific total synthesis of a "slow reacting substance" of anaphylaxis: leukotriene C. *J. Am. Chem. Soc.*, 102:1436–1439.

9. Fernandez, H. N., and Hugli, T. E. (1976): Partial characterization of human C5a anaphylatoxin I: chemical description of the carbohydrate and polypeptide portions of human C5a. *J. Immunol.*, 117:1688–1694.

10. Fjellstedt, T. A., Allen, R. H., Duncan, B. K., and Jakoby, W. B. (1973): Enzymatic conjugation of epoxides with glutathione. *J. Biol. Chem.*, 248:3702–3707.

11. Habig, W. H., Pabst, M. J., and Jakoby, W. B. (1974): Glutathione S-transferases: the first enzymatic step in mercapturic acid formation. *J. Biol. Chem.*, 249:7130–7139.

12. Hammarström, S., Murphy, R. C., Samuelsson, B., Clark, D. A., Mioskowski, C., and Corey, E. J. (1979): Structure of leukotriene C: identification of the amino acid part. *Biochem. Biophys. Res. Commun.*, 91:1266–1272.

13. Hammarström, S., Samuelsson, B., Clark, D. A., Goto, G., Marfat, A., Mioskowski, C., and Corey, E. J. (1980): Stereochemistry of leukotriene C-1. *Biochem. Biophys. Res. Commun.*, 91:946–953.

14. Hugli, T. E., Vallota, E. H., and Müller-Eberhard, H. J. (1975): Purification and partial characterization of human and porcine C3a anaphylatoxin. *J. Biol. Chem.*, 250:1472–1478.

15. Jakoby, W. B., and Habig, W. H. (1980): Glutathione transferases. In: *Enzymatic Basis of Detoxification*, edited by W. B. Jakoby, Vol. 2, pp. 63–94. Academic Press, New York.

16. Jakschik, B. A., and Lee, L. H. (1980): Enzymatic assembly of slow reacting substances. *Nature (Lond.)*, 287:51–52.

17. Javitt, N. B. (1976): Biochemical probes for the study of binding and conjugation of glutathione S-transferases. In: *Glutathione: Metabolism and Function*, edited by I. M. Arias and W. B. Jakoby, pp. 309–315. Raven Press, New York.

18. Ketley, J. N., Habig, W. H., and Jakoby, W. B. (1975): Binding of non-substrate ligands to the glutathione S-transferases. *J. Biol. Chem.*, 250:8670–8673.

19. Kirsch, R., Fleischner, G., Kamisaka, K., and Arias, I. M. (1975): Structural and functional studies of ligands, a major renal organic anion-binding protein. *J. Clin. Invest.*, 55:1009–1019.

20. Lewis, R. A., Austen, K. F., Drazen, J. M., Clark, D. A., Marfat, A., and Corey, E. J. (1980): Slow reacting substances of anaphylaxis identification of leukotrienes C-1 and D from human and rat sources. *Proc. Natl. Acad. Sci. USA*, 77:3710–3714.

21. Lewis, R. A., Drazen, J. M., Austen, K. F., Clark, D. A., and Corey, E. J. (1980): Identification of the C(6)-S-conjugate of leukotriene A with cysteine as a naturally occurring slow reacting substance of anaphylaxis (SRS-A): importance of the 11-cis-geometry for biological activity. *Biochem. Biophys. Res. Commun.*, 96:271–277.

22. Lewis, R. A., Drazen, J. M., Corey, E. J., and Austen, K. F. (1981): Structural and functional characteristics of the leukotriene components of slow reacting substance of anaphylaxis (SRS-A). In: *SRS-A and Leukotrienes*, edited by P. J. Piper, pp. 101–117. Wiley, New York.

23. Morris, H. R., Taylor, G. W., Piper, P. J., Samhoun, M. N., and Tippins, J. R. (1980): Slow reacting substances (SRSs): the structure identification of SRSs from rat basophil leukemia (RBC-1) cells. *Prostaglandins*, 19:185–201.

24. Murphy, R. C., Hammarström, S., and Samuelsson, B. (1979): Leukotriene C: a slow-reacting substance from murine mastocytoma cells. *Proc. Natl. Acad. Sci. USA*, 76:4275–4279.
25. Ohl, V. S., and Litwack, G. (1977): Selective inhibition of glutathione-S-transferases by 17-β-estradiol disulfate. *Arch. Biochem. Biophys.*, 180:186–190.
26. Orange, R. P., Moore, E. G., and Gelfand, E. W. (1980): The formation and release of slow reacting substance of anaphylaxis (SRS-A) by rat and mouse peritoneal mononuclear cells induced by ionophore A23187. *J. Immunol.*, 124:2264–2267.
27. Örning, L., and Hammarström, S. (1980): Inhibition of leukotriene C and leukotriene D biosynthesis. *J. Biol. Chem.*, 255:8023–8026.
28. Örning, L., Wedegran, D., Hammarström, S., and Samuelsson, B. (1980): Leukotriene D: a slow reacting substance from rat basophil leukemia cells. *Proc. Natl. Acad. Sci. USA*, 77:2014–2017.
29. Parker, C. W., Fischman, C. M., and Wedner, H. J. (1980): Relationship of biosynthesis of slow reacting substance to intracellular glutathione concentrations. *Proc. Natl. Acad. Sci. USA*, 77:6870–6873.
30. Parker, C. W., Jakschik, B. A., Huber, M. G., and Falkenhein, S. F. (1979): Characterization of slow reacting substance as a family of thiolipids derived from arachidonic acid. *Biochem. Biophys. Res. Commun.*, 89:1186–1192.
31. Parker, C. W., Koch, D., Huber, M. M., and Falkenhein, S. F. (1980): Formation of the cysteinyl form of slow reacting substance (leukotriene E) in human plasma. *Biochem. Biophys. Res. Commun.*, 97:1038–1046.
32. Rådmark, O., Malmsten, C., and Samuelsson, B. (1980): Leukotriene A4: enzymatic conversion to leukotriene C4. *Biochem. Biophys. Res. Commun.*, 96:1679–1687.
33. Rankin, J. A., Hitchcock, M., Merrill, W. W., and Askenase, P. W. (1981): Slow reacting substance (SRS): IgE dependent release from alveolar macrophages (AM). *Fed. Proc.*, 40:1014 (abstr. 4444).
34. Regal, J. F., and Pickering, R. J. (1981): C5a-induced tracheal contractions: effect of an SRS-A antagonist and inhibitors of arachidonate metabolism. *J. Immunol.*, 126:313–316.
35. Rouzer, C. A., Scott, W. A., Cohn, Z. A., Blackburn, P., and Manning, J. M. (1980): Mouse peritoneal macrophages release leukotriene C in response to a phagocytic stimulus. *Proc. Natl. Acad. Sci. USA*, 77:4928–4932.
36. Rouzer, C. A., Scott, W. A., Griffith, O. W., Hamill, A. L., and Cohn, Z. A. (1981): Inhibition of the synthesis of leukotriene C with buthionine sulfoximine. *Clin. Res.*, 29:492A.
37. Steel, L., and Kaliner, M. (1981): Prostaglandin generating factor of anaphylaxis: identification and isolation. *J. Biol. Chem.* (*in press*).
38. Steinhof, M. M., Lee, L. H., and Jakschik, B. A. (1980): Enzymatic formation of prostaglandin D2 by rat basophilic leukemia cells and normal rat mast cells. *Biochim. Biophys. Acta*, 618:28–34.
39. Stimler, N. P., Brocklehurst, W. E., Bloor, C. M., and Hugli, T. E. (1981): C5a-stimulated release of SRS-A from guinea pig lung. *Fed. Proc.*, 40:1115 (abstr. 5026).
40. Wallach, D. F. H., and Ullrey, D. (1962): Studies on the surface and cytoplasmic membranes of Ehrlich ascites carcinoma cells. I. The hydrolysis of ATP and related nucleotides by microsomal membranes. *Biochim. Biophys. Acta*, 64:526–539.

Leukotrienes and Other Lipoxygenase Products,
edited by B. Samuelsson and R. Paoletti.
Raven Press, New York © 1982.

Leukotrienes: Their Metabolism, Structure, and Role in Allergic Responses

Charles W. Parker

*Department of Internal Medicine, Division of Allergy and Immunology, and
Howard Hughes Medical Institute Research Laboratory, Washington University
School of Medicine, St. Louis, Missouri 63110*

Slow-reacting substance (SRS) was first described by Feldberg and Kellaway in 1938 as a smooth muscle contracting activity (8) released during the perfusion of cat and guinea pig lungs with cobra venom. In 1940 Kellaway and Trethewie demonstrated the production of a substance with similar properties during stimulation of perfused sensitized guinea pig lungs with antigen (17). Brocklehurst clearly distinguished the SRS activity from histamine by the slow onset and sustained nature of the response as well as the absence of inhibition by antihistamines (4). Subsequent progress was impeded by the lability of the SRS molecule and the general problem of lack of precision and specificity inherent in bioassay measurements. However, the spectrum of SRS activity on various smooth muscle preparations was determined, the molecule was shown to be produced *de novo* in response to a variety of immunological and nonimmunological stimuli, and it was defined as a relatively polar acidic lipid more stable at alkaline than at acid pH. In 1975, at the time our laboratory entered the SRS field, there was still no definitive information available in regard to the mechanism of biosynthesis and structure of SRS despite 35 years of research (24). We reasoned that the most important initial problems were to: (a) find a reasonably homogeneous tissue (preferably a stable cell line maintainable *in vitro*) in which SRS could be produced in substantial quantities; (b) improve existing purification procedures for SRS; and (c) attempt to identify possible biosynthetic precursors.

IDENTIFICATION OF RBL-1 CELLS AS A SOURCE OF SRS

Since SRS had been reported to be produced during immunoglobulin E (IgE)-mediated reactions in primate lungs (12), and mast cells and basophils are the major effector cells in this form of allergy, we conducted our initial studies in a line of rat basophilic leukemia (RBL-1) cells which had been utilized for studies of the receptor for IgE (18). This cell line contains histamine, and the cells closely resemble normal basophils in their general appearance and staining properties with dyes. Moreover, they can be grown in considerable quantity in tissue culture in the absence of contaminating cells. Although IgE and anti-IgE antibodies were ineffective, the divalent cation ionophore A23187 generated substantial amounts of SRS-like activity from these cells (13,14). The activity produced was similar or identical to various immunologically and nonimmunologically generated SRSs described in the literature with respect to its chromatographic behavior, character of the guinea pig ileal response, spectrum of other smooth muscle-contracting activities, failure of enzymes which degrade other mediators of immediate hypersensitivity to affect the response, and susceptibility of the spasmogen to inhibition by either arylsulfatase or the selective SRS antagonist FPL-55712. Thus it was apparent that we were studying a molecule that was similar or identical to previously described SRSs and that RBL-1 cells would be very useful as a source of SRS.

Using the RBL-1 cell system we went on to provide the initial evidence that: (a) SRS was a metabolite of arachidonic acid (AA) produced through the lipoxygenase pathway; (b) the 5-lipoxygenase product, 5-hydroxyeicosatetraenoic acid (5-HETE) was produced in substantial quantities concomitantly with SRS, suggesting involvement of a 5-lipoxygenase in SRS synthesis; (c) SRS was heterogeneous and that each of the various species had a sulfur-containing side chain bound in thioether linkage to a fatty acid core; (d) the side chains in the most prominent species of SRS contained glutathione or cysteinyl-glycine rather than cysteine, although an SRS-containing cysteine was eventually produced; and (e) cleavage of SRS produced an acidic lipid with ultraviolet absorbancy near 280 nm which comigrated with 5-HETE chromatographically, supporting the role of the lipoxygenase pathway.

The initial portions of this chapter review this and subsequent work by our laboratory on SRS metabolism and structure in some detail. This is followed by a brief discussion of some of our other studies with SRS and related molecules.

IDENTIFICATION OF ARACHIDONATE AS AN SRS PRECURSOR

Having established that RBL-1 cells were suitable for studying SRS biosynthesis, we went on to study the possible involvement of long-chain fatty acids in SRS synthesis. One approach was to add an inhibitor of fatty acid metabolism before the ionophore and determine if the yield of spasmogenic activity was altered. A second approach was to produce SRS in the presence of radiolabeled and unlabeled fatty acids and determine if the yield of SRS was altered or radioactivity comigrated with SRS after extensive chromatographic purification. From this work, convincing evidence was obtained that AA was an SRS precursor (13,15). A marked decrease (by as much as 90%) in the SRS response was observed at low micromolar concentrations of 5,8,11,14 eicosatetraynoic acid (ETYA), a cyclo-oxygenase and lipoxygenase inhibitor. In addition, exogenous unlabeled AA considerably enhanced (up to five-fold increase) the SRS response to A23187 but was only modestly stimulatory when A23187 was absent, even at high AA concentrations (15). Other long-chain fatty acids, including 8,11,14-eicosatrienoic acid, which is identical to AA except for the absence of the five double bonds, had little or no enhancing activity, and several of the C-18 fatty acids were significantly inhibitory.

The much greater effect of AA in the presence than in the absence of A23187 was of considerable interest in itself as it suggested that the ionophore was doing more than simply providing free AA for SRS formation. It was later found by Jakschik et al. (16) and confirmed by our laboratory *(in preparation)* that the initial enzyme in the SRS pathway (the 5-lipoxygenase, see below) is markedly stimulated by Ca^{++}. The strongest evidence that AA was an SRS precursor came from radioactive AA incorporation studies. SRS was generated by A23187 in RBL-1 cells which had been preincubated either for 16 hr or for 1 to 60 min with ^3H- or ^{14}C-labeled AA (15). The SRS was then extensively purified by column and preparative thin-layer chromatography (TLC). Regardless of the preincubation time, significant amounts of radioactivity (usually 0.1 to 0.5%) still comigrated with SRS at the completion of the purification procedure. Moreover, the purified material continued to show comigration of radioactivity and spasmogenic activity in a variety of additional chromatographic systems, establishing convincingly that the label was in the molecule. The radiochemical purity of these preparations was later confirmed by high-pressure liquid chromatography (HPLC). The overall yields of SRS activity during the SRS purification were usually of the order of 50%, making it highly likely that the most active SRS species were being recovered. In contrast to AA, the other long-chain fatty acids that were evaluated did not incorporate radioactivity into SRS. These studies also established that there were at least two chromatographically distinguishable species of SRS with somewhat different spasmogenic activities, as determined by the ratio of radioactivity to spasmogenic activity in the guinea pig ileal system. It was apparent even at this stage of the work in 1977 that SRS was considerably more potent as a spasmogenic agent than histamine in this particular smooth muscle system (15).

Although we were successful in obtaining radiochemically pure SRS from RBL-1 cells, the pathway involved was clearly a quantitatively minor one from the point of view of overall AA metabolism in these cells. Only 0.1 to 0.5% of the arachidonic acid radioactivity originally added to the cells copurified with SRS. Even allowing for the moderate losses of SRS activity usually experienced during purification, the level of incorporation was clearly considerably less than for several other AA metabolites produced by these cells, such as prostaglandin D_2 (PGD_2). This, together with the marked lability of the SRS molecule and the considerable difficulty of isolating SRS from complex tissue such as lung, where the proportion of the label in SRS is even lower, undoubtedly explains why several laboratories were initially unable to confirm our observations. The first apparent confirmation that radioactive AA is incorporated into SRS was by Bach et al. (1) in rat peritoneal mononuclear cells, but their observations were preliminary and were apparently retracted in early 1979 (2). They did conclude that ETYA inhibited SRS synthesis (1) in accord with our RBL-1 cell results (13,15). On the other hand, based on studies with indomethacin and a newly developed thromboxane synthetase inhibitor, they suggested that the cyclo-oxygenase instead of the lipoxygenase pathway was involved (1).

As discussed in the succeeding paragraphs, our own studies strongly implicated the lipoxygenase pathway, and this was later confirmed by a number of other groups.

IDENTIFICATION OF SRS AS A LIPOXYGENASE PRODUCT

To determine which pathway of AA metabolism is involved in SRS biosynthesis, the effect of indomethacin, a selective inhibitor of the cyclo-oxygenase pathway, was studied in the RBL-1 cell system. SRS formation was unaffected or enhanced by indomethacin under conditions in which PGD_2 formation was markedly decreased. This strongly suggested that SRS was a lipoxygenase product (15,23,24). Moreover, when SRS biosynthesis was maximally stimulated by A23187 with radiolabeled AA present, a substantial percentage of the radiolabel was incorporated into a product cochromatographing with 5-HETE which had just been described in rabbit neutrophils (3). We soon found that A23187 stimulated 5-HETE formation in human neutrophils since human neutrophils were already known to make SRS (5), and 8,11,14-eicosatetraenoic acid did not appear to be an SRS precursor. In 1977 I suggested that SRS was likely to be a 5-lipoxygenase product (23).

CHARACTERIZATION OF SRS AS A THIOETHER

In addition to our screening studies for biosynthetic labeling of SRS with radiolabeled fatty acids, a variety of sugars, amino acids, and metabolic intermediates were also evaluated, and ^{35}S-cysteine and ^{35}S-methionine were shown to be incorporated into the SRS molecule. RBL-1 cells were grown for 16 hr in tissue culture medium with ^{35}S-labeled cysteine or methionine, washed, and SRS generated with A23187. Using our usual chromatographic purification procedure for SRS in combination with reverse-phase (RP) HPLC on C-18 reverse-phase columns, three distinct SRS species, each containing ^{35}S radioactivity, were obtained. When the cells were preincubated with a ^{35}S precursor for 60 min or less before stimulation with A23187, incorporation of radioactivity was barely detectable, suggesting (although not proving) that the incorporation involved a metabolite of cysteine instead of cysteine itself. When SRS preparations labeled with ^{14}C-AA or ^{35}S-cysteine were compared, behavior of the radioactivity on RP-HPLC columns and other chromatographic systems was indistinguishable. Moreover, when the spasmogenic activity of the SRS was inactivated by boiling in dilute acid, the chromatography patterns of the ^{35}S and ^{14}C radiolabels were altered in parallel with one another and in stoichiometric relationship to the loss of spasmogenic activity, confirming that the labels were indeed present in the biologically active molecule (33).

These observations were reported briefly in 1978 (24) and in more detail in April and early May of 1979 (25,26) at national meetings. Most of the basic information was summarized in the meeting abstracts, although more detail and some entirely new information was brought forth in the presentations themselves. A more detailed report was published in the summer of 1979 (26).

The ^{35}S incorporation studies by our laboratory represented the first convincing evidence that the SRS molecules contain sulfur (26,28,33). Indirect evidence for the presence of sulfur based on inactivation of SRS by aryl sulfatase (21) and apparent augmentation of SRS biosynthesis by exogenous thiols (22) turned out to be invalid in that the inactivation by aryl sulfatase is primarily due to a contaminating peptidase which degrades the oligopeptide side chain of SRS rather than to hydrolysis of a sulfate ester linkage (29), whereas the enhancing effect of the various thiols is probably due to protection of SRS from degradation by similar enzymes (37). Spark source microanalytical data for SRS published in 1974 (21) were also invalid in that, in retrospect, the samples show no resemblance to any known SRS species.

The evidence from our laboratory that each of the SRS species contained sulfur, taken together with the earlier evidence that SRS was a lipoxygenase product, was of considerable interest since it was known that thiols form thioether adducts with lipid hydroperoxides (9). However, verification of a similar reaction in SRS biosynthesis obviously required structural confirmation. Although gas chromatography-mass spectroscopy (GC-MS) of the undegraded SRS molecule before and after derivatization was unsuccessful, we were able to identify reagents which could chemically cleave SRS. In early May, 1979 (25) we reported that the major SRS species could be degraded by reagents that cleave thioether bonds including cyanogen bromide, Raney nickel, and sodium metal in liquid ammonia as evidenced by separation of the ^{14}C and ^{35}S labels by TLC at the end of the chemical reactions. The studies with cyanogen bromide were particularly informative as sulfoxides and sulfones are not normally affected by this reagent, in contrast to sodium (Na) in liquid ammonia and Raney nickel, which are not at all specific for thioether bonds (25). As radioactivity was also incorporated from [2,3^{14}C]cysteine or [3,3'^{3}H]cystine SRS, it was apparent that at least a portion of the alkyl side chain of the cysteine molecule was also being incorporated into SRS. As discussed below, chromatographic studies of the Na metal cleavage product indicated that the side chain was not cysteine but one of its metabolites (25–27).

Our results with respect to incorporation of radiolabeled AA and ^{35}S into SRS were confirmed by Samuelsson and his colleagues for a single species of SRS from mouse mastocytoma cells (36). However, the side chain was stated to be cysteine; as we first briefly reported in September (30) and later in greater detail (27), amino acid analysis and chromatographic studies indicated that the sulfur-containing side chain in the major form of SRS from RBL-1 cells was not cysteine but glutathione or cysteinyl-glycine. This was especially clear for the glutathionyl species of SRS, which not only had the appropriate amino acid composition but also produced a free side chain which showed the appropriate chromatographic behavior for glutathione after cleavage with Na metal in liquid ammonia. The results with cysteinyl-glycyl SRS were less definitive initially because this species of SRS was variably contaminated with glutathionyl SRS. However, with relatively minor adjustments in the chromatographic purification procedure for separating the two major SRS species, a product giving the expected amino acid composition for cysteinyl-glycyl SRS was obtained. Chromatographic studies with the side chain of this SRS after

Na metal cleavage product confirmed the presence of cysteinyl-glycine. Based on chemical inactivation studies, we also suspected the presence of a reducible function in the SRS molecule, but because of the small amounts of SRS available definitive studies were not possible. Now that the most important SRS molecules have been synthesized chemically and are available in substantial quantities, the presence of a reducible function other than the conjugated double bond system itself appears unlikely. Subsequent studies by two other laboratories confirmed the existence of the cysteinyl-glycine and glutathionyl forms of SRS by amino acid sequence and composition analysis. Hammarström et al. (11) reported that the predominant species of SRS from mouse mastocytoma cells had a glutathionyl side chain. We later reported that the cysteinyl form of SRS could be demonstrated in RBL-1 cell supernatants, particularly when the incubation of the cells with A23187 was extended to 30 to 60 min (31). This was later confirmed by Lewis et al. (19).

CHARACTERIZATION OF THE LIPID PORTION OF THE SRS MOLECULE

We also reported in early May that when either of the major SRS species was labeled with ^{14}C from [^{14}C]AA and degraded with Na metal the major radioactive product was an acidic lipid which migrated very similarly to 5-HETE in three TLC systems. An unsubstituted C-20 fatty acid also was present as determined by mass spectroscopy, presumably due to loss of the OH group from the 5-HETE during leavage. As 5-lipoxygenases introduce molecular oxygen at the 5-position of the AA molecule, the presence of 5-HETE after Na metal cleavage of SRS produced the first direct structural confirmation that a 5-lipoxygenase is indeed involved in SRS synthesis. At this stage of our studies, GC-MS studies on the major Na metal cleavage product had not been completed; later, however, we showed that after reduction with hydrazine in an oxygen atmosphere and esterification and silylation, derivatized 5-hydroxyarachidic acid could be identified, establishing conclusively that both of the major SRS species have a fatty acid core with 20 carbons and an oxygen group at the 5-position (27).

Samuelsson and his colleagues soon confirmed and extended these observations (20,36). Treatment of the SRS molecule with Raney nickel produced 5-hydroxy-arachidic acid, which was identified by mass spectroscopy. Digestion of intact SRS with soybean lipoxygenase led to an increase in ultraviolet absorbancy at 308 nm, indicating the formation of a tetraene; taken together with the results of ozonolysis and the original 280 nm absorption maximum in undigested SRS, this indicated that the double bonds were at the 7-, 9-, 11-, and 14-positions, and the sulfur-containing side chain was at the 6-position. Although important new information was provided in their presentation, several of their major observations were confirmatory of our own previously presented work, and in several respects their data were more preliminary than our own (28,33). The only thioether cleavage reagent that was evaluated, Raney nickel, is highly nonspecific. Moreover, the side chain was stated to be cysteine (36), which we later showed to be incorrect (see above). No data were given on the important question of SRS yields during purification, and no attempt was made to correlate changes in decreased SRS spasmogenic activity with altered chromatographic behavior of radiolabeled SRS to prove that the radioactivity was indeed incorporated into the SRS molecule. The full structure of SRS, including identification of double-bond geometry as 7(*trans*), 9(*trans*), 11(*cis*), and 14(*cis*) was later established by several groups by organic synthesis, as described elsewhere in this volume.

The structural studies by the group in Sweden and ourselves indicating that SRS contains an oxygen atom at the 5-position strongly supported the possible role of a 5-lipoxygenase in SRS synthesis. Nonetheless, there are other possible mechanisms for introducing oxygen into unsaturated fatty acids, and the possibility that a new pathway of AA metabolism was involved had to be excluded. As the oxygen introduced by lipoxygenase is from molecular oxygen, we produced SRS biosynthetically in an atmosphere containing $^{18}O_2$, purified the SRS to homogeneity, and degraded it in the usual way with Na metal and hydrazine. ^{18}O was identified at the 5-position by mass spectroscopy at an isotope abundance similar to that in the O_2 on the original reaction mixture, establishing that the 5-oxygen in the SRS was derived from molecular oxygen (27). We later demonstrated that SRS synthesis was markedly stimulated when unlabeled or labeled synthetic 5-hydroperoxyeicosatetraenoic acid (5-HPETE) was added to RBL-1 cell suspensions in the absence or presence of A23187 (7). It was also found that radiolabel from [1-^{14}C]5-HPETE was incorporated into SRS, and that under conditions of maximal stimulation or SRS biosynthesis by A23187 in RBL-1 cells 5-HETE and 5,12-dihydroxy-6,8,10,14-eicosatetraenoic acid (5,12-DHETE) increased in parallel with SRS as verified by chromatography and mass spectroscopy (7). Moreover, the partially selective lipoxygenase inhibitors nordihydroguaiaretic acid and pyrogallol inhibited the formation of all three products to about the same extent. Taken together, these observations provided strong additional evidence that a 5-lipoxygenase is involved in SRS biosynthesis. Moreover, since 5-HPETE by itself produced a marked SRS response, they further indicated that the major effect of A23187 is to increase the availability of 5-HPETE rather than to stimulate other enzymes in the SRS pathway.

FURTHER STUDIES OF SRS METABOLISM

We had originally suggested that the cysteinyl-glycyl form of SRS might arise from glutathionyl SRS by the action of γ-glutamyltranspeptidase, a widely distributed enzyme which cleaves the γ-glutamyl linkage in glutathione (30). Alternatively, the possibility had been considered that cysteinyl-glycine, which is a normal component of tissues but present at much lower concentrations intracellularly than glutathione, itself was being conjugated to the activated fatty acid molecule to produce cysteinyl-glycyl (cys-gly) SRS directly. Subsequent studies in our laboratory with RBL-1 cells demonstrated that glutathionyl SRS is produced first during the first few minutes of stimulation by A23187 and is sequentially degraded to cys-gly SRS, which is the predominant species at 15 min, and finally to cysteinyl SRS, which has about 10% of the spasmogenic activity of cys-gly SRS in the guinea pig ileal system (31). These transformations were established by adding purified [1-^{14}C]- or [^{35}S]glutationyl or cys-gly SRS to RBL-1 cells, either in the presence or absence of A23187, and analyzing the products chromatographically, spectrally, and by amino acid analysis. Studies with glutathionyl and cys-gly SRS labeled with [1,2-^{14}C]glycine confirmed these conversions and established that both carbons of the glycyl residue were being removed. Interestingly, cysteinyl SRS itself proved relatively stable to further degradation by RBL-1 cells, although it was slowly inactivated. Glutathionyl SRS was also converted to cys-gly SRS by partially purified γ-glutamyltranspeptidase, indicating that the glutamyl residue is bound in γ-peptide linkage as would be expected if the side chain is indeed glutathione. Depending on the digestion conditions, a peptidase that is present as a contaminant in the enzyme preparation could complete the conversion to cysteinyl SRS (31). Commercial preparations of limpet aryl sulfatase also cleave the glycine moiety and to a lesser extent LTC$_4$, presumably due to a contaminating protease; this helps explain the partial inactivation of these particular species of SRS by this enzyme (29). Degradation of the sulfur-containing side chain also occurs over a several-hour period at 37°C in human plasma, with changes being seen at plasma dilutions of as much as 1:2,500 (32), indicating that enzymatic degradation of the SRS side chain is likely to occur *in vivo*. This has been confirmed in studies in the urine of mice injected intravenously with SRS labeled at various positions, although it is also apparent that other metabolic alterations of the SRS molecule are occurring.

INHIBITION OF SRS BIOSYNTHESIS

The inhibition of SRS biosynthesis by ETYA and partially selective lipoxygenase inhibitors has already been discussed. It remains to be established that these or other recently identified lipoxygenase inhibitors would be sufficiently selective and nontoxic for long-term treatment of human asthma, in which SRS is thought to be an important mediator. Nonetheless, as the biosynthesis of other 5-lipoxygenase products which may also contribute to allergic inflammation would be inhibited, the potential value of lipoxygenase inhibitors in the treatment of asthma deserves careful evaluation.

As glutathionyl SRS is the initial species of SRS that is formed when SRS is produced, agents which lower intracellular glutathione concentrations might also be of value in controlling SRS formation, at least on a temporary basis. We have found that when intracellular glutathione concentrations in RBL-1 cells are decreased to 10% or less of their original value by exposing cells to unsaturated compounds which react enzymatically and nonenzymatically with glutathione, SRS biosynthesis is inhibited almost completely (34). Moreover, at least some inhibition of SRS synthesis is observed, even with relatively modest reductions of glutathione levels, suggesting that physiological fluctuations in glutathionyl levels in SRS-producing cells might help regulate SRS production. Although the sustained reduction of intracellular glutathione concentrations must be presumed to be undesirable, there is preliminary evidence to indicate that rats tolerate low glutathione levels for up to 2 weeks without apparent untoward symptoms after treatment with inhibitors of glutathione synthesis (10), so this approach may merit more careful evaluation.

NONIMMUNOLOGICALLY GENERATED SRS FROM NORMAL RAT MAST CELLS AND OTHER SOURCES

Our laboratory has also studied the role of AA in SRS biosynthesis in A23187-stimulated normal rat peritoneal mast cells and peripheral blood myelogenous leukemia cells rich in apparently mature basophils. In each system SRS was produced in substantial quantities in response to A23187, and ETYA inhibited and AA increased SRS release (39–41). Moreover, when either cell type was stimulated with A23187 in the presence of ^{14}C- or ^{3}H-AA and the SRS was purified, incorporation of radioactivity into SRS could be shown. The behavior of these SRSs on HPLC columns was indistinguishable from that of the major RBL-1 cells' SRS product. Thus although too little of these SRSs has been available for mass spectroscopy or amino acid analysis, there is every reason to believe that they are similar or identical to other forms of SRS. In contrast to our results, several laboratories have reported that very little or no SRS is obtained from highly purified rat peritoneal mast cells. However, on a per cell basis, our yields of SRS from highly purified (> 95%) mast cells are at least equal to those from polymorphonuclear leukocytes or mononuclear cells when A23187 is used as the stimulus.

Although it is clear that mast cells produce SRS, the cell source of SRS in IgE-mediated reactions in lung and other tissues *in vivo* remains to be established. In contrast to what is observed with A23187, when highly purified mast cells are stimulated with anti-IgE antibody the SRS response is small and inconsistent even though histamine is being released. Despite the lack of a convincing SRS response in immunologically activated mast cells, we believe that the demonstrated capacity of normal rat mast cells to produce SRS is likely to be of significance for allergic responses, as SRS is normally produced during IgE-mediated responses, and mast cells and basophils contain large numbers of IgE receptors. It is possible that some cofactor analogous to phosphatidyl serine (which activates histamine release in antigen-stimulated mast cells) that is necessary for SRS generation is lacking after mast cells have been extensively purified. On the other hand, other cell types, particularly macrophages, which in some instances apparently bear IgE receptors (6), must be considered as possible sources, either alone or in collaboration with mast cells. There is a recent preliminary report that rat alveolar macrophages that have been preincubated with mouse IgE anti-DNP antibody produce SRS in response to DNP antigen (35). However, our own results in a similar system have thus far failed to confirm this, although the macrophages do respond to A23187. We continue to feel that mast cells are probably the major source of SRS in IgE-mediated reactions in lung.

IMMUNOLOGICALLY GENERATED SRS FROM LUNG

Although the above studies provided structural information on nonimmunologically produced forms of SRS, it was still necessary to study the SRSs produced during allergic responses, as it was by no means certain that they would be identical or even closely related. This question was investigated in our laboratory with SRS from antigen-stimulated lung fragments from guinea pigs immunized 4 to 6 weeks previously. As in the other systems, SRS biosynthesis was enhanced by indomethacin and AA, unaffected or inhibited by other long-chain fatty acids, and inhibited by ETYA, although the inhibition by ETYA was considerably less impressive than in RBL-1 cell suspensions, possibly because of less-effective penetration of the reagent (38). When SRS was released from lung fragments that had been prelabeled with [I¹⁴C]AA and purified chromatographically, a radiolabeled species of SRS containing 0.03% of the original radioactivity was isolated which was similar or identical to cys-gly SRS from RBL-1 cells in molar spasmogenic activity in the guinea pig ileum, chromatographic behavior on HPLC, ultraviolet absorption spectrum in the presence and absence of soybean lipoxygenase, and amino acid composition. Moreover, after cleavage with Na metal in liquid ammonia and further reduction with hydrazine, 5-hydroxyarachidic acid was identified by mass spectroscopy. Despite the extended purification procedure that was used, 50% of the original SRS activity was recovered, making it highly probable that the most active species was being characterized. Thus it is almost certain that the immunologically and nonimmunologically generated forms of SRS are the same. Since the guinea pig lung studies were carried out using a 20-min incubation time, it is not surprising that cys-gly SRS was the predominant SRS species: it is also the major SRS species in RBL-1 cells incubated with A23187 for similar time periods. It seems highly likely that if other incubation periods had been used the other SRS species would have been identified.

REFERENCES

1. Bach, M. K., Brashler, J. R., and Gorman, R. R. (1977): On the structure of slow reacting substance of anaphylaxis: evidence of biosynthesis from arachidonic acid. *Prostaglandins*, 14:21–38.
2. Bach, M. K., Brashler, J. R., Brooks, C. D., and Neerken, A. J. (1979): Slow reacting substances: comparison of some properties of human lungs SRS-A and two distinct fractions from ionophore-induced rat mononuclear cells SRS. *J. Immunol.*, 122:160–165.
3. Borgeat, P., Hamberg, M., and Samuelsson, B. (1976): Transformation of arachidonic acid and homo-γ-linolenic acid by rabbit polymorphonuclear leukocytes: monohydroxy acids from novel lipoxygenase. *J. Biol. Chem.*, 251:7816–7820.
4. Brocklehurst, W. E. (1960): The release of histamine and formation of a slow-reacting substance (SRS-A) during anaphylactic shock. *J. Physiol. (Lond.)*, 151:416–435.
5. Conroy, M. C., Orange, R. P., and Lichtenstein, L. M. (1976): Release of slow reacting substance of anaphylaxis (SRS-A) from human leukocytes by the calcium ionophore A23187. *J. Immunol.*, 116:1677–1681.
6. Dessaint, J. P., Torpier, G., Capron, M., Brazin, H., and Capron, A. (1979): Cytophilic binding of IgE to the macrophage. I. Binding characteristics of IgE on the surface of macrophages in the rat. *Cell Immunol.*, 46:12–23.
7. Falkenhein, S. F., MacDonald, H., Huber, M. M., Koch, D., and Parker, C. W. (1980): Effect of the 5-hydroperoxide of eicosatetraenoic acid and inhibitors of the lipoxygenase pathway on the formation of slow reacting substance by rat basophilic leukemia cells: direct evidence that slow reacting substance is a product of the lipoxygenase pathway. *J. Immunol.*, 125:163–168.
8. Feldberg, W., and Kellaway, C. H. (1938): Liberation of histamine and formation of lysolecithin-like substances by cobra venom. *J. Physiol. (Lond.)*, 94:187–226.

9. Gardner, H. W., Kleiman, R., Weisleder, D., and Inglett, G. E. (1977): Cysteine adds to lipid hydroperoxide. *Lipids*, 12:655–660.
10. Griffith, O. W., and Meister, A. (1979): Glutathione: interorgan translocation, turnover, and metabolism. *Proc. Natl. Acad. Sci. USA*, 65:5606–5610.
11. Hammarström, S., Murphy, R. C., Samuelsson, B., Clark, D. A., Mioskowski, C., and Corey, E. J. (1979): Structure of leukotriene C: identification of the amino acid part. *Biochem. Biophys. Res. Commun.*, 91:1266–1272.
12. Ishizaka, T., Ishizaka, K., Orange, R. P., and Austen, K. F. (1971): Pharmacologic inhibition of the antigen-induced release of histamine and slow reacting substance of anaphylaxis (SRS-A) from monkey lung tissues mediated by human IgE. *J. Immunol.*, 106:1267–1273.
13. Jakschik, B. A., and Parker, C. W. (1976): Probable precursor role for arachidonic acid in slow reacting substance (SRS) biosynthesis. *Clin. Res.*, 24:575A (abstr.).
14. Jakschik, B. A., Kulczycki, A., MacDonald, H. H., and Parker, C. W. (1977): Release of slow reacting substance (SRS) from rat basophilic leukemia (RBL-1) cells. *J. Immunol.*, 119:618–622.
15. Jakschik, B. A., Falkenhein, S., and Parker, C. W. (1977): Precursor role of arachidonic acid in slow reacting substance from rat basophilic leukemia cells. *Proc. Natl. Acad. Sci. USA*, 74:4577–4581.
16. Jakschik, B. A., Sun, F. F., Lee, L., and Steinhoff, M. M. (1980): Calcium stimulation of a novel lipoxygenase. *Biochem. Biophys. Res. Commun.*, 95:103–110.
17. Kellaway, C. H., and Trethewie, E. R. (1940): The liberation of a slow-reacting smooth muscle-stimulating substance in anaphylaxis. *Q. J. Exp. Physiol.*, 30:121–145.
18. Kulczycki, A., Jr., Isersky, C., and Metzger, H. (1974): The interaction of IgE with rat basophilic leukemia cells. I. Evidence for specific binding of IgE. *J. Exp. Med.*, 139:600–616.
19. Lewis, R. A., Drazen, J. M., Austen, K. F., Clark, D. A., and Corey, E. J. (1980): Identification of the C(6)-S-conjugate of leukotriene with cysteine as a naturally occurring slow reacting substance of anaphylaxis (SRS-A): importance of the 11-cis-geometry for biological activity. *Biochem. Biophys. Res. Commun.*, 96:271–277.
20. Murphy, R. C., Hammarstrom, S., and Samuelsson, B. (1979): Leukotriene C: a slow-reacting substance from murine mastocytoma cells. *Proc. Natl. Acad. Sci. USA*, 76:4275–4279.
21. Orange, R. P., Murphy, R. C., and Austen, K. F. (1974): Inactivation of slow reacting substance of anaphylaxis (SRS-A) by arylsulfatases. *J. Immunol.*, 113:316–322.
22. Orange, R. P., and Chang, P. L. (1975): The effect of thiols on immunologic release of slow reacting substance of anaphylaxis. *J. Immunol.*, 115:1072–1077.
23. Parker, C. W. (1977): Aspirin sensitive asthma. In: *Asthma: Physiology, Immunopharmacology, and Treatment*, edited by Lichtenstein, Austen, and Simon, pp. 301–313. Academic Press, New York.
24. Parker, C. W. (1979): Prostaglandins and slow reacting substance. *J. Allergy Clin. Immunol.*, 63:1–14.
25. Parker, C. W., Huber, M. M., and Falkenhein, S. (1979): Evidence that slow reacting substance (SRS) is a fatty acid thioether formed through a previously undefined pathway of arachidonate metabolism. *Clin. Res.*, 27:473A (abstr.).
26. Parker, C. W., Huber, M. M., and Falkenhein, S. F. (1979): Incorporation of ^{35}S into slow reacting substance (SRS). *Fed. Proc.*, 38:1167A (abstr.).
27. Parker, C. W., Huber, M. M., Hoffman, M. K., and Falkenhein, S. F. (1979): Characterization of the two major species of slow reacting substance from rat basophilic leukemia cells as glutathionyl thioethers of eicosatetraenoic acid oxygenated at the 5 position: evidence that peroxy groups are present and important for spasmogenic activity. *Prostaglandins*, 18:673–686.
28. Parker, C. W. (1980): Incorporation of ^{14}C-AA into SRS. In: *Proceedings of the Symposium on SRS-A and Leukotrienes*. Wiley, London (*in press*).
29. Parker, C. W., Koch, D. A., Huber, M. M., and Falkenhein, S. F. (1980): Arylsulfatase inactivation of slow reacting substance: evidence for proteolysis as a major mechanism when ordinary commercial preparations of the enzyme are used. *Prostaglandins*, 20:887–908.
30. Parker, C. W. (1980): Pulmonary effects of inhibitors. In: *Prostaglandin Synthetase Inhibitors in Clinical Medicine*, pp. 30–43. Proceedings of the International Prostaglandins Meeting, Paris, September, 1979. Alan R. Liss, New York.
31. Parker, C. W., Falkenhein, S. F., and Huber, M. M. (1980): Sequential conversions of the glutathionyl side chain of slow reacting substance (SRS) to cysteinyl-glycine and cysteine in rat basophilic leukemia cells stimulated with A23187. *Prostaglandins*, 20:863–886.

32. Parker, C. W., Koch, D. A., Huber, M. M., and Falkenhein, S. F. (1980): Incorporation of ra-
    diolabel from [^{14}C]5-hydroxy-eicosatetraenoic acid into slow reacting substances. *Biochem. Bio-
    phys. Res. Commun.*, 94:1037–1043.
33. Parker, C. W. (1981): Chemical nature of slow reacting substance. In: *Advances in Inflammation
    Research*. Raven Press, New York *(in press)*.
34. Parker, C. W., Fischman, C. M., and Wedner, H. J. (1980): Relationship of slow reacting sub-
    stance biosynthesis to intracellular glutathione levels. *Proc. Natl. Acad. Sci. USA*, 77:6876–6883.
35. Rankin, J. A., Hitchcock, M., Merrill, W. W., and Askenase, P. W. (1981): Slow reacting sub-
    stance (SRS): IgE dependent release from alveolar macrophages (AM). *Fed. Proc.*, 409:1014.
36. Samuelsson, B., Borgeat, P., Hammarström, S., and Murphy, R. C. (1979): Introduction of a
    nomenclature: leukotrienes. *Prostaglandins*, 17:785–787.
37. Sok, D., Pai, J., Atrach, V., and Sih, C. J. (1980): Characterization of slow reacting substances
    (SRSs) of rat basophilic leukemia (RBL-1) cells: effect of cysteine on SRS profile. *Proc. Natl.
    Acad. Sci. USA*, 77:6481–6485.
38. Watanabe-Kohno, S., and Parker, C. W. (1980): Role of arachidonic acid in the biosynthesis of
    slow reacting substance of anaphylaxis (SRS-A) from sensitized guinea pig lung frag-
    ments: evidence that SRS-A is very similar or identical structurally to nonimmunologically induced
    forms of SRS. *J. Immunol.*, 125:946–955.
39. Yecies, L. D., Johnson, S. M., Wedner, H. J., and Parker, C. W. (1979): Slow reacting substances
    (SRS) from ionophore A23187 stimulated peritoneal mast cells of the normal rat. II. Evidence
    for a precursor role of arachidonic acid and further purification. *J. Immunol.*, 122:2090–2095.
40. Yecies, L. D., Watanabe, S., and Parker, C. W. (1979): Slow reacting substance from ionophore
    A23187 stimulated basophilic leukemia cells and peritoneal mast cells in the rat. I. Purification
    and comparison during sequential Sephadex LH-20 and thin layer chromatography. *Life Sci.*,
    25:1909–1916.
41. Yecies, L. D., Wedner, H. J., Johnson, S. M., Jakschik, B. A., and Parker, C. W. (1979): Slow
    reacting substance (SRS) from ionophore A23187 stimulated peritoneal mast cells of the normal
    rat. I. Conditions of generation and initial characterization. *J. Immunol.*, 122:2083–2089.

Leukotrienes and Other Lipoxygenase Products,
edited by B. Samuelsson and R. Paoletti.
Raven Press, New York © 1982.

Modulation of Leukotriene Formation By Various Acetylenic Acids

Barbara A. Jakschik, Denise M. DiSantis, *S. K. Sankarappa, and *Howard Sprecher

*Department of Pharmacology, Washington University Medical School, St. Louis, Missouri 63110; and *Department of Physiological Chemistry, Ohio State University, Columbus, Ohio 43210*

Various enzymatic pathways will oxygenate arachidonic acids and form different products. Prostaglandins and thromboxanes are formed by the cyclo-oxygenase pathway, whereas polyunsaturated hydroxy fatty acids are generated by lipoxygenases. Two mammalian lipoxygenases have been described, 12-lipoxygenase in platelets (5) and 5-lipoxygenase in leukocytes (1). 5-Lipoxygenase is the first step in the biosynthesis of leukotrienes (1,14). A number of tissues and cells synthesize prostaglandins (PGs) as well as lipoxygenase products. Rat basophilic leukemia (RBL-1) cells, the cell used in this study, produce leukotrienes (LT), mainly LTD_4 (13,16,19), as well as PGD_2 (8). It is of interest to determine how these enzymes, which utilize the same substrate, arachidonic acid, vary in their requirements for binding and catalysis. Hopefully, such knowledge will make it possible to design specific antagonists with differential inhibition of one pathway of arachidonic acid metabolism and not the others.

Utilizing a cell-free enzyme system obtained from RBL-1 cells, we investigated factors which modulate leukotriene formation. We found that calcium enhances the 5-lipoxygenase step (9), which differentiates this enzyme from cyclo-oxygenase as well as 12-lipoxygenase. Studying the substrate requirements for 5-lipoxygenase, we tested several polyenoic fatty acids that vary in chain length as well as number and position of double bonds. We found that this 5-lipoxygenase has a stringent substrate requirement for polyenoic acids with double bonds at carbons 5, 8, and 11 (10). This substrate requirement further differentiates 5-lipoxygenase from cyclo-oxygenase and 12-lipoxygenase (15).

In this study we investigated the effect of a large number of acetylenic acids (that vary in chain length and in position and number of triple bonds) on arachidonic acid metabolism by RBL-1 cells. The results further differentiate cyclo-oxygenase and leukotriene-forming enzymes.

MATERIALS

The acetylenic fatty acids were prepared by total organic synthesis as described previously (18,20) except for 5,8,11,14-eicosatetraynoic acid, which was kindly supplied by Roche Laboratories. Arachidonic acid was obtained from Nu-Chek Prep. and [^{14}C]arachidonic acid (55 Ci/mole) from New England Nuclear. Leukotriene A_4 was kindly supplied by Dr. J. Rokach (Frosst Merck, Canada) and prostaglandin standards by Dr. J. Pike (Upjohn Co., Kalamazoo, MI). Thin-layer chromatography (TLC) was performed on silica gel plates from MCG Manufacturing Chemists, Inc. (Cincinnati, OH).

METHODS

RBL-1 cells were grown in suspension cultures as described previously (21). The cells were washed with 50 mM phosphate buffer, 1 mM EDTA, resuspended at 5×10^7/ml, and homogenized in 35 mM phosphate buffer, 1 mM EDTA, as described previously (21).

Determination of 5-Lipoxygenase Activity

The 10,000 $\times$ g supernatant of the RBL-1 homogenate was used for determining 5-lipoxygenase activity. The acetylenic acids (0 to 180 μM) were preincubated with 0.5 ml of the 10,000 $\times$ g supernatant, 1.5 mM Ca^{++}, for 10 min on ice. Preincubation at 37°C or room temperature was not possible as the enzyme activity was destroyed under these conditions. The preparation was incubated with [^{14}C]arachidonic acid (220,000 cpm, 4μM) for 15 min at 37°C. The protein was precipitated with 2 volumes acetone and the supernatant extracted with chloroform (pH 3.4). The extract was concentrated and applied to silica gel thin-layer plates. Chromatography was performed with solvent system A9 (ethyl acetate:2,2,4-trimethylpentane:acetic acid:water (110:50:20:100). This solvent system separates prostaglandins (4) as well as 5-HETE and 5,12-DHETE (9). Radioautography was performed with Kodak X-Omat RAX-5 film (3 days). Quantitation was obtained scraping the appropriate bands and liquid scintillation counting. The sum of the counts per minute in 5-HETE and 5,12-DHETE bands was designated as the 5-lipoxygenase products and the sum in the bands comigrating with $PGF_{2\alpha}$, PGE_2, and PGD_2 as cyclo-oxygenase products. Each experiment was performed in duplicate three to five times.

Determination of SRS Activity

For the generation of slow-reacting substance (SRS), a mixture of LTC_4 and LTD_4 (12), the whole homogenate was used as less SRS activity is obtained with the 10,000 $\times$ g supernatant (12). Preincubation of the homogenate with the acetylenic acids (0 to 180 μM) and 1.5 mM Ca^{++} was performed on ice for 10 min. Reduced glutathione (GSH) (1 mM) and arachidonic acid (4 μM) were then added and the mixture was incubated at 37°C for 15 min. SRS activity was monitored on the superfused (Krebs Henseleit solution) guinea pig ileum in the presence of various inhibitors, including pyrilamine maleate, 8 μg/ml, and indomethacin, 1 μg/ml (8). No inhibition was observed when the acetylenic acids were added to controls (no acetylenic acid present during the incubation) just before the sample was applied to the guinea pig ileum. Quantitation was obtained by determining the area under the curve produced by the contraction. Each experiment was repeated four to eight times.

Evaluation of Data

Dose-response curves were obtained for each acetylenic acid and the results expressed as percent of control. The concentration of acetylenic acid required to cause 50% reduction in activity (IC_{50}) as compared to the control was determined.

RESULTS AND DISCUSSION

Table 1 shows the effect of the various $\Delta 5$ acetylenic acids on arachidonic acid metabolism. Relatively high concentrations are required for inhibition of cyclooxygenase as well as 5-lipoxygenase, and no important differences in the effect on the two enzymes were observed. This differs greatly from the specific requirement of a $\Delta 5$ double bond in polyenoic acids to be a substrate for the 5-lipoxygenase (10). These results may indicate differential requirements of 5-lipoxygenase at the catalytic side ($\Delta 5$ double bond) and the binding side as the affinity of the $\Delta 5$ acetylenic acids for the enzyme is relatively low.

The $\Delta 4$ acetylenic acids are an interesting group of compounds (Table 2). None of them inhibits cyclo-oxygenase or 5-lipoxygenase. However, all inhibit SRS formation, and the potency increases with the carbon chain length of the compound

TABLE 1. *Effect of $\Delta 5$ acetylenic acids on arachidonic acid metabolism*

Compound	IC_{50} (μM)		
	Cyclo-oxygenase	5-Lipoxygenase	SRS activity
18:3 (5a,8a,11a)	89 ± 7	74 ± 4	42 ± 8
18:4 (5a,8a,11a,14a)	62 ± 18	> 100	41 ± 8
19:3 (5a,8a,11a)	37 ± 7	54 ± 6	49 ± 6
19:4 (5a,8a,11a,14a)	54 ± 14	> 100	32 ± 6
20:3 (5a,8a,11a)	57 ± 7	87 ± 13	34 ± 4
20:4 (5a,8a,11a,14a)	29 ± 4	52 ± 4	17 ± 4
21:4 (5a,8a,11a,14a)	65 ± 8	> 100	52 ± 11

TABLE 2. *Comparison of $\Delta 4$ acetylenic acids*

Compound	IC_{50} (μM)		
	Cyclo-oxygenase	5-Lipoxygenase	SRS activity
16:3 (4a,7a,10a)	> 60	> 60	31 ± 8
19:4 (4a,7a,10a,13a)	> 55	> 55	26 ± 2
20:4 (4a,7a,10a,13a)	> 50	> 50	14 ± 2
21:4 (4a,7a,10a,13a)	> 50	> 50	7 ± 1

(11). The Δ4 acetylenic acids therefore seem to be selective inhibitors of glutathione transferase, the enzyme that catalyzes the addition of glutathione to LTA_4. This is further substantiated by the observation that these compounds also block the formation of SRS (LTC_4 and LTD_4) from the immediate precursor LTA_4 (Fig. 1). These findings make the Δ4 acetylenic acids valuable in the investigation of the role of LTC_4 and LTD_4 in biological processes.

The Δ7 acetylenic acids (Table 3) are rather potent inhibitors of cyclo-oxygenase as well as 5-lipoxygenase. Only one compound, 21:3 (7a,10a,13a), shows a moderate amount of differential inhibition, with the IC_{50} for cyclo-oxygenase four-fold higher than that for 5-lipoxygenase. There are a number of compounds which

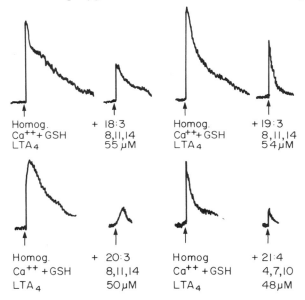

FIG. 1. Inhibition of SRS formation by acetylenic acids. RBL-1 homogenate, 1.5 mM Ca^{++}, was preincubated on ice with the acetylenic acids at concentrations indicated. Then 1 mM GSH and LTA_4 were added and the mixture incubated at 37°C for 15 min. The incubation mixture was applied to the guinea pig ileum to measure SRS activity.

TABLE 3. *Action of Δ7 acetylenic acids*

	IC_{50}(μM)		
Compound	Cyclo-oxygenase	5-Lipoxygenase	SRS activity
20:3 (7a,10a,13a)	25 ± 9	10 ± 3	10 ± 2
20:4 (7a,10a,13a,16a)	5 ± 1	15 ± 6	10 ± 1
21:3 (7a,10a,13a)	14 ± 2	3 ± 0.4	16 ± 3
22:4 (7a,10a,13a)	3 ± 0.2	6 ± 0.7	9 ± 1

preferentially inhibit cyclo-oxygenase, e.g., 20:2 (10a,13a) (Table 5, below), 17:3 (8a,11a,14a), 18:3 (8a,11a,14a), 20:2 (8a,11a) (Table 4, below), and 20:4 (6a,9a,12a,15a) (Table 5, below).

The Δ8 acetylenic acids tested are a large group of compounds with varying chain length and number of triple bonds. They allow a more detailed study of the effect of chain length and number of triple bonds. As can be seen from Table 4 and Figs. 2 and 3, the potency increases with the chain length up to 20 carbons. A minimum of 19 carbons must be present to exhibit some inhibition of 5-lipoxygenase (Fig. 1). No such requirement for the carbon chain length was observed for cyclo-oxygenase. Compounds 17:3 (8a,11a,14a) and 18:3 (8a,11a,14a) are potent cyclo-oxygenase inhibitors. A similar requirement for carbon chain length

TABLE 4. *Comparison of Δ8 acetylenic acids*

Compound	IC$_{50}$(μM)		
	Cyclo-oxygenase	5-Lipoxygenase	SRS activity
17:3 (8a,11a,14a)	5 ± 1	> 80	46 ± 8
18:3 (8a,11a,14a)	5 ± 2	> 80	13 ± 2
19:3 (8a,11a,14a)	22 ± 9	52 ± 10	14 ± 1
20:2 (8a,11a)	16 ± 2	> 70	21 ± 6
20:3 (8a,11a,14a)	14 ± 5	25 ± 5	10 ± 3
20:4 (8a,11a,14a,17a)	24 ± 3	> 70	11 ± 1
21:3 (8a,11a,14a)	32 ± 6	36 ± 3	27 ± 2
21:4 (8a,11a,14a,17a)	8 ± 3	13 ± 3	7 ± 1
22:3 (8a,11a,14a)	17 ± 2	12 ± 3	14 ± 2

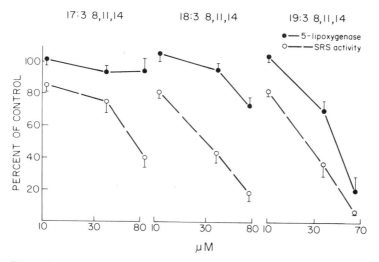

FIG. 2. Effect of carbon chain length on inhibition. For experimental details, see *Methods*. Bars denote SEM (mean of three to eight experiments).

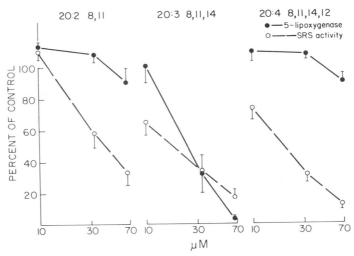

FIG. 3. Effect of number of triple bonds on inhibition. For experimental details, see *Methods*. Bars denote SEM (mean of three to eight experiments).

TABLE 5. *Action of Δ6 and Δ10 acetylenic acids*

Compound	$IC_{50}(\mu M)$		
	Cyclo-oxygenase	5-Lipoxygenase	SRS activity
18:3 (6a,9a,12a)	< 55	> 55	22 ± 4
18:4 (6a,9a,12a,15a)	< 55	> 55	50 ± 3
19:4 (6a,9a,12a,15a)	53 ± 3	> 53	41 ± 4
20:3 (6a,9a,12a)	9 ± 1	15 ± 2	16 ± 2
20:4 (6a,9a,12a,15a)	11 ± 2	> 70	15 ± 3
21:3 (6a,9a,12a)	62 ± 17	32 ± 6	15 ± 3
22:4 (6a,9a,12a,15a)	31 ± 3	33 ± 2	14 ± 2
20:2 (10a,13a)	20 ± 6	> 70	52 ± 7
22:3 (10a,13a,16a)	10 ± 4	14 ± 1	11 ± 0.6

was observed with the Δ6 acetylenic acids (Table 3). The number of triple bonds present also influenced the degree of inhibition of 5-lipoxygenase; a minimum of three triple bonds seem to be necessary (Fig. 2). Compounds 20:2 (8a,11a) and 20:2 (10a,13a) (Table 5) did not have an inhibitory effect on 5-lipoxygenase. Again, such a requirement was not noticed with cyclo-oxygenase. Compounds 20:2 (8a,11a) and 20:2 (10a,13a) are potent cyclo-oxygenase inhibitors. Comparison of the various Δ8 acetylenic acids also showed that with a carbon chain length of 20 the optimum number of double bonds is three (20:3 Δ8). Increasing the number to four (20:4 Δ8) causes a loss of the inhibitory action on 5-lipoxygenase. A similar effect of the number of double bonds present is observed with Δ6 compounds (Table 5).

Compound 20:3 (6a,9a,12a) is a good inhibitor of 5-lipoxygenase, but 20:4 (6a,9a,12a,15a) is not. Again, such a variation in the number of triple bonds does not affect the action of acetylenic acid analogs on cyclo-oxygenase. Compounds 20:4 Δ8 and 20:4 Δ6 are good cyclo-oxygenase inhibitors.

The effect of some of the acetylenic acids utilized in this study have also been tested by some other investigators for their effect on platelet cyclo-oxygenase and lipoxygenase (2,3,6,22–24). Our results with regard to the effect on cyclooxygenase vary in some instances. This may be due to the fact that the other investigators used platelets as an intact cell preparation, whereas we used a broken cell preparation from RBL-1 cells which did not allow preincubation with the acetylenic acids at 37°C. Orning and Hammarström (17) tested 20:3 (5a,8a,11a) on RBL-1 cells and obtained an IC_{50} which was lower than ours. Again, this could be due to different incubation conditions.

As can be seen from Tables 1 through 5, all the acetylenic acids tested had inhibitory activity on SRS formation, even though some of these compounds did not inhibit 5-lipoxygenase. In general, an increase of potency was observed with an increase of the carbon chain length. It appears that these acetylenic acids directly inhibit the glutathione transferase which synthesizes LTC_4 from LTA_4. This conclusion is substantiated by the finding that these acetylenic acids block the formation of SRS from LTA_4 (Fig. 1). The observation that such a large number of acetylenic acids block the glutathione transferase suggests that this inhibition is rather nonspecific for this enzyme. It agrees with the observation that glutathione transferases from other tissues have a broad substrate specificity (7).

Even though glutathione transferase is inhibited by a large number of acetylenic acids, the Δ4 compounds are specific blockers of SRS formation as compared to arachidonic acid metabolism by other enzymes, as the Δ4 acetylenic acids do not inhibit cyclo-oxygenase or 5-lipoxygenase. The Δ4 acetylenic acids tested, except for 20:4 (4a,7a,11a,13a), also do not have an effect on 12-lipoxygenase in platelets (Sams et al., *this volume*). Therefore these compounds are valuable tools for studying the function of LTC_4 and LTD_4 in cellular mechanisms.

SUMMARY

A series of acetylenic acids varying in chain length as well as in number and position of triple bonds were tested for their effect on arachidonic acid metabolism via cyclo-oxygenase and 5-lipoxygenase in homogenates of RBL-1 cells. A number of trends which differentiate 5-lipoxygenase from cyclo-oxygenase were observed: (a) In general, the potency of the acetylenic acids to inhibit 5-lipoxygenase increased with the carbon chain length up to 20 carbons. (b) A minimum of three triple bonds is necessary for the inhibitory action. (c) With the 20-carbon Δ8 and Δ6 acetylenic acids, activity decreased when the number of triple bonds increased from three to four. These structure activity relationships were not observed with cyclo-oxygenase. The Δ4 acetylenic acids proved to be an interesting group of compounds. They did not inhibit cyclo-oxygenase or 5-lipoxygenase but were good inhibitors of SRS synthesis. The potency of the $\Delta 4$ acetylenic acids increased with chain length, and 21:4 (4a,7a,10a,13a) was the most potent compound inhibiting SRS formation. The $\Delta 4$ acetylenic acids are therefore useful tools for investigating the function and action of LTC_4 and LTD_4.

ACKNOWLEDGMENTS

This work was supported by NIH grants 2RO1 HL21874-04, 5-P30-CA16217-03, and HV-72945 (H.S.).

REFERENCES

1. Borgeat, P., and Samuelsson, B. (1979): Arachidonic acid metabolism in polymorphonuclear leukocytes: unstable intermediate in the formation of dihydroxy acids. *Proc. Natl. Acad. Sci. USA*, 76:3213–3217.
2. Dutilh, C. E., Haddeman, E., Jouvenaz, G. H., Hoor, F. T., and Nugteren, D. H. (1979): Study of the two pathways for arachidonate oxygenation in blood platelets. *Lipids*, 14:241–246.
3. Goetz, J. M., Sprecher, H., Cornwell, D. G., and Paganamala, R. V. (1976): Inhibition of prostaglandin biosynthesis by triynoic acids. *Prostaglandins*, 12:187–192.
4. Hamberg, M., and Samuelsson, B. (1966): Prostaglandins in human seminal plasma. *J. Biol. Chem.*, 241:257–263.
5. Hamberg, M., and Samuelsson, B. (1974): Prostaglandin endoperoxides: novel transformation of arachidonic acid in human platelets. *Proc. Natl. Acad. Sci. USA*, 71:3400–3404.
6. Hammarström, S. (1977): Selective inhibition of n-8 lipoxygenase by 5,8,11-eicosatriynoic acid. *Biochim. Biophys. Acta*, 487:517–519.
7. Jakoby, W. B. (1978): The glutathione S-transferases: a group of multifunctional detoxification proteins. *Adv. Enzymol.*, 46:383–414.
8. Jakschik, B. A., Lee, L. H., Shuffer, G., and Parker, C. W. (1978): Arachidonic acid metabolism in rat basophilic leukemia (RBL-1) cells. *Prostaglandins*, 16:733–748.
9. Jakschik, B. A., Sun, F. F., Lee, L. H., and Steinhoff, M. M. (1980): Calcium stimulation of a novel lipoxygenase. *Biochem. Biophys. Res. Commun.*, 95:103–110.
10. Jakschik, B. A., Sams, A. R., Sprecher, H., and Needleman, P. (1980): Fatty acid structural requirements for leukotriene biosynthesis. *Prostaglandins*, 20:401–410.
11. Jakschik, B. A., DiSantis, D. M., and Sprecher, H. (1981): Delta four acetylenic acids—inhibitors of SRS formation. *Biochim. Biophys. Res. Commun.*, 102:624–629.
12. Jakschik, B. A., Harper, T., and Murphy, R. C. (1981): Manuscript in preparation.
13. Morris, H. R., Taylor, C. W., Piper, P. J., Samhoun, M. N., and Tippins, J. R. (1980): Slow reacting substances: the structure identification of SRSs from rat basophil leukemia (RBL-1) cells. *Prostaglandins*, 19:185–201.
14. Murphy, R. C., Hammarström, S., and Samuelsson, B. (1979): Leukotriene C: a slow-reacting substance from murine mastocytoma cells. *Proc. Natl. Acad. Sci. USA*, 76:4275–4279.
15. Needleman, P., Wyche, A., LeDuc, L., Sankarappa, S. K., Jakschik, B. A., and Sprecher, H. (1981): Fatty acid as sources of potential "magic bullets" for the modification of platelet and vascular function. *Prog. Lipid Res. (in press)*.
16. Orning, L., Hammarström, S., and Samuelsson, B. (1980): Leukotriene D: a slow reacting substance from rat basophilic leukemia cells. *Proc. Natl. Acad. Sci. USA*, 77:2014–2017.
17. Orning, L., and Hammarström, S. (1980): Inhibition of leukotriene C and D biosynthesis. *J. Biol. Chem.*, 255:8023–8026.
18. Osbond, J. M., Philpott, P. G., and Wickens, J. C. (1961): Essential fatty acids. I. Synthesis of linoleic, α-linolenic, arachidonic and docosa-4,7,10,13,16-pentaenoic acid. *J. Chem. Soc.*, II:2779–2787.
19. Parker, C. W., Huber, M. M., Hoffman, M. K., and Falkenhein, S. F. (1979): Characterization of the two major species of slow reacting substance from rat basophilic leukemia cells as glutathionyl thioethers of eicosatetraenoic acids oxygenated at the 5 position: evidence that peroxy groups are present and important for spasmogenic activity. *Prostaglandins*, 18:673–686.

20. Sprecher, H. (1978): The organic synthesis of unsaturated fatty acids. In: *Progress in the Chemistry of Fats and Other Lipids*, edited by R. T. Hohman, Vol. 15, pp. 219–254. Pergamon Press, London.

21. Steinhoff, M. M., Lee, L. H., and Jakschik, B. A. (1980): Enzymatic formation of prostaglandin D_2 by rat basophilic cells and normal rat mast cells. *Biochim. Biophys. Acta*, 618:28–34.

22. Sun, F. F., McGuire, J. C., Merton, D. R., Pike, J. E., Sprecher, H., and Kunau, W. H. (1981): Inhibition of platelet arachidonic acid 12-lipoxygenase by acetylenic acid compounds. *Prostaglandins*, 21:333–343.

23. Tobias, L. D., and Hamilton, J. G. (1979): The effect of 5,8,11,14-eicosatetraynoic acid on lipid metabolism. *Lipids*, 14:181–193.

24. Wilhelm, T. E., Sankarappe, S. H., Van Rollins, M., and Sprecher, H. (1981): Selective inhibitors of platelet lipoxygenase: 4,7,10,13-icosatetraynoic acid and 5,8,11,14-henicosatetraynoic acid. *Prostaglandins*, 21:323–332.

Leukotrienes and Other Lipoxygenase Products,
edited by B. Samuelsson and R. Paoletti.
Raven Press, New York © 1982.

Structure, Function, and Metabolism of Leukotriene Constituents of SRS-A

*¶Robert A. Lewis, *¶K. Frank Austen, **Jeffrey M. Drazen,
§¶Nicholas A. Soter, *¶Joanne C. Figueiredo, and †E. J. Corey

*Departments of *Medicine and §Dermatology, Harvard Medical School, and
¶Department of Rheumatology and Immunology, Brigham and Women's Hospital, Boston,
Massachusetts 02115; **Department of Physiology, Harvard School of Public Health,
and Departments of Medicine, Harvard Medical School and Brigham and Women's
Hospital, Boston, Massachusetts 02115; and †Department of Chemistry, Harvard
University, Cambridge, Massachusetts 02138*

The biologically active moiety referred to for 40 years as "slow reacting substance of anaphylaxis"(SRS-A) is now known to comprise three related metabolites of arachidonic acid: leukotrienes (LT) C_4, D_4, and E_4 (22,23,29,30,37). LTC_4, the first member of this family of molecules to be completely defined chemically, was identified by analysis of a nonimmunologically generated product (30). However, the finding that SRS-A was derived from acute immunologic reactions (3,19) provided a compelling impetus to seek LTC_4 and LTD_4 as products of mast-cell-dependent hypersensitivity reactions in fragments of guinea pig (29) and human (22) lung and immune complex-, neutrophil-, and complement-dependent reactions in the rat peritoneal cavity (22,23). The determination of whether the triggering mechanism for SRS-A generation in complex tissue systems is immunoglobulin E (IgE)-mast-cell-dependent (32,34) or IgGa-neutrophil-complement-dependent (31,35) does not define the predominant cell source of the leukotrienes. Moreover, with dispersed human lung cells, there is evidence for a major contribution by nonmast cell types responding to the activation of mast cells (25). This finding is consistent with the capacity of mammalian mononuclear cells to produce the leukotrienes in response to nonimmunologic stimuli (1,43).

That SRS-A, as a product of allergic reactions, might have relevance to bronchial asthma was suggested by the exquisite spasmogenic response of human bronchiolar tissue to crude material (4) and was supported by the later findings that partially purified SRS-A had a preferential action on peripheral as compared to central airways of the guinea pig *in vivo* (8) and *in vitro* (11). The unique capacity of this mediator class to preferentially impair pulmonary compliance compared to conductance was confirmed by the intravenous administration of synthetic LTC_4 and LTD_4 to guinea pigs (9) and by recent demonstrations in humans of dose-dependent and selective decrements in flow rates at 30% of vital capacity after inhalation of LTC$_4$ (Weiss, Drazen, Lewis, Weller, Corey, and Austen, *unpublished data*). The earlier demonstration that SRS-A augmented venular permeability in guinea pig skin (31) has been confirmed with synthetic LTC$_4$, LTD$_4$, and LTE$_4$ (9,23) and extended to the demonstration of wheal and prolonged flare cutaneous responses in the human (28), a finding which again has implications for allergic or nonallergic inflammatory diseases. The detection of the potent neutrophil-attracting chemotactic factor LTB$_4$ (2) in the synovial fluids of rheumatoid arthritis patients (20), along with various cyclo-oxygenase products (42) and material reacting antigenically with antibody to LTD$_4$ (Lewis, Coblyn, Corey, Levine, and Austen, *unpublished data*), emphasizes the critical need to consider the interactions of the oxidative products of arachidonic acid (21) in inflammatory diseases.

A number of interactions have been described between the leukotrienes and the arachidonic acid metabolites arising via the cyclo-oxygenase pathway, including positive or negative modulation of specific product biosynthesis or action on target tissues and even direct recruitment of products. The cyclo-oxygenase products prostaglandin (PG) F$_{2\alpha}$ and PGE$_1$ augment and suppress, respectively, IgE-dependent generation of SRS-A from human lung tissue (45). SRS-A generation in guinea pig lung and rat peritoneal cavity by IgG$_1$ and IgGa class antibody-dependent mechanisms is inhibited by pretreatment of each tissue with PGI$_2$ (5,12). The SRS-A leukotrienes apparently directly stimulate production of PGE$_2$, PGI$_2$, and thromboxane A$_2$ (TXA$_2$) from rat peritoneal macrophages (14) and TXA$_2$ from guinea pig lung (13). At the target tissue level, PGD$_2$ acts synergistically with LTB$_4$ to produce local cutaneous neutrophil infiltration in the rhesus monkey (26) and the human (28). As a precondition to understanding these interactions, the structural basis by which leukotrienes exert their tissue effects was explored.

STRUCTURAL BASIS OF SRS-A LEUKOTRIENE FUNCTION

Leukotriene D$_4$ (Fig. 1) is structurally composed of an eicosanoid (C$_{20}$) backbone (I), a C-5 hydroxyl (II), and a C-6 sulfidopeptide (III). Each of these structural components is itself comprised of one or more structural domains. The eicosanoid is comprised of a terminal (C-1) carboxylic acid (IA), a conjugated (7,9-*trans*,11-*cis*) triene (IB), and an unconjugated olefinic bond that forms a *cis*,*cis*-1,4-diene with the ω end of the triene (IC). The C-5 hydroxyl (II) is of *S*-chirality. The C-6 thiopeptide has several domains, as diagrammed for LTD$_4$ (Fig. 1): C-6 with *R*-chirality (IIIA), sulfur with a valence of 2 (IIIB), linked to 2-L-amino (IIIC) proprionyl (IIID) amino acetyl (glycyl) chain (IIIE) with a free carboxylic acid (IIIF).

Within this scheme, LTC$_4$ can thus be described as a substitution at IIIC, LTE$_4$ as a deletion of IIIC, and LTB$_4$ as a deletion of IIIA-F and introduction of a 12-*S*-hydroxyl into the previous domain of the *cis*,*cis*-1,4-diene (IC), thereby altering the position and stereochemistry of the triene (IB) to 6-*cis*,8,10-*trans* (6,26) (Fig. 1).

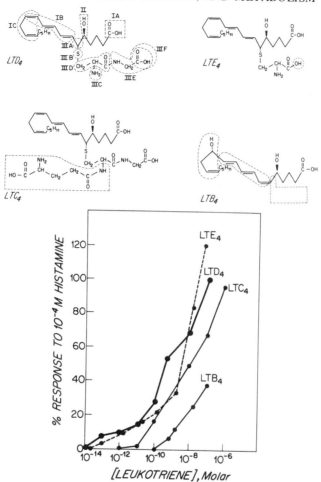

FIG. 1. Natural leukotrienes of the 5-lipoxygenase pathway: structure and function on guinea pig pulmonary parenchymal strips. (Adapted in part from refs. 9,10, and 23).

Biological activity of the SRS-A leukotrienes on guinea pig nonvascular smooth muscle assays is essentially eliminated by displacement of the sulfido-peptide (III) on the eicosanoid backbone (I), with or without displacement of the hydroxyl group (II). The concentrations of LTC_4, LTD_4, and LTE_4 that evoke a half-maximal response (EC_{50}) of pulmonary parenchymal strips are 1×10^{-8} M, 6×10^{-10} M, and 3×10^{-9} M, respectively (9,23). The two diastereomeric analogs in which the LTC_4 sulfidopeptide is attached at C-12 rather than at C-6 are inactive on the pulmonary parenchymal strip and ileum at concentrations up to 0.2 μM. The C-14-S-glutathionyl,C-15-hydroxyl analog of LTC_4 and the C-14-S-cysteinylglycyl,C-15-hydroxyl analog of LTD_4 are also only minimally active at 0.2 μM concentrations. The 11-hydroxyl,12-S-peptidyl analogs of LTC_4 and LTD_4, although more

active than any of the other peptide-displacement analogs, were still less than 0.1%
as active as their parent compounds (10). Displacement of the thiopeptide in these
analogs alters the molecule in several ways, including a decrement in the hydrophobicity of the ω portion of the molecule, displacement of the thiopeptide with
all of its subdomains (IIIA-F) from the terminal carboxylic acid of the eicosanoid,
and alteration of the stereochemistry of the triene (IB), the *cis,cis*-1,4-diene (IC),
or both.

Functional evaluation of an additional set of analogs suggests that displacement
of the thiopeptide impairs function because of the effect on hydrophobicity of C-7
to C-20. The ratio of the concentration of the parent compound to that of an analog
required to elicit the same standard contractile response of the guinea pig parenchymal strip or ileum is used to define the percent activity of the analog relative
to the parent structure in each bioassay. The 14,15-dihydro analogs of LTC_4 and
LTD_4, which alter only the *cis,cis*-1,4-diene (IC), are, respectively, 19% and 66%
as active as their parent compounds (10). Saturation of two (C-9,11) olefinic bonds
of the triene, as well as the unconjugated one at C-14, produces a pair of C-7 *cis-
trans* stereoisomers for LTC_4 and LTD_4, which have biological activities on the
smooth muscle assays intermediate between those of the parent compounds and
those of the S-peptide displacement analogs. The most active of these analogs (7-
trans, hexahydro-LTC_4) has 19% of the activity of its parent compound on the
pulmonary parenchymal strip and 35% on the ileum, and the least active (7-*trans*
hexahydro-LTD_4) has 2% and 0.5% on pulmonary parenchyma and ileum, respectively (10). In contrast, the product of 15-lipoxygenase action on LTD_4, which
leaves the triene (IB) intact but alters the unconjugated diene by displacing the C-
14 olefin to C-13 and interferes with hydrophobicity by hydroxylating C-15, has
less than 0.1% of LTD_4 activity on the pulmonary parenchyma. These data suggest
that activation of the nonvascular contractile tissues requires interaction of the ω
portion of the leukotriene with hydrophobic moieties on the recognition unit and
that the stereochemical requirements for this interaction are not strictly defined.

Each of the subdomains of the S-peptide (IIIA-F) has some determinant role on
the biological efficacy of the SRS-A leukotrienes for the contractile assay tissues.
There is no S-peptide in the structure of LTB_4, and the molecule is inactive on the
guinea pig ileum at concentrations up to 2 μM. This inactivity cannot be attributed
solely to the introduction of the second (C-12) hydroxyl in the hydrophobic ω
portion of the eicosanoid because LTB_4 is 1 log less active than the 15-lipoxygenase
product of LTD_4 on a molar basis.

Two diastereomeric LTD_4 sulfoxides that vary structurally from LTD_4 at the IIIB
domain differ in their bioactivities on the pulmonary parenchymal strip and ileum
by 2 logs, such that one is 10% and the other less than 0.1% as active as LTD_4
(24). The 5(*S*)-hydroxy-6(*R*)-S-cysteinylglycine isomer of LTD_4 (6-epi-LTD_4, varied in the IIIA domain) has an EC_{50} of 0.4% of its parent compound. The substitution
of D-cysteine for L-cysteine (IIID domain) decreases activity 1 log on the pulmonary
parenchyma and nearly 2 logs on the ileum, and the substitution of D-penicillamine

for L-cysteine produces a 3-log decrement in activity in each bioassay (24) (Fig. 2). These data indicate that the conformational relationship between the peptide and eicosanoid moieties of LTD$_4$ is critical for spasmogenic activity of the agonist. That these geometric and structural changes can significantly decrease LTD$_4$ activity also argues for the presence of a specific receptor for LTD$_4$ on the target tissues, although the additional criteria of saturability and a dose-dependent relationship of binding to functional response are not yet available.

Evaluation of the role of the C-5 hydroxyl for smooth muscle spasmogenic activity is limited by virtue of the few relevant analogs available. The 7-*trans* hexahydro LTD$_4$ analog, which elicits 2% of the LTD$_4$ effect on pulmonary parenchyma and 0.5% on the ileum, is without measurable activity after 5-dehydroxylation (II domain). Further, the 5(R)-hydroxy,6(R)-cysteinylglycine isomer of LTD$_4$ (5-epi-LTD$_4$, II domain variation) has an EC$_{50}$ of 0.4% of its parent compound on the

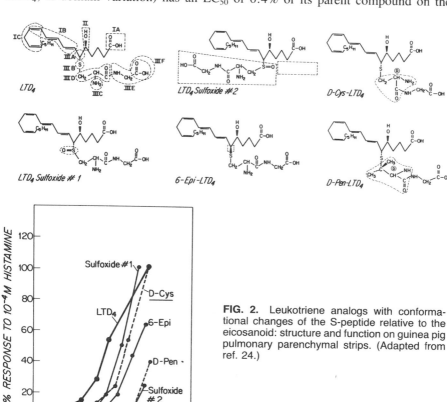

FIG. 2. Leukotriene analogs with conformational changes of the S-peptide relative to the eicosanoid: structure and function on guinea pig pulmonary parenchymal strips. (Adapted from ref. 24.)

pulmonary parenchymal strips (Fig. 3) and an activity of 0.5% of LTD_4 on the ileum (24). These data suggest a role for the C-5 hydroxyl and emphasize the influence of conformation on agonist function.

With regard to the recognition unit of a putative receptor, it appears that the C-1 carboxyl (IA) and probably also the peptidic carboxyl (IIIF) need not be ionized for functional activity on the guinea pig pulmonary parenchyma, and each may serve separately as a proton donor for hydrogen bonding or they may interchangeably bind to a recognition unit for a single free carboxylate domain. On the pulmonary parenchyma, the C-1 monoamide of LTD_4 (IA domain variation) is equiactive with its parent compound (Fig. 4), and on the ileum it retains 25% activity (24). Whereas the monoamide of the LTD_4 glycine carboxyl (IIIF domain) has an EC_{50} of 10% that of LTD_4 on the pulmonary parenchymal strips, the dimethylamide, which has

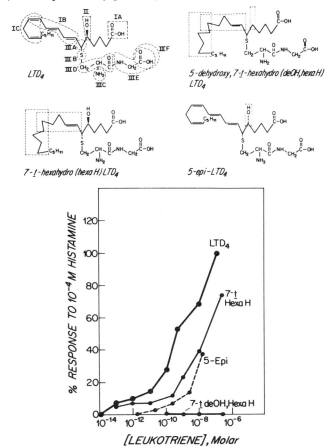

FIG. 3. Leukotriene analogs with primary variation at the C-5 hydroxyl: structure and function on guinea pig pulmonary parenchymal strips. (Adapted from ref. 24.)

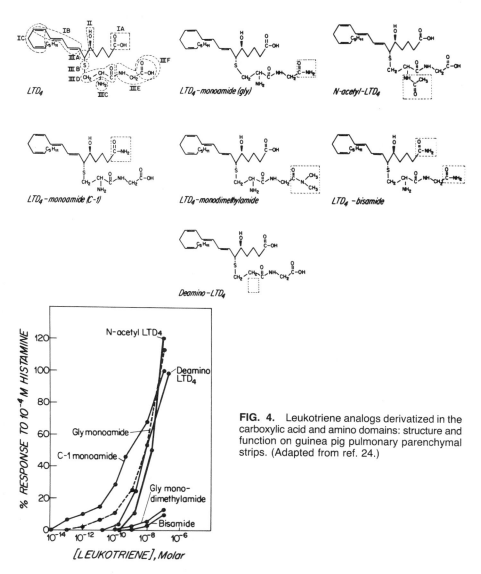

FIG. 4. Leukotriene analogs derivatized in the carboxylic acid and amino domains: structure and function on guinea pig pulmonary parenchymal strips. (Adapted from ref. 24.)

a 3-log reduction in potency (Fig. 4) does not have the possibility of hydrogen bonding by that domain. That at least one carboxylic acid group is critical to function is implied by the minimal activity of the LTD$_4$-bisamide on the pulmonary strips (24) (Fig. 4).

The absence of the N-terminal amino group (deamino LTD$_4$, IIIC domain variation) or its N-acetylation reduces activity to 3 to 5% of its parent's activity on the pulmonary parenchymal strip (Fig. 4) and to 4 to 20% on the ileum. Although a contribution of the free amino group to full agonist activity is apparent, the data obviate a requirement for an obligatory ammonium charge in the peptide.

In summarizing these data on the SRS-A leukotriene analogs, we hypothesize that the pulmonary parenchyma and probably the ileum of the guinea pig possess true receptors for LTD_4 and that these receptors comprise a loose hydrophobic binding site for accepting the ω portion of the eicosanoid and a polar activating site. The latter site appears to accommodate the C-1 carboxyl and the peptide amino and carboxyl domains, with only one carboxylate possibly required in an ionized form.

The existence of a true LTD_4 receptor on the guinea pig peripheral airway is also implied by the two patterns of pulmonary parenchymal tissue response curves to this agonist. Of 99 tissue strips demonstrating standard dose-response curves to histamine, approximately one-third exhibited a biphasic response to increasing doses of LTD_4 with the initial response at 10^{-14} to 10^{-12} M, and two-thirds exhibited a monophasic response beginning at higher concentrations. The presence of a low-concentration, high-affinity LTD_4 receptor in a substantial number of pulmonary parenchymal tissues is suggested.

STRUCTURAL BASIS OF LTB$_4$ FUNCTION

LTB_4 is approximately 100-fold more potent than 5-L-hydroxy-6,8,10,14-eicosatetraenoic acid (5-HETE) when these 5-lipoxygenase products are compared for molar concentrations eliciting a half-maximal chemotactic response (EC_{50}) (16,17). The *in vitro* chemotactic activity of 5-HETE depends substantially on its ionizable carboxylic acid but not on an ionizable C-5 hydroxyl. The methyl ester has barely detectable chemotactic activity at a concentration of 15 μM which is over 1 log greater than the EC_{50} of the free acid (16); O-acetylation of the C-5 hydroxyl of 5-HETE does not reduce its chemotactic potency.

The structural basis for chemotactic function of LTB_4 is different from that of 5-HETE, suggesting a different receptor. Acetylated LTB_4 is 10 to 33% as active as LTB_4 and, assuming that both hydroxyl groups were acetylated, it is suggested that hydrogen bonding to the putative receptor is critical to agonist function. The maintainance of full activity by the methyl ester of LTB_4 implies that the neutrophil receptor for LTB_4 could interact with a nonionized C-1 carboxylic acid by hydrogen bonding.

A clear point of structural dependence of LTB_4 activity resides in the stereochemistry of its triene domain. LTB_4 analogs, which are varied in this domain—including the 6,8,10-*trans*, the 6,10-*trans*,8-*cis*, and the 6,8-*trans*,10-*cis* stereoisomers—are all less potent, by a factor of at least 2 logs, for elicitation of responses at the EC_{50} level than native or synthetic 5S,12R-dihydroxy,6,14-*cis*, 8,10-*trans* eicosatetraenoic acid (LTB_4) (26). The all-*trans* isomers are the least active of all. The LTB_4 recognition unit on the human neutrophil thus requires a fairly specific configuration of the LTB_4 C-1 to C-12 domain, including the stereochemistry of the 6-, 8-, and 10-olefinic bonds and the presence of both hydroxyl groups to optimize hydrogen bonding via one or both hydroxyl groups. Schematic overlays on LTB_4 of the structures of each of three double bond stereoisomer analogs are shown in Fig. 5, with their triene hydrogens in a single plane. Increased displacement of the C-12 hydroxyl from its preferred configuration in LTB_4 to the all-*trans* isomers correlates with a loss of chemotactic activity (26).

LEUKOTRIENE CASCADE

LTC_4, the initial SRS-A component produced via the 5-lipoxygenase pathway, is metabolized by γ-glutamyl transpeptidase to LTD_4 (36), and then via an as yet undefined peptidase (40,44) to LTE_4 (23). This cascade generates a series of biologically active molecules and is not viewed as a major detoxification pathway. LTD_4 is 20 to 50 times as potent as LTC_4 and 5 to 10 times as active as LTE_4 in contracting guinea pig pulmonary parenchymal strips (23), whereas LTC_4 and LTD_4 are equipotent as spasmogens for isolated human bronchial smooth muscle strips (7,18). *In vivo*, the three leukotrienes of SRS-A have comparable activity in altering venular permeability in guinea pig skin (9,23) and in eliciting a wheal and prolonged cutaneous flare in the human (28). LTE_4 is unselective as compared to LTC_4 or LTD_4 for peripheral versus central airway effects when given intravenously to the unanesthetized guinea pig (Drazen, Venugopalan, Austen, Brion, and Corey, *unpublished data*). As LTC_4 and LTD_4 have a differential effect on guinea pig peripheral versus central airways when given intravenously, whereas LTE_4 has equivalent effects on both, it is unlikely that guinea pig pulmonary tissue rapidly converts the former to the latter.

It has been known for several years that, of the various human peripheral leukocytes, eosinophils are the most active inactivators of SRS-A biological activity, as measured on the guinea pig ileum (46); these cells either metabolize LTC_4 to a product other than LTE_4 or make it otherwise biologically unavailable. Direct metabolism via eosinophil arylsulfatase B per se, the first mechanism proposed for SRS-A inactivation (33,47), is unproven (39). Although eosinophil arylsulfatase B

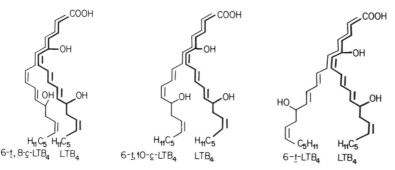

FIG. 5. Superimposition of the structure of LTB_4 with those of three 6,8,10-olefinic stereoisomers.

is inhibited by partially purified (47) SRS-A, and homogeneous enzyme is inhibited by nanomolar concentrations of synthetic leukotrienes by a mechanism which recognizes the nature and position of the sulfidopeptide (48), there is no inactivation of spasmogenic activity. Further, ^{3}H-labeled LTC_4 is recovered equally from C_{18} reverse-phase high-performance liquid chromatography (RP-HPLC) as a single unaltered peak after incubation with purified eosinophil arylsulfatase B or with buffer alone.

As analyzed by RP-HPLC in six experiments with up to 95% recovery of triene (assessed by the integrated 280 nm absorbance of the fractions and confirmed by the appropriate spasmogenic activity for the leukotriene species), the half-life ($t_{1/2}$) for conversion of 2 μg LTC_4 to LTD_4 and 2 μg LTD_4 to LTE_4 in 5 ml human plasma is 22 and 12 min, respectively. The triene recovery indicates the stability of LTE_4 and suggests cellular mechanisms for detoxification. Although the functional integrity of the active site of purified arylsulfatase was not established, after addition of 4 units/ml undiluted human plasma, the rate of conversion of LTC_4 was not altered by the addition of this enzyme.

FUNCTIONAL EFFECTS OF THE LEUKOTRIENES IN PRIMATES

To study the functional effects of LTC_4 in primates, three 4-kg *Macaca fascicularis* monkeys were anesthetized with ketamine and azopromazine and then intubated; a retrocardiac liquid-filled esophageal catheter was inserted for continuous monitoring of pleural pressure changes, and the monkeys were placed in a body plethysmograph for monitoring of pulmonary volume and flow (38). LTC_4, dissolved in 1.5 ml normal saline, was administered by aerosol from a Collison nebulizer, driven by oxygen at 15 pounds/square inch, by five inflations at each dose to approximately two-thirds total lung capacity. The relationship of dose to response was not consistent among the three animals or in a single animal from one day to another. Nonetheless, significant decrements of 30 to 60% in dynamic lung compliance occurred in six experiments after aerosolization of 1 to 5 μg LTC_4 from the reservoir, whereas changes in airway resistance were of lesser magnitude and more transient. The maximal effect on dynamic lung compliance was observed at 15 to 30 min after the administration of LTC_4 (Fig. 6).

An analogous result was obtained after inhalation of 2 to 30 μg LTC_4 by five humans. Each person inhaled 10 breaths of LTC_4 dissolved in normal saline from a 1-ml reservoir of a DeVilbiss nebulizer and was tested by forced exhalations from 60% of vital capacity to residual volume while in a computerized variable-volume body plethysmograph. Falls of 15 to 50% in flow rates at 30% of vital capacity were demonstrated reproducibly in each subject at approximately the same reservoir concentrations of LTC_4 on repeated testing, although intersubject variation was as great as 10-fold for concentrations evoking an equivalent effect (Weiss et al., *unpublished observations*).

A single rhesus monkey was repeatedly tested by intradermal injections of 10 to 500 ng LTB_4, LTC_4, LTD_4, and five double-bond stereoisomers of LTB_4. Punch

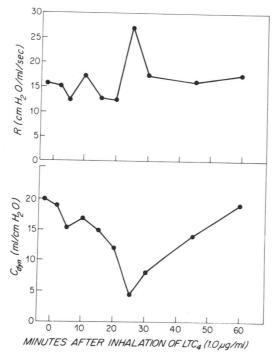

FIG. 6. Time-dependent responses of pulmonary functions of a representative *Macaca fascicularis* monkey after inhalation of 1 μg LTC₄. **Upper panel:** Airway resistance (R). **Lower panel:** Dynamic lung compliance (C$_{dyn}$).

biopsies of injection sites after 4 hr, cut in 1 μm thick sections and stained with Giemsa's reagent, showed dilation of superficial and deep venules as well as of some deep arterioles. No cellular infiltration was observed at sites prepared with up to 500 ng LTC₄ or LTD₄ or 100 ng of the stereoisomers of native LTB₄ (26). In contrast, 100 ng LTB₄ per site induced a neutrophil infiltrate of 100 to 300 cells per 40 high power fields, evaluated from a 4 mm diameter biopsy, and 10 ng evoked an infiltrate of 5 to 50 cells per biopsy section (26). Natural LTB₄, generated by divalent cation ionophore treatment of human neutrophils (17), and synthetic (5S,12R-6,14-*cis*, 8,10-*trans* DHETE) LTB₄ elicited approximately equivalent responses (Fig. 7).

Three humans injected with 500 ng LTB₄ per site exhibited a neutrophil infiltrate after 4 hr as evaluated in Giemsa-stained 1 μm thick sections of 4-mm punch biopsies. Clinically, LTB₄ induced a transient early wheal and flare which dissipated in 30 to 60 min, and tenderness with minimal induration at 4 to 6 hr. Sites injected with 500 ng LTC₄, LTD₄, or LTE₄ exhibited an early wheal and flare with central pallor and surrounding halos of erythema; the flare persisted for up to 4 hr but did not become tender at any time. Biopsies of LTC₄ or LTD₄ injection sites at 2 hr

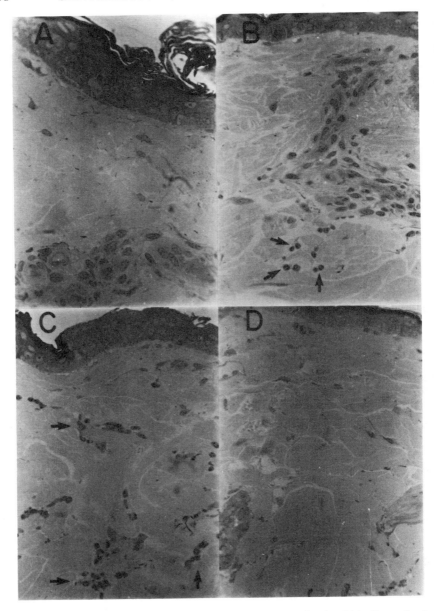

FIG. 7. Histological responses in rhesus monkey skin 4 hrs after local injections of natural LTB$_4$, synthetic LTB$_4$, and synthetic analog. **A**: Control, uninjected. **B**: Neutrophil-derived, purified LTB$_4$, 100 ng. **C**: Synthetic LTB$_4$, 100 ng, **D**: Synthetic 6t-LTB$_4$, 100 ng. Note infiltrating neutrophilis (*arrows*). 1 μm Epon-embedded specimens, Giemsa. **A** × 165.4, **B, C** × 100, **D** × 78.1.

demonstrated vasodilation of superficial and deep venules as well as of small dermal arterioles, but no cellular infiltration (28).

PGD_2, a vasodilating (15) arachidonic acid metabolite generated by mast cells (27,41), elicited, at 1 μg per site, a transient wheal and flare in the human. In another test, 1 μg PGD_2 was injected in combination with 100 ng LTB_4 in the monkey or 500 ng LTB_4 in humans, and biopsies of the injected sites were compared to those of sites injected with LTB_4 or PGD_2 alone. The combination of agents was found to act synergistically to elicit neutrophil infiltration in the monkey and in two of the three human subjects (26,28).

ACKNOWLEDGMENTS

This work was supported in part by grants AI-07722, AI-10356, HL-17382, HL-19777, HL-00549, AI-00399, and RR-05669 from the National Institutes of Health and a grant from the National Science Foundation.

REFERENCES

1. Bach, M. K., Brashler, J. R., and Gorman, R. R. (1977): On the structure of slow reacting substance of anaphylaxis: evidence of biosynthesis from arachidonic acid. *Prostaglandins*, 14:21–38.
2. Borgeat, P., and Samuelsson, B. (1979): Metabolism of arachidonic acid in polymorphonuclear leukocytes. *J. Biol. Chem.*, 254:7865–7868.
3. Brocklehurst, W. E. (1953): Occurrence of an unidentified substance during anaphylactic shock in cavy lung. *J. Physiol. (Lond.)*, 120:16P-17P.
4. Brocklehurst, W. E. (1960): The release of histamine and formation of a slow-reacting substance (SRS-A) during anaphylactic shock. *J. Physiol. (Lond.)*, 151:416–435.
5. Burka, J. F., and Garland, L. G. (1976): A possible modulatory role for prostacyclin (PGI) in IgGa-induced release of slow reacting substance of anaphylaxis in rats. *Br. J. Pharmacol.*, 61:697–699.
6. Corey, E. J., Marfat, A., Goto, G., and Brion, F. (1980): Leukotriene B: total synthesis and assignment of stereochemistry. *J. Am. Chem. Soc.*, 102:7984–7988.
7. Dahlén, S. E., Hedqvist, P., Hammarström, S., and Samuelsson, B. (1980): Leukotrienes are potent constrictors of human bronchi. *Nature*, 288:484–486.
8. Drazen, J. M., and Austen, K. F. (1974): Effects of intravenous administration of slow-reacting substance of anaphylaxis, histamine, bradykinin, and prostaglandin $F_2α$ on pulmonary mechanics in the guinea pig. *J. Clin. Invest.*, 53:1679–1685.
9. Drazen, J. M., Austen, K. F., Lewis, R. A., Clark, D. A., Goto, G., Marfat, A., and Corey, E. J. (1980): Comparative airway and vascular activities of leukotrienes C-1 and D in vivo and in vitro. *Proc. Natl. Acad. Sci. USA*, 77:4354–4358.
10. Drazen, J. M., Lewis, R. A., Austen, K. F., Toda, M., Brion, F., Marfat, A., and Corey, E. J. (1981): Contractile activities of structural analogs of leukotrienes C and D: necessity of a hydrophobic region. *Proc. Natl. Acad. Sci. USA*, 78:3195–3198.
11. Drazen, J. M., Lewis, R. A., Wasserman, S. I. Orange, R. P., and Austen, K. F. (1979): Differential effects of a partially purified preparation of slow reacting substance of anaphylaxis on guinea pig tracheal spirals and parenchymal strips. *J. Clin. Invest.*, 63:1–5.
12. Engineer, D. M., Jose, P. J., Piper, P. J., and Tippins, J. R. (1978): Modulation of slow reacting substance of anaphylaxis and histamine release by prostacyclin and thromboxanes. *J. Physiol. (Lond.)*, 281:42P (abstr.).
13. Engineer, D. M., Piper, P. J., and Sirois, P. (1977): Release of prostaglandins and rabbit aorta contracting substance (RCS) from guinea pig lung by slow reacting substance of anaphylaxis (SRS-A). *Br. J. Pharmacol.*, 59:444P (abstr.).
14. Feuerstein, N., Foegh, M., and Ramwell, P. W. (1981): Leukotrienes C4 and D4 induce prostaglandin and thromboxane release from rat peritoneal macrophages. *Br. J. Pharmacol.*, 72:389–391.

15. Flower, R. J., Harvey, E. A., and Kingston, W. P. (1976): Inflammatory effects of prostaglandin D2 in rat and human skin. *Br. J. Pharmacol.*, 56:229–233.
16. Goetzl, E. J., Brash, A. R., Tauber, A. I., Oates, J. A., and Hubbard, W. C. (1980): Modulation of human neutrophil function by monohydroxy-eicosatetraenoic acids. *Immunology*, 39:491–501.
17. Goetzl, E. J., and Pickett, W. C. (1981): Novel structural determinants of the human neutrophil chemotactic activity of leukotriene B. *J. Exp. Med.*, 153:482–487.
18. Hanna, C. J., Bach, M. K., Pare, P. D., and Schellenberg, R. R. (1981): Slow-reacting substances (leukotrienes) contract human airway and pulmonary vascular smooth muscle in vitro. *Nature*, 290:343–344.
19. Kellaway, C. H., and Trethewie, E. R. (1940): The liberation of a slow-reacting smooth-muscle-stimulating substance in anaphylaxis. *Q. J. Exp. Physiol.*, 30:121–145.
20. Klickstein, L. B., Shapleigh, C., and Goetzl, E. J. (1980): Lipoxygenation of arachidonic acid as a source of polymorphonuclear leukocyte chemotactic factors in synovial fluid and tissue in rheumatoid arthritis and spondyloarthritis. *J. Clin. Invest.*, 66:1166–1170.
21. Lewis, R. A., and Austen, K. F. (1981): Mediation of local homeostasis and inflammation by leukotrienes and other mast cell–dependent compounds. *Nature*, 293:103–108.
22. Lewis, R. A., Austen, K. F., Drazen, J. M., Clark, D. A., Marfat, A., and Corey, E. J. (1980): Slow reacting substances of anaphylaxis: identification of leukotrienes C-1 and D from human and rat sources. *Proc. Natl. Acad. Sci. USA*, 77:3710–3714.
23. Lewis, R. A., Drazen, J. M., Austen, K. F., Clark, D. A., and Corey, E. J. (1980): Identification of the C(6)-S-conjugate of leukotriene A with cysteine as a naturally occurring slow reacting substance of anaphylaxis (SRS-A): importance of the 11-cis geometry for biological activity. *Biochem. Biophys. Res. Commun.*, 96:271–277.
24. Lewis, R. A., Drazen, J. M., Austen, K. F., Toda, M., Brion, F., Marfat, A., and Corey, E. J. (1981): Contractile activities of structural analogs of leukotrienes C and D: role of the polar substituents. *Proc. Natl. Acad. Sci. USA*, 78:4579–4583.
25. Lewis, R. A., Drazen, J. M., Corey, E. J., and Austen, K. F. (1981): Structural and functional characteristics of the leukotriene components of slow reacting substance of anaphylaxis (SRS-A). In: *SRS-A and the Leukotrienes*, edited by P. J. Piper, pp. 101–117. Wiley, London.
26. Lewis, R. A., Goetzl, E. J., Drazen, J. M., Soter, N. A., Corey, E. J., and Austen, K. F. (1981): Functional characterization of synthetic leukotriene B and its stereochemical isomers. *J. Exp. Med.*, 154:1243–1248.
27. Lewis, R. A., Holgate, S. T., Roberts, S. J., II, Oates, J. A., and Austen, K. F. (1981): Preferential generation of prostaglandin D2 by rat and human mast cells. In: *Biochemistry of the Acute Allergic Reactions, 4th International Symposium*, edited by E. L. Becker, A. S. Simon, and K. F. Austen, pp. 239–254. Alan R. Liss, New York.
28. Lewis, R. A., Soter, N. A., Corey, E. J., and Austen, K. F. (1981): Local effects of synthetic leukotrienes on monkey and human skin. *Clin. Res.*, 29:492A (abstr.).
29. Morris, H. R., Taylor, G. W., Piper, P. J., and Tippins, J. R. (1980): Structure of slow reacting substance of anaphylaxis from guinea pig lung. *Nature*, 285:104–105.
30. Murphy, R. C., Hammarström, S., and Samuelsson, B. (1979): Leukotriene C: a slow-reacting substance from murine mastocytoma cells. *Proc. Natl. Acad. Sci. USA.* 76: 4275–4279.
31. Orange, R. P., and Austen, K. F. (1969): Slow reacting substance of anaphylaxis. *Adv. Immunol.*, 10:106–144.
32. Orange, R. P., Austen, W. G., and Austen, K. F. (1971): Immunologic release of histamine and slow-reacting substance of anaphylaxis from human lung. I. Modulation by agents influencing cellular levels of cyclic 3',5'-adenosine monophosphate. *J. Exp. Med.*, 134:136s-148s.
33. Orange, R. P., Murphy, R. C., and Austen, K. F. (1974): Inactivation of slow reacting substance of anaphylaxis (SRS-A) by arylsulfatases. *J. Immunol.*, 113:316–321.
34. Orange, R. P., Stechschulte, D. J., and Austen, K. F. (1970): Immunochemical and biologic properties of rat IgE. II. Capacity to mediate the immunologic release of histamine and slow-reacting substance of anaphylaxis (SRS-A). *J. Immunol.*, 105:1087–1095.
35. Orange, R. P., Valentine, M. D., and Austen, K. F. (1968): Inhibition of the release of slow-reacting substance of anaphylaxis (SRS-A) in rats prepared with homologous antibody. *Proc. Soc. Exp. Biol. Med.*, 127:127–132.
36. Orning, L., and Hammarström, S. (1980): Inhibition of leukotriene C and leukotriene D biosynthesis. *J. Biol. Chem.*, 255:8023–8026.
37. Orning, L., Hammarström, S., and Samuelsson, B. (1980): A slow reacting substance from rat basophilic leukemia cells. *Proc. Natl. Acad. Sci. USA*, 77:2014–2017.

38. Pare, P. D., Michoud, M. C., and Hogg, J. C. (1976): Lung mechanics following antigen challenge of Ascaris suum-sensitive rhesus monkeys. *J. Appl. Physiol.*, 41:668–676.
39. Parker, C. W., Koch, D. A., Huber, M. M., and Falkenhein, S. F. (1980): Arylsulfatase inactivation of slow reacting substance: evidence for proteolysis as a major mechanism when ordinary commercial preparations of the enzyme are used. *Prostaglandins*, 20:887–907.
40. Parker, C. W., Koch, D., Huber, M. M., and Falkenhein, S. F. (1980): Formation of the cysteinyl form of slow reacting substance (leukotriene E4) in human plasma. *Biochem. Biophys. Res. Commun.*, 97:1038–1046.
41. Roberts, L. J., Lewis, R. A., Oates, J. A., and Austen, K. F. (1979): Prostaglandin, thromboxane, and 12-hydroxy-5,8,10,14-eicosatetraenoic acid production by ionophore-stimulated rat serosal mast cells. *Biochim. Biophys. Acta*, 575:185–192.
42. Robinson, D. R., Dayer, J. M., and Krane, S. M. (1979): Prostaglandins and their regulation in rheumatoid inflammation. *Ann. NY Acad. Sci.*, 332:279–294.
43. Rouzer, C. A., Scott, W. A., Cohn, Z. A., Blackburn, P., and Manning, J. M. (1980): Mouse peritoneal macrophages release leukotriene C in response to phagocytic stimulus. *Proc. Natl. Acad. Sci. USA*, 77:4928–4932.
44. Sok, D. E., Pai, J. K., Atrache, V., and Sih, C. J. (1980): Characterization of slow reacting substances (SRSs) of rat basophilic leukemia (RBL-1) cells: effect of cysteine on SRS profile. *Proc. Natl. Acad. Sci. USA*, 77:6481–6485.
45. Tauber, A. I., Kaliner, M., Stechschulte, D. J., and Austen, K. F. (1973): Immunologic release of histamine and slow reacting substance of anaphylaxis from human lung. V. Effects of prostaglandins on release of histamine. *J. Immunol.*, 111:27–32.
46. Wasserman, S. I., Goetzl, E. J., and Austen, K. F. (1975): Inactivation of human SRS-A by intact human eosinohils and by eosinophil arylsulfatase. *J. Allergy Clin. Immunol.*, 55:72 (abstr.).
47. Wasserman, S. I., Goetzl, E. J., and Austen, K. F. (1975): Inactivation of slow-reacting substance of anaphylaxis by human eosinophil arylsulfatase. *J. Immunol.*, 114:645–649.
48. Weller, P. F., Lewis, R. A., Austen, K. F., and Corey, E. J. (1981): The interaction of purified human eosinophil arylsulfatase B with synthetic leukotrienes. *Fed. Proc.*, 40:1023 (abstr.).

Leukotrienes and Other Lipoxygenase Products,
edited by B. Samuelsson and R. Paoletti.
Raven Press, New York © 1982.

Biological Activity of Leukotriene C_4 in Guinea Pigs: In Vitro and In Vivo Studies

G. C. Folco, C. Omini, T. Viganò, G. Brunelli, G. Rossoni, and F. Berti

Institute of Pharmacology and Pharmacognosy, University of Milan, 20129 Milan, Italy

The recent discovery that some arachidonic acid metabolites derived from the 5-lipoxygenase pathway are responsible for slow-reacting substance of anaphylaxis (SRS-A) activity has greatly stimulated research in the field of asthma and other allergic diseases (22–26). In fact these compounds, which were first detected in leukocytes, possess as a common feature a conjugated triene and have been named leukotrienes (25).

Among them, leukotrienes C, D, and E cause alterations that are likely to occur in immediate-type hypersensitivity reactions, where the prominent effects are peripheral airway constriction and hypotension (9,10). In this regard several investigators have demonstrated that leukotrienes C_4 (LTC$_4$), D_4 (LTD$_4$), and E_4 (LTE$_4$) constrict guinea pig and human airway with a potency that is several orders of magnitude higher than that of histamine (19). Furthermore, a preferential peripheral airway site of action has been shown for LTC$_4$, LTD$_4$, and LTE$_4$ together with their ability to affect vascular permeability in guinea pig skin (10).

Studies performed *in vivo* indicate that functional modifications of pulmonary mechanics are associated with a fall of mean systemic arterial pressure, a phenomenon attributed by Drazen et al. (10) to a direct cardiovascular action of the leukotrienes.

SRS-A and LTC$_4$ are also able to activate arachidonic acid metabolism and promote formation and release of thromboxane A_2 (TXA$_2$) from isolated guinea pig lungs: this effect is prevented by pretreatment with cyclo-oxygenase inhibitors and with FPL-55712 (12). These observations prompted us to consider whether the biological activity of LTC$_4$ is merely direct or also mediated by other arachidonic acid metabolites. In fact, as far as bronchoconstriction is concerned, very early evidence by Berry and Collier indicated that aspirin-like drugs prevent the increased airway resistance induced in the guinea pig by SRS-A (2).

Finally, drugs which are currently employed in the treatment of bronchial obstructive disease have been considered in view of their ability to inhibit pulmonary and cardiovascular actions of LTC$_4$.

153

METHODS

In Vitro Studies

In *in vitro* studies lungs from male guinea pigs (350 to 450 g) were perfused through the pulmonary artery as previously described (11). The pulmonary effluent superfused a bank of tissues including strips of rabbit aorta (RbA), bovine coronary artery (BCA), rabbit mesenteric artery (RbMA), rat colon (RC), and rat stomach (RSS) in order to detect arachidonate metabolite-like activity. These tissues were therefore treated with a mixture of receptor antagonists and with indomethacin to increase their specificity (14).

Some experiments were carried out perfusing, through the pulmonary artery, guinea pig lung vessels free from parenchyma and, through the mesenteric artery, guinea pig isolated mesenteric vascular bed in order to detect formation and release of thromboxanes and prostacyclin-like material. Further *in vitro* experiments were performed using helical strips of guinea pig trachea and lung parenchymal strips prepared for recording isotonic contractile activity: Resting tension was 1 and 0.5 g, respectively. The perfusates and the incubating media were collected for assay of TXB_2 and 6-keto-prostaglandin $F_{1\alpha}$ (6-keto-$PGF_{1\alpha}$) as already described (15–20).

In Vivo Studies

Guinea pigs (males 400 to 500 g) anesthetized with ethyl urethane (1.2 g/kg i.p.) were prepared for recording pulmonary mechanics according to Drazen et al. (10). Dynamic compliance (C_{Dyn}), lung resistance (R_L), transpulmonary pressure (TPP), tidal volume (V), and respiratory air flow ($\overset{\circ}{V}$) were registered using Hewlett-Packard instruments. The animals were also arranged for extracorporeal circulation following the blood-bathed organ technique (24) to detect the presence of PGI_2 and TXA_2-like activity: assay tissues were BCA and RbA pretreated overnight, as already indicated for the *in vitro* experiments. Drugs were given intravenously, and LTC_4 was also tested for its biological activity through the aerosol route.

RESULTS

In Vitro Studies

When LTC_4 (0.8 to 1.6 pmoles) is injected into isolated guinea pig lungs, it brings about activation of arachidonic acid metabolism with formation of PGI_2 and TXA_2 in a dose-dependent manner. This release is prevented by pretreating the lungs with aspirin and FPL-55712 (Table 1). The potency of LTC_4 in triggering this mechanism is, on a molar basis, approximately 5,000 times higher than that of bradykinin (5 nmoles) and even higher when it is compared with histamine (50 nmoles).

In view of these results, the contributions of lung vasculature and lung parenchyma to the formation of PGI_2 and TXA_2 were examined. As reported in Fig. 1,

TABLE 1. *LTC₄, bradykinin, and histamine activation of arachidonic acid metabolism in isolated guinea pig lung*

Treatment	TXB_2 (ng/ml)	6-Keto-PGF$_{1\alpha}$ (ng/ml)
Saline	2.0 ± 0.3	0.50 ± 0.08
LTC₄ (0.8 pmole)	18.6 ± 1.0	2.20 ± 0.16
LTC₄ (1.6 pmole)	27.9 ± 2.0	3.20 ± 0.11
Aspirin (3 × 10⁻⁵ M)	1.7 ± 0.1	0.40 ± 0.05
FPL-55712 (10⁻⁷ M)	1.8 ± 0.2	0.53 ± 0.09
A + L[a]	2.2 ± 0.3	0.58 ± 0.08
F + L[a]	2.3 ± 0.3	0.70 ± 0.11
Bradykinin (5 nmole)	15.7 ± 1.4	4.56 ± 0.44
Histamine (50 nmole)	11.5 ± 0.7	1.75 ± 0.22

The mean values ± SEM are derived from at least 10 experiments. LTC₄, bradykinin, and histamine were given as a bolus, whereas aspirin and FLP-55712 were perfused for 15 min before challenge of the lung with LTC₄.
[a]A = aspirin (3 × 10⁻⁵ M). L = LTC₄ (1.6 pmole). F = FPL-55712 (10⁻⁷ M).

the lung vascular tissue of the guinea pig, free from parenchyma, was challenged with LTC₄ (1.6 pmoles). This substance was found to significantly augment the tissue's ability to release PGI₂ in the perfusate as monitored by the tissues in cascade and by the quantitative determination of 6-keto-PGF$_{1\alpha}$. Furthermore, as might be expected from vascular tissue, no formation of TXA₂ was detected. Bradykinin (5 nmoles) caused release of PGI₂ as well, a phenomenon sensitive to indomethacin (30 μM).

Using a different vascular region (e.g., guinea pig mesentery), a similar path of activation of the eicosanoid system was shown by LTC₄ and bradykinin. In fact, as reported in Fig. 2, the biological evidence indicates preferential formation of PGI₂, as no modification of the basal tonus of the RbA was observed. The slight relaxation and the decrease of spontaneous motility of RC suggest that no primary prostaglandins (PGE and PGF$_\alpha$) are present in the vascular outflow.

The experiments carried out with strips of lung parenchyma (which, as indicated by Gryglewski et al. (16), should represent the main source of TXA₂ in the lung tissue) demonstrate that LTC₄ (0.8 to 2.4 nM) and histamine (50 μM) increase the tonus of the parenchymal strip together with a significant release of TXA₂ into the incubating medium (Fig. 3). Together with TXA₂, a barely detectable amount of PGI₂ is formed during challenge with LTC₄ and histamine. This formation may be due to the presence of vascular and bronchial smooth muscle in the parenchyma. Pretreatment with indomethacin (30 μM) reduces the contraction of the parenchymal tissue and completely prevents generation of TXA₂. However, with respect to histamine a different pattern of indomethacin activity can be observed. In fact, whereas the parenchymal strip contraction is not affected by the cyclo-oxygenase inhibitor, the TXA₂ production is totally abolished.

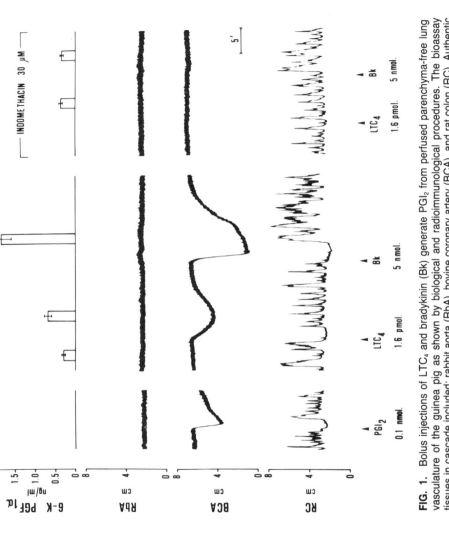

FIG. 1. Bolus injections of LTC₄ and bradykinin (Bk) generate PGI₂ from perfused parenchyma-free lung vasculature of the guinea pig as shown by biological and radioimmunological procedures. The bioassay tissues in cascade included: rabbit aorta (RbA), bovine coronary artery (BCA), and rat colon (RC). Authentic PGI₂ was injected directly over the cascade tissues. Columns represent the mean values ± SEM of 6-keto-PGF₁α of at least 10 experiments.

INDOMETHACIN 30 μM

5'

6-K PGF₁d
ng/ml
2.0
1.5
1.0
0.5
0

RbA
cm
8
4
0

BCA
cm
8
4
0

RC
cm
8
4
0

PGI₂
0.1 nmol.

LTC₄
1.6 pmol.

Bk
5 nmol.

LTC₄
1.6 pmol.

Bk
5 nmol.

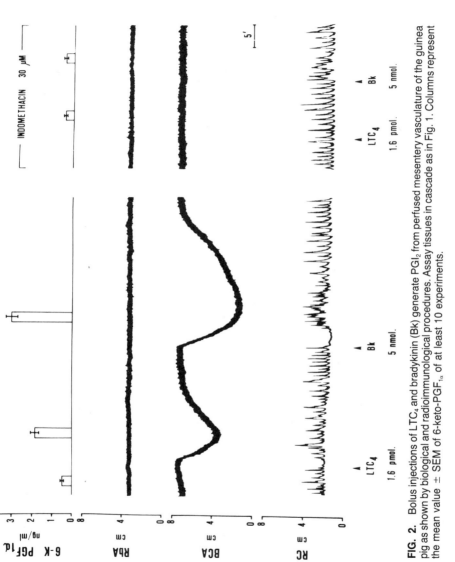

FIG. 2. Bolus injections of LTC_4 and bradykinin (Bk) generate PGI_2 from perfused mesentery vasculature of the guinea pig as shown by biological and radioimmunological procedures. Assay tissues in cascade as in Fig. 1. Columns represent the mean value ± SEM of 6-keto-$PGF_{1\alpha}$ of at least 10 experiments.

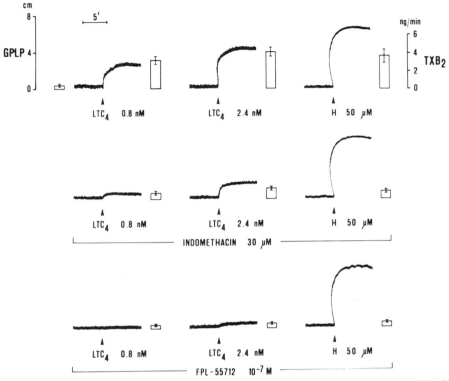

FIG. 3. LTC_4 and histamine (H) induce contraction of guinea pig lung parenchymal strip (GPLP) and generate TXA_2. Effect of indomethacin and FPL-55712. Columns represent the mean values $\pm$ SEM of TXB_2 (at least 10 experiments) measured in the bathing fluid during organ contraction.

The leukotriene receptor antagonist FPL-55712 (10^{-7} M) was also tested in this set of experiments and, at variance with indomethacin, the ability of LTC_4 to increase the tonus of the lung strip and TXA_2 generation is completely affected. In contrast, the contractile activity of histamine is fully retained, whereas TXA_2 concentration in the bathing fluid is reduced to basal level (Fig. 3). The ability of LTC_4 and histamine to activate arachidonic acid metabolism has been observed in tracheal smooth muscle of the guinea pig. As shown in Fig. 4, LTC_4 (10 to 20 nM) and histamine (50 μM) induce contraction of tracheal tissue accompanied by generation of PGI_2, as indicated by the radioimmunological determination of 6-keto-$PGF_{1\alpha}$ in the bathing fluid. In these experiments, the bioassay on RSS for PGE_2-like activity indicates that this primary prostaglandin occurs quite unsteadily in the tracheal incubating medium. However, the fact that indomethacin (30 μM) potentiates the contractile response of guinea pig tracheas to exogenous LTC_4 and histamine suggests that cyclo-oxygenase products with relaxing activity are formed.

Considering the observation of Levi and Burke (18) that SRS-A sensitizes the heart to the tachyarrythmic effect of histamine, a possible sensitizing effect of LTC_4

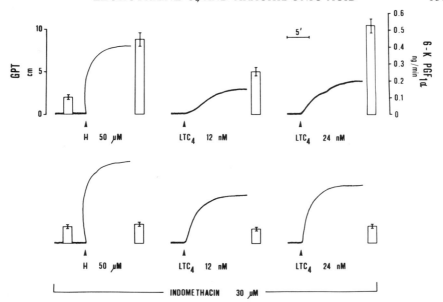

FIG. 4. LTC$_4$ and histamine (H) induce contraction of helical strips of guinea pig trachea (GPT) and generate PGI$_2$. Effect of indomethacin. Columns represent the mean values ± SEM of 6-keto-PGF$_{1\alpha}$ (at least eight experiments) measured in the bathing fluid during organ contraction.

on histamine- and bradykinin-induced TXA$_2$ generation from isolated guinea pig lung has been investigated. As shown in Fig. 5, when LTC$_4$ is perfused at a concentration (16 pM) which does not significantly activate arachidonic acid metabolism per se, the ability of histamine and bradykinin to form TXA$_2$ is enhanced. On the other hand, formation of TXA$_2$ induced by injection of arachidonic acid (17 nmoles) is not enhanced on perfusion with LTC$_4$. The sensitizing activity shown by LTC$_4$ seems to be specific for this compound as threshold doses of histamine (10 μM) and bradykinin (1 μM) do not sensitize LTC$_4$ (0.8 nmoles) in producing TXA$_2$. In view of the potential role of TXA$_2$ in the pathogenesis of broncho-obstructive disease and pulmonary hypertension, the pharmacological control of its release from the lungs following challenge with LTC$_4$ has been examined. Activation of β-adrenoceptors perfusing the lungs with salbutamol and fenoterol (0.4 μM) antagonizes the capacity of LTC$_4$ to increase TXA$_2$ formation (Fig. 6). This protecting activity, which is rapid in onset and readily reversible upon stopping perfusion with the β$_2$-adrenoceptor agonists, is likely to be related to activation of adenylate cyclase and to intracellular accumulation of 3′-5′-adenosine cyclic monophosphate (cyclic AMP) *(unpublished data)*.

Blockade of muscarinic receptors with atropine (1.3 μM) reduces the ability of LTC$_4$ (0.8 pmoles) and histamine (50 nmoles) to generate TXA$_2$ and PGI$_2$ from perfused guinea pig lungs. As shown in Fig. 7, atropine at the concentration used seems to be more effective against histamine than against LTC$_4$. This may be explained considering the antihistaminic effect of the antimuscarinic compound.

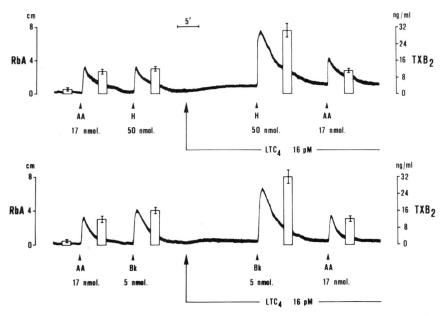

FIG. 5. LTC$_4$ sensitizes the ability of histamine (H) and bradykinin (Bk), but not that of arachidonic acid (AA), to generate TXA$_2$ from perfused guinea pig lungs as shown by biological (RbA contraction) and radioimmunological procedures. Columns represent the mean values ± SEM of TXB$_2$ of at least 10 experiments.

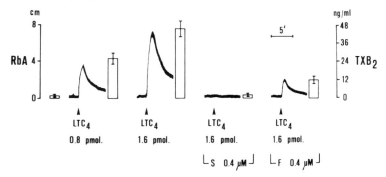

FIG. 6. Salbutamol (S) and fenoterol (F) antagonize the ability of LTC$_4$ to generate TXA$_2$ from perfused guinea pig lungs as shown by biological (RbA contraction) and radioimmunological procedures. Columns represent the mean values ± SEM of TXB$_2$ of at least eight experiments.

In Vivo Studies

When LTC$_4$ (0.01 to 1.0 μg/kg) was given intravenously to anesthetized spontaneously breathing guinea pigs, a dose-related change of respiratory mechanics and lowering of systemic arterial pressure occurred. In particular, the decrease in C_{Dyn} and the increase in R_L were consistent with a predominant peripheral site of

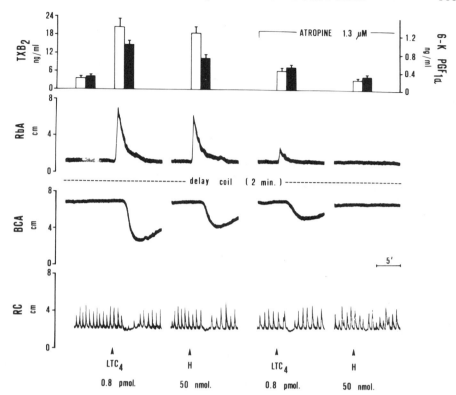

FIG. 7. Atropine reduces the ability of LTC₄ and histamine (H) to generate PGI₂ and TXA₂ from perfused guinea pig lungs as shown by biological and radioimmunological procedures. Bioassay tissues in cascade as in Fig. 1. Open and closed columns represent the mean values ± SEM of TXB₂ and 6-keto-PGF₁ₐ, respectively, of at least 10 experiments.

action of the lipoxygenase product. The bronchoconstriction due to LTC₄ (0.25 to 1.0 µg/kg) was associated with the appearance in the arterial blood of TXA₂- and PGI₂-like activity as monitored by the contraction of RbA and relaxation of BCA (Table 2). A similar pattern of activity was observed with bradykinin (1.0 µg/kg) (Fig. 8).

Pretreatment of the animals with aspirin (10 mg/kg) fully antagonized the cardiovascular and pulmonary modifications due to LTC₄ and bradykinin (Table 2, Fig. 8). When the effect of histamine was examined, more rapid bronchoconstriction was observed, associated with a marked release of TXA₂-like activity in the circulation. However, the systemic arterial pressure was minimally affected (biphasic effect) by the autacoid, and BCA was not sufficiently sensitive to detect PGI₂-like activity in the blood. The early contraction exhibited by this tissue might be due to a rise in the circulation of cyclic prostaglandin endoperoxide and TXA₂. Furthermore, as previously reported (24), acetylcholine induced bronchoconstriction

TABLE 2. *LTC₄ induction of bronchoconstriction and systemic arterial pressure decrease in anesthetized guinea pigs; correlation with blood arachidonic acid metabolites*

Treatment	Lung resistance (cm H$_2$O/ml/sec)	Dynamic compliance (ml/cm H$_2$O)	Blood pressure (mm Hg)	RbA[a] Tension increase (mg)		BCA[b] Tension decrease (mg)	
Saline	0.18 ± 0.06	0.65 ± 0.01	72 ± 1.4	—	—	—	—
LTC₄ (0.25 μg/kg)	0.35 ± 0.015[c]	0.35 ± 0.01[c]	54 ± 2.4[c]	155 ± 6		172 ± 9	
LTC₄ (1 μg/kg)	0.58 ± 0.034[c]	0.17 ± 0.01[c]	39 ± 2.5[c]	972 ± 37		811 ± 49	
Aspirin (10 mg/kg)	0.17 ± 0.004	0.66 ± 0.01	73 ± 2.0	—	—	—	—
A + L[d]	0.18 ± 0.006	0.62 ± 0.02	71 ± 1.8	—	—	—	—

The mean values ± SEM are derived from at least 10 experiments. The compounds were injected intravenously and parameters were recorded approximately 6 min after their administration, when peak effect was evident.

[a]Rabbit aorta (RbA): resting tension during blood superfusion = g 1.5 ± 0.025.
[b]Bovine coronary artery (BCA): resting tension during blood superfusion = g 5.5 ± 0.050.
[c]Significant difference from saline, $p < 0.05$.
[d]A = aspirin (10 mg/kg). L = LTC₄ (1 μg/kg).

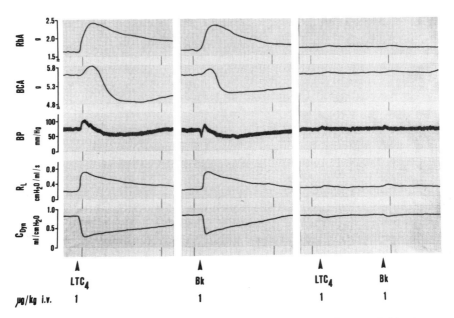

FIG. 8. Protective activity of aspirin (10 mg/kg i.v.) against changes in systemic blood pressure (BP), lung resistance (R$_L$), and dynamic compliance (C$_{Dyn}$) induced by LTC₄ and bradykinin (Bk) in anesthetized guinea pigs. Rabbit aorta (RbA) and bovine coronary artery (BCA), superfused by extracorporeal circulating blood, indicate the appearance of TXA₂ and PGI₂-like activity.

and a fall in systemic arterial pressure which was fast in onset, short-lasting, and totally independent from stimulation of arachidonic acid metabolism. In fact, during cholinergic stimulation the tonus of blood-bathed RbA and BCA was not affected.

When LTC_4 was given by aerosol (0.001% for 5 min) there was significant bronchoconstriction associated with a sustained rise of TXA_2-like activity in the blood. However, differently from intravenously injected LTC_4, inhaled LTC_4 increased blood pressure of the guinea pig, and BCA did not reveal the presence of PGI_2-like activity in the blood. In these experiments the effects of LTC_4 were almost completely antagonized by aspirin.

The capacity of atropine and salbutamol to control the effects of LTC_4 on pulmonary mechanics and arterial pressure was investigated in another set of experiments. Atropine (2.5 mg/kg) significantly reduced the increase in R_L and the fall of C_{Dyn} due to LTC_4 administration (Fig. 9). Parallel to this protection, atropine prevented the increase of circulating TXA_2-like activity, whereas PGI_2-like activity seemed to be less affected. The systemic arterial pressure, which is lowered by atropine per se, did not show a further fall (Fig. 9). A similar pattern of protecting activity was observed with salbutamol (5 μg/kg), in agreement with the results from the *in vitro* studies.

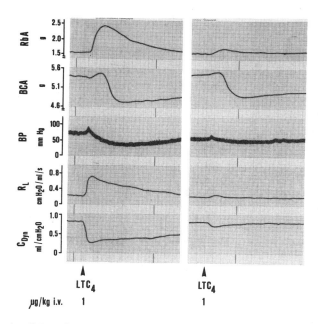

FIG. 9. Atropine (2.5 mg/kg i.v.) reduces the effect of LTC_4 on systemic arterial blood pressure (BP), lung resistance (R_L), and dynamic compliance (C_{Dyn}) in anesthetized guinea pigs. Rabbit aorta (RbA) and bovine coronary artery (BCA), superfused by extracorporeal circulating blood, indicate the appearance of TXA_2- and PGI_2-like activity.

DISCUSSION

The myotropic substance described by Kellaway and Trethewie (17) as a compound released during lung anaphylaxis was recently found to consist of a mixture of leukotrienes (22). One of these, LTC_4, possesses potent biological activity (8). The present findings support previous observations (4) and particularly point out that guinea pig cardiopulmonary modifications due to exogenous LTC_4 are mediated by arachidonic acid metabolites of the cyclo-oxygenase pathway. In fact, bronchoconstriction and the fall in systemic arterial pressure following LTC_4 administration are prevented by cyclo-oxygenase inhibitors. The capacity of LTC_4 to activate arachidonic acid metabolism is clearly evident in isolated guinea pig lung, where generation and release of TXA_2 and PGI_2 has been proved. Concerning the possible source of this lipidic material in the airway system, the evidence from our experiments indicate that pulmonary vessels mainly form PGI_2, whereas TXA_2 seems to be generated from parenchymal tissues. This is in line with the well-documented ability of the vascular endothelium to synthesize PGI_2 preferentially (21), whereas guinea pig alveolar interstitial cells probably provide TXA_2 (16). Bradykinin and histamine show a similar behavior in generating the cyclo-oxygenase products even if indomethacin affects their contractile activity differently.

In this respect, comparing LTC_4 with histamine on lung parenchymal strips, it is evident that although the contraction induced by histamine is not affected by indomethacin and is totally independent from TXA_2 generation, that of LTC_4 is partially present in spite of complete cyclo-oxygenase inhibition and is totally suppressed by the receptor antagonist FPL-55712. It is worth noting that this compound also prevents the formation of TXA_2 induced by LTC_4 and histamine, an effect which might be explained by its ability to affect cyclic AMP phosphodiesterase activity (PDE) (6). In fact, PDE inhibitors (e.g., theophylline and papaverine) are known to lower the formation of TXA_2 due to histamine and SRS-A in perfused guinea pig lungs (28). Although the site of action of LTC_4 is mainly at the periphery of the airway system, a contractile activity at a higher level of the respiratory tract has been reported (8). As in the case for histamine, contracting guinea pig trachea exposed to LTC_4 releases PGI_2 and probably other bronchodilating prostaglandins. This may be interpreted as a "self defense" mechanism to control "overdone" bronchoconstriction.

The significance of PGI_2 formation is difficult to understand in terms of its activity on tracheal smooth muscle. In fact, evidence has been presented in favor of contractile activity of PGI_2 (23) as well as in favor of relaxing action (13). Furthermore, the amount of PGI_2 detected during contraction with LTC_4 is very low and probably does not affect the tonus of the preparation.

Another point which deserves some consideration is the ability of LTC_4, at a concentration which does not activate arachidonic acid metabolism per se to sensitize histamine and bradykinin in generating TXA_2 in guinea pig lungs. This is in line with early findings that SRS-A sensitizes the smooth muscles of the gut to the effect of histamine (5) and potentiates and extends the chronotropic and arrythmogenic effect of the released autacoid during cardiac anaphylaxis (18). However, even if in the heart TXA_2 does not seem to explain the sensitizing effect of SRS-A (18), products of lipoxygenase different from SRS-A should be taken into account. In fact, as already demonstrated in guinea pig lungs (1), the anaphylactic release of mediators is increased by indomethacin, and this has been attributed to augmented synthesis of fatty acid hydroperoxides (HPETE).

However, the picture that arises from studies in guinea pig lungs points to the fact that the effects of primary mediators of anaphylaxis alter the function of the respiratory system directly and through activation of the eicosanoid system ($PGF_{2\alpha}$, TXA_2), triggering a vicious cycle which amplifies enormously the original signals. In this context of intricate relationships, LTC_4 joins the various mediators mobilized in the lungs that mimic SRS-A activity. The exact mechanism through which LTC_4 promotes the generation of TXA_2 and PGI_2 and sensitizes the activity of histamine and bradykinin in guinea pig lung is difficult to explain. Nevertheless, as the conversion of exogenous arachidonic acid is not affected by LTC_4, the mechanism of the phenomenon may occur at the phospholipase level or at the early steps coupled with the availability of arachidonic acid. In this respect, it has been observed that LTC_4 activity in the lungs is antagonized by a very well known phospholipase A_2 inhibitor, dexamethasone *(unpublished observations)*.

Other drugs currently employed in the treatment of bronchial asthma (e.g., β_2-adrenoceptor agonists and atropinics) antagonize the capacity of LTC_4 to stimulate arachidonic acid metabolism *in vitro* and protect the guinea pigs from broncho-constriction caused by this lipoxygenase product. As far as salbutamol and fenoterol are concerned, accumulation of cyclic AMP may explain their protecting activity not only through inhibition of TXA_2 formation but also through their relaxing effect. Our *in vitro* observations fit with the results of Chignard and Vargaftig (7), who found a decrease of TXA_2 formation in platelets incubated with arachidonic acid in the presence of β-adrenoceptor agonists.

The possibility that atropinics might interfere in some way with cyclo-oxygenase or thromboxane synthetase has been ruled out (3). It is therefore conceivable that atropine might somewhat prevent either lysosomal phospholipase A_2 release or the activity of this enzyme, hence making arachidonic acid less available.

Regarding the activity of LTC_4 on pulmonary mechanics and vasculature, me-diation of arachidonic acid metabolites (e.g., TXA_2 and PGI_2, respectively) seems to be predominant. In fact, TXA_2 is a powerful bronchoconstrictor and causes, in artificially ventilated guinea pigs, a long-lasting dyspnea (27). On the other hand, the decrease in blood pressure due to LTC_4 seems to occur because of formation of PGI_2 by different vascular areas: The *in vitro* experiments indicate that lung and mesenteric vasculature respond to challenge of LTC_4 with PGI_2 generation.

The relevance of these observations to human pathology is still questionable, and the possibility of measuring the formation of lipoxygenase products in specific tissues during asthma and other immediate hypersensitivity reactions will be of great help in clarifying the physiopathological role of leukotrienes.

REFERENCES

1. Adcock, J. J., Garland, L. G., Moncada, S., and Salmon, J. A. (1978): The mechanism of enhancement by fatty acid hydroperoxides of anaphylactic mediator release. *Prostaglandins*, 16:179–187.
2. Berry, P. A., and Collier, H. O. J. (1964): Bronchoconstrictor action and antagonism of a slow reacting substance from anaphylaxis of guinea-pig isolated lungs. *Br. J. Pharmacol.*, 23:201–216.
3. Berti, F., Folco, G. C., Giachetti, A., Malandrino, S., Omini, C., and Viganò, T. (1980): Atropine inhibits thromboxane-A₂ generation in isolated lungs of the guinea-pig. *Br. J. Pharmacol.*, 68:467–472.
4. Borgeat, P., and Sirois, P. (1981): Leukotrienes: a major step in the understanding of immediate hypersensitivity reactions. *J. Med. Chem.*, 24:121–126.
5. Chakravarty, N. (1959): A method for the assay of "slow reacting substance." *Acta Physiol. Scand.*, 46:298–313.
6. Chasin, M., and Scott, C. (1978): Inhibition of cyclic nucleotide phosphodiesterase by FPL-55712, an SRS-A antagonist. *Biochem. Pharmacol.*, 27:2065–2067.
7. Chignard, M., and Vargaftig, B. B. (1978): Why do some β-adrenergic agonists inhibit generation of thromboxane-A₂ in incubates of platelets with arachidonic acid? *Biochem. Pharmacol.*, 27:1603–1606.
8. Dahlén, S.-E., Hedqvist, P., Hammarström, S., and Samuelsson, B. (1980): Leukotrienes are potent constrictors of human bronchi. *Nature*, 288:484–486.
9. Drazen, J. M., and Austen, K. F. (1974): Effects of intravenous administration of slow reacting substance of anaphylaxis, histamine, bradykinin, and prostaglandin PGF₂α on pulmonary mechanics in the guinea pig. *J. Clin. Invest.*, 53:1679–1685.
10. Drazen, M. J., Austen, K. F., Lewis, R. A., Clark, D. A., Goto, G., Marfat, A., and Corey, E. J. (1980): Comparative airway and vascular activities of leukotrienes-C₁ and D in vivo and in vitro. *Proc. Natl. Acad. Sci. USA*, 77:4354–4358.
11. Engineer, D. M., Niederhauser, U., Piper, P. J., and Sirois, P. (1978): Release of mediators of anaphylaxis: inhibition of prostaglandin synthesis and the modification of release of slow reacting substance of anaphylaxis and histamine. *Br. J. Parmacol.*, 62:61–66.
12. Folco, G. C., Hansson, G., and Granström, E. (1981): Leukotriene-C₄ stimulates TXA₂ formation in isolated sensitized guinea pig lungs. *Biochem. Pharmacol.*, 30:2491–2493.
13. Gardiner, P. J., and Collier, H. O. J. (1980): Specific receptors for prostaglandins in airways. *Prostaglandins*, 19:819–841.
14. Gilmore, N., Vane, J., and Wyllie, J. H. (1968): Prostaglandins released by the spleen. *Nature*, 218:1135–1140.
15. Granström, E., and Kindahl, H. (1978): Radioimmunoassay of prostaglandins and thromboxanes. *Adv. Prostaglandin Thromboxane Res.*, 5:119–210.
16. Gryglewski, R. J., Dembinska-Kiec, A., and Grodzinska, L. (1976): Generation of prostaglandin and thromboxane-like substances by large airways and lung parenchyma. In: *Prostaglandins and Thromboxanes*, edited by F. Berti, B. Samuelsson, and G. P. Velo, pp. 165–178. Plenum Press, New York.
17. Kellaway, C. H., and Trethewie, E. R. (1940): The liberation of a slow reacting smooth muscle-stimulating substance in anaphylaxis. *Q. J. Exp. Physiol.*, 30:121–145.
18. Levi, R., and Burke, J. A. (1980): Cardiac anaphylaxis: SRS-A potentiates and extends the effects of released histamine. *Eur. J. Pharmacol.*, 62:41–49.
19. Lewis, R. A., Drazen, M. J., Austen, F. K., Clark, D. A., and Corey, E. J. (1980): Identification of the C(6)-S-conjugate of leukotriene A with cysteine as a naturally occurring slow reacting substance of anaphylaxis (SRS-A): importance of the 11-*cis*-geometry for biological activity. *Biochem. Biophys. Res. Commun.*, 96:271–277.
20. Maclouf, J. (1982): Radioimmunoassay for 6-keto-PGF₁α: *Methods Enzymol. (in press)*.
21. Moncada, S., and Vane, R. J. (1981): Prostacyclin. In: *The Prostaglandin System. Endoperoxides, Prostacyclin and Thromboxanes*, edited by F. Berti and G. P. Velo, pp. 203–221. Plenum Press, New York.
22. Murphy, R. C., Hammarström, S., and Samuelsson, B. (1979): Leukotriene C: a slow-reacting substance from murine mastocytoma cells. *Proc. Natl. Acad. Sci. USA*, 76:4275–4279.
23. Omini, C., Moncada, S., and Vane, J. R. (1977): The effects of prostacyclin (PGI₂) on tissues which detect prostaglandins (PG's). *Prostaglandins*, 14:625–632.

24. Rossoni, G., Omini, C., Viganò, T., Mandelli, V., Folco, G. C., and Berti, F. (1980): Bronchoconstriction by histamine and bradykinin in guinea pig: relationship to thromboxane A_2 generation and effect of aspirin. *Prostaglandins*, 20:547–557.
25. Samuelsson, B., and Hammarström, S. (1980): Nomenclature for leukotrienes. *Prostaglandins*, 19:645–648.
26. Samuelsson, B., Borgeat, P., Hammarström, S., and Murphy, R. C. (1980): Leukotrienes: a group of biologically active compounds. *Adv. Prostaglandin Thromboxane Res.*, 6:1–17.
27. Svensson, J., Strandberg, K., Tuvemo, T., and Hamberg, M. (1977): Thromboxane-A_2: effect on airway and vascular smooth muscle. *Prostaglandins*, 14:425–436.
28. Viganò, T., Peleska, J., Varin, L., Omini, C., Folco, G. C., and Berti, F. (1980): La stimolazione del recettore β-adrenergico inibisce la formazione di TXA_2 nel polmone isolato di cavia. In: *Proceedings: XX Meeting of the Italian Pharmacological Society, Verona, Italy, 2–3 July*, abstract 31.

Leukotrienes and Other Lipoxygenase Products,
edited by B. Samuelsson and R. Paoletti.
Raven Press, New York © 1982.

Actions of Leukotrienes on Vascular, Airway, and Gastrointestinal Smooth Muscle

Priscilla J. Piper, L. G. Letts, Marwa N. Samhoun, J. R. Tippins, and M. A. Palmer

Department of Pharmacology, Institute of Basic Medical Sciences,
Royal College of Surgeons of England, Lincoln's Inn Fields,
London WC2A 3PN, England

The leukotrienes (LTs) (23) are recently described lipoxygenase metabolites of arachidonic acid. They are a family of substances which includes the slow-reacting substances (14–16), and they display potent biological actions in various systems. The type of pharmacological action varies among the leukotrienes. For example, the dihydroxy acid LTB_4 causes chemotaxis and chemokinesis (7), responses which are involved in inflammation. On the other hand, the peptidolipids LTC_4 and LTD_4 induce potent contractions of smooth muscle (5). Availability of pure, naturally occurring or synthetic LTs has allowed extensive study of their pharmacology and confirmation of observations originally made with the slow-reacting substance of anaphylaxis (SRS-A; usually partially purified), such as the actions on smooth muscle and sensitization of tissues to other agonists (3). In addition to causing constriction of gastrointestinal and airway smooth muscle (4,20) LTC_4 and LTD_4 caused potent vasoconstriction in the microvasculature of guinea pig skin (18) and the coronary circulation (11). The actions of the LTs show variation between species; the release of thromboxane A_2 (TXA_2) by LTs B_4, C_4, and D_4 (19,24), for example, has been demonstrated only in guinea pig lung tissue; LTD_4 caused exudation of plasma in guinea pig skin but not in rabbit (18,25).

METHODS

Superfusion of Assay Tissues

Isolated tissues, guinea pig ileum smooth muscle (GPISM), rat stomach strip (RSS), rabbit aorta (RbA), and rabbit celiac artery (RbCA) were superfused in series with Tyrode solution at 5 ml min^{-1}. The specificity of the tissues was increased by the infusion of a combination of antagonists to histamine, 5-hydroxytryptamine, acetylcholine, and norepinephrine into the Tyrode solution (21).

Perfusion of Isolated Lungs

Lungs from guinea pigs (Dunkin Hartley, 250 to 350 g) or rats (Wistar, 200 to 250 g) were perfused with Tyrode solution (5 ml min^{-1}) via the pulmonary artery.

The effluent from the lungs then superfused a bank of assay tissues. Indomethacin ($1 \ \mu g \ ml^{-1}$) was infused into guinea pig lungs to prevent stimulation of release of TXA_2 and other cyclo-oxygenase products by LTC_4 and LTD_4.

Perfusion of Isolated Working Hearts

Using a modification of the Langendorff preparation, spontaneously beating hearts from guinea pigs (as above) were perfused via retrograde cannulation of the aorta under constant pressure (50 mm Hg) with Ringer-Locke solution maintained at 37°C and gassed with oxygen. A drop counter provided a continuous record of coronary flow; flow was also measured every minute. A thread hooked to the apex of the left ventricle and attached to an isometric muscle transducer allowed continuous monitoring of the rate and force of contraction. Rat and rabbit hearts were perfused by the same method.

In another series of experiments, the coronary circulation of spontaneously beating guinea pig hearts was perfused at constant flow (5 to 10 ml · min^{-1}) and the effluent superfused over assay tissues: RbA, RbCA, GPISM, RSS, and colon (RC). Pressure in the coronary circulation was measured with an EM 750 transducer.

Parenchymal Strips

Strips of parenchyma from guinea pig, rat, or rabbit lung were prepared by the method of Lulich et al. (13) and superfused with Tyrode solution at 5 ml min^{-1}.

Guinea Pig Blood Pressure In Vivo

Guinea pigs (Dunkin Hartley 400 to 600 g, either sex) were anesthetized with urethane 2.0 g · kg^{-1}. The trachea, jugular vein, and carotid artery were cannulated. Sometimes the second carotid was cannulated and the cannula positioned so that its tip was in the arch of the aorta. Blood pressure was recorded from the carotid artery with an EM 750 transducer. When necessary the animal was artificially ventilated.

Guinea Pig Trachea

Tracheas from guinea pigs (GPT) (Dunkin Hartley 500 g) were cut according to the method of Emmerson and McKay (6) and suspended in Tyrode solution bubbled with oxygen in a 15-ml organ bath. Changes in muscle length were recorded with a Harvard smooth muscle transducer. In another series of experiments, GPT was superfused above assay tissues sensitive to cyclo-oxygenase products (see above).

Tracheal rings were incubated at 37°C with LTD_4 (2×10^{-7} M) in Tyrode solution for 15 min. The supernatant was extracted twice with equal volumes of diethyl ether at pH 3, and the cyclo-oxygenase products released were quantitated by radioimmunoassay (10).

RESULTS

Activation of LTC_4

Guinea Pig Ileum

Leukotrienes C_4 and D_4 contracted guinea pig ileum or GPISM, but LTD_4 was more potent by approximately one order of magnitude and caused a faster contraction than LTC_4 (20). When bolus doses of LTC_4 and LTD_4 were superfused over a series of six strips of smooth muscle from guinea pig ileum, the profile of biological activity of LTC_4 changed over the 1 min taken to superfuse the tissues, whereas that of LTD_4 remained constant. The relative height of contraction in response to LTC_4 increased whereas the duration decreased, and the response to LTC_4 became more like that seen with LTD_4, suggesting that LTC_4 was converted to LTD_4 by γ-glutamyl transpeptidase (γ-GT) present on the assay tissues. In the presence of δ-D-glutamyl(O-carboxy)phenylhydrazine (2.5×10^{-4} molar), an inhibitor of γ-GT (8), the relative height of contraction in response to LTC_4 was reduced on the first tissue and progressively decreased until the sixth tissue. There was no increase in the activity of LTC_4 during superfusion over the tissues, and responses to LTD_4 were unchanged (Fig. 1). (Morris, Jones, Piper, Samhoun, and Taylor, *unpublished observations*).

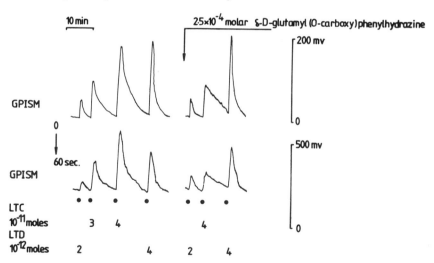

FIG. 1. Inhibition of increase in activity of LTC_4 by δ-D-glutamyl (O-carboxy)phenylhydrazine. Strips of guinea pig ileum smooth muscle (GPISM) were superfused in series. In the first panel, the height of contraction of the upper GPISM to LTC_4 4.0×10^{-11} mole was almost equivalent to that to LTD_4 4×10^{-12} mole, but on the lower GPISM (60 sec later) the response to LTC_4 was considerably greater than LTD_4 4.0×10^{-12} mole. In the second panel, in the presence of δ-D-glutamyl (O-carboxy)phenylhydrazine 2.5×10^{-4} molar, the response of the first GPISM to LTC_4 was reduced, that of the second GPISM was further reduced relative to LTD_4, and the rate of recovery was slower. The contractions to LTD_4 were unchanged.

Pulmonary Circulation

Leukotrienes C_4 and D_4 (0.5 to 2 × 10^{-9} M) were given either by bolus injection or infusion (1.0 to 2.0 × 10^{-9} M, 3 to 5 min) into the pulmonary circulation (IA) of guinea pig or rat isolated lungs and the responses of GPISM superfused by the effluent from the lungs compared with the responses to infusions of LTC_4, LTD_4 given directly to the assay tissues (DIR). Doses of prostaglandin (PG) E_2 were given IA and DIR to ensure that inactivation of PGE_2 in the pulmonary circulation was within the normal range (90 to 98%) (22). As shown in Fig. 2 the height of responses of GPISM to LTC_4 increased after passage through the pulmonary circulation in guinea pig lung. The responses to LTD_4 were either unchanged or slightly reduced. Responses to LTC_4 or LTD_4 given DIR remained constant. This apparent activation of LTC_4 also occurred in rat lung when LTC_4 was given by infusion (Fig. 2) but not when given by bolus injection. The potentiated LTC_4-induced

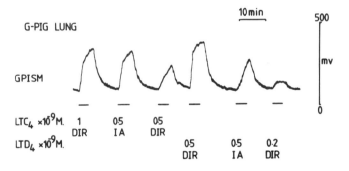

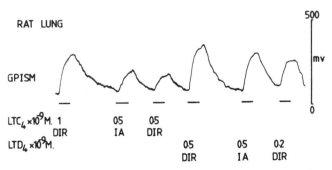

FIG. 2. Infusion of LTC_4 and LTD_4 into the pulmonary circulation of rat or guinea pig isolated perfused lung. In the upper panel GPISM was superfused with the effluent from guinea pig isolated lung perfused with Tyrode solution 5 ml min⁻¹. In the lower panel GPISM was superfused with the effluent from rat perfused lung. Doses of LTC_4 or LTD_4 were given either DIR or IA. In guinea pig and rat lung the responses of GPISM to LTC_4 given IA were greater than those to the same dose given DIR. The responses to LTD_4 given IA were smaller than those to the same dose given DIR.

contraction of the GPISM was sensitive to FPL 55712 (1 μg ml^{-1}) and unchanged by δ-D-glutamyl(O-carboxy)phenylhydrazine 2.5 × 10^{-4} molar given IA 15 min before and during infusion of LTC$_4$.

Coronary Circulation

When infused into the coronary circulation, LTC$_4$ caused a marked increase in coronary arterial pressure (see below), and the activity of LTC$_4$ on GPISM showed a similar increase (Fig. 3). LTD$_4$ given IA also increased coronary arterial pressure, but its biological activity (as shown by contraction of GPISM) was unchanged by passage through the coronary vessels.

Sensitization to Other Agonists

Strips of GPISM (not bathed with mepyramine) superfused by effluent from either guinea pig lung or heart were contracted by infusions of histamine or PGE$_2$ prior to IA infusion of LTC$_4$ (0.5 to 2.0 × 10^{-9} M). After infusion of LTC$_4$ either DIR or IA, the sensitivity of GPISM to histamine had noticeably increased (Fig. 4). A similar potentiation occurred to responses to PGE$_2$.

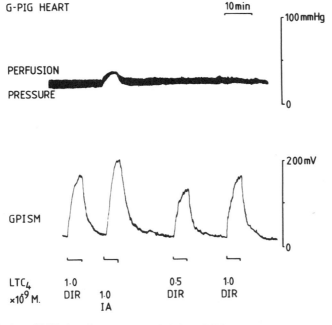

FIG. 3. Infusion of LTC$_4$ into the coronary circulation. GPISM was superfused with the effluent from a guinea pig heart perfused with Tyrode solution 5 ml min^{-1}. Pressure in the coronary artery was recorded. When LTC$_4$ 1 × 10^{-9} M (5 min) was infused into the coronary artery IA, there was a sustained increase in arterial pressure and the contraction of GPISM was greater than that in response to the same dose given DIR.

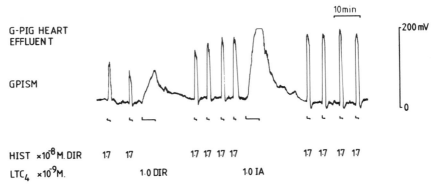

FIG. 4. Potentiation of responses of GPISM to histamine (HIST). GPISM was superfused with the effluent from a guinea pig heart perfused with Tyrode solution 5 ml min⁻¹. Stable contractions of GPISM in response to infusions (2 min) of histamine 1.7×10^{-8} M were obtained. After infusion of LTC₄ 1×10^{-9} M (5 min) DIR, the responses to histamine were increased. LTC₄ 1×10^{-9} M was then infused (5 min) into the coronary artery, and the contraction to GPISM was greater than that seen in response to the same dose of LTC₄ given DIR. The height of responses to histamine were still further increased after the IA dose of LTC₄.

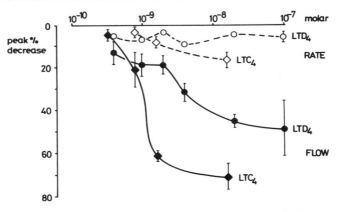

FIG. 5. Logarithmic dose-response curves of peak percentage reduction in coronary flow *(closed symbols)* of guinea pig working hearts in response to LTC₄ and LTD₄. The curve indicating the response to LTC₄ is steeper than that showing the response to LTD₄, showing a maximum decrease of 71 ± 6% at 2.0×10^{-8} M. The maximum reduction after LTD₄ was 45 ± 2% at 1.6×10^{-7} M. The heart rate was reduced ·following LTC₄ administration, but LTD₄ had less effect.

Cardiac Actions of LTC₄ and LTD₄

When LTC₄ (4.0×10^{-10} to 2.0×10^{-8} M) or LTD₄ (1.6×10^{-10} to 8.0×10^{-8} M) was infused (5 min) into the coronary circulation of spontaneously beating guinea pig hearts perfused under constant pressure, it reduced coronary flow in a dose-dependent manner. However, there were differences in response to the leukotrienes. LTC₄ is more active than LTD₄ (Fig. 5), causing a maximum percentage reduction in flow of 71 ± 6 at 2.0×10^{-8} M, whereas the maximum percentage reduction produced by LTD_4 (1.6×10^{-8} M) was $45 \pm 2\%$. The log-dose response curves for the peak percentage reduction in flow (Fig. 5) show that LTC_4 has a steeper curve than LTD_4. LTD_4 also caused a reduction in coronary flow in spontaneously beating rat or rabbit isolated perfused hearts. There was some reduction in the spontaneous rate of beating and force of contraction of the hearts, but neither LTC_4 nor LTD_4 induced cardiac arrhythmias.

In experiments where guinea pig hearts were perfused at constant flow, LTC_4 or LTD_4 (1 to 2×10^{-9} M) caused dose-related increases of the coronary perfusion pressure. The vasoconstriction caused by LTC_4 1.3 ng ml^{-1} was sufficient to obscure the record of pulse pressure from the peristaltic pump.

The flow rate began to return toward control values during the 5-min infusions of LTC_4 or LTD_4 in hearts perfused under constant pressure, but usually the increase in pressure was maintained in hearts perfused at constant flow.

On a comparative basis, the results indicate that LTC_4 has a greater efficacy in reducing coronary flow, whereas LTD_4 has a greater efficacy in reducing the force of contraction ($37 \pm 11\%$ versus $24 \pm 7\%$ for similar LT-induced reductions in coronary flow). LTD_4 had a faster onset of action. Indomethacin (0.3 to 1.4×10^{-5} M) significantly decreased the peak reduction in coronary flow in response to 1.6×10^{-9} M LTC_4 from $46 \pm 3\%$ to $21 \pm 3\%$ ($N = 4,5$) but not in response to 2×10^{-8} M LTD_4 ($48 \pm 2\%$ to $43 \pm 3\%$, $N = 4,5$). The fall in flow in response to LTD_4 was more rapid than that in response to LTC_4, and the fall in flow during the first minute in response to LTD_4 was significantly reduced by indomethacin ($45 \pm 2\%$ to $31 \pm 5\%$, $N = 4,5$). The decreases in contractility of the working hearts to LTC_4 and LTD_4 were reduced by indomethacin, but only the LTD_4 reduction was significant ($37 \pm 11\%$ to $9 \pm 7\%$, $N = 4,5$). FPL 55712 (3.8×10^{-6} M) significantly blocked the peak LTC_4-induced decrease in flow ($46 \pm 3\%$ to $7 \pm 3\%$, $N = 4,3$) as well as the initial phase of LTD_4-induced reduction in flow ($48 \pm 2\%$ to $27 \pm 6\%$, $N = 4,4$). There was no significant difference in the LTD_4-induced reduction of flow after a 5-min infusion. FPL 55712 completely antagonized the decreases in contractility due to LTC_4 and LTD_4.

When the effluent from guinea pig heart was superfused over RbA, RbCA, RSS, and GPISM to investigate the possible release of mediators following infusion of leukotrienes, only GPISM and RSS contracted. These responses were blocked by FPL 55712. Guinea pig chopped lung was agitated above the assay tissues to ensure that RbA and RbCA were sensitive to TXA_2 (17). LTC_4 and LTD_4 (up to 1×10^{-9} M), given DIR, caused contraction of RSS and GPISM but had no action on RbA or RbCA. In other experiments rabbit isolated pulmonary, carotid, renal, or femoral arteries were unaffected by LTC_4 or LTD_4.

Cardiovascular Actions of Leukotrienes In Vivo

When bolus injections of LTC_4 or LTD_4 (1.5×10^{-10} to 1.5×10^{-9} mole) were given into the jugular vein of anesthetized guinea pigs, they produced a rapid fall in carotid arterial blood pressure. Following the initial drop in blood pressure, there was sometimes a reflex rise, particularly with LTC_4. Recovery of the total hypotensive effect took 7 to 10 min. Administration of LTC_4 or LTD_4 into the arch of the aorta also produced a long-lasting hypotensive effect.

Administration of indomethacin 1 mg kg^{-1} 15 min prior to LTC_4 or LTD_4 caused a persistent small elevation in blood pressure, shortened the duration of the hypotensive effect, and blocked the reflex rise in blood pressure.

Actions of Leukotrienes on Parenchyma

Parenchymal strips from guinea pig, rat, rabbit, and human lung were contracted by bolus injections of LTC_4 and LTD_4. In all cases LTC_4 and LTD_4 were approximately equipotent (19), but the sensitivity of the lung varied between species. Doses of LTC_4 and LTD_4 required to contract GPP were 0.05 to 5.0 × 10^{-12} moles, but doses 1,000 to 10,000 times higher were required to contract rat, rabbit, or human lung strips. In confirmation of the findings of Sirois et al. (24), GPP was contracted by LTB_4 (produced from rabbit polymorphonuclear leukocytes) but in lower doses (3 × 10^{-12} to 3 x 10^{-11} moles) than reported by the above authors, LTE_4 (2 × 10^{-11} to 5 x 10^{-10} mole) also contracted GPP. The LTE_4-induced contractions of GPP were of longer duration than those produced by the other LTs. As with LTC_4 and LTD_4, the LTB_4-induced contractions of GPP were reduced by imidazole. Contractions of rabbit, rat, or human parenchymal strips to LTs were not inhibited by indomethacin.

LTC_4- or LTD_4-induced contractions of GPP were completely abolished by FPL 55712 (1 μg ml^{-1}). Responses of GPP to low doses of LTs were completely blocked by indomethacin (5 μg ml^{-1}), whereas responses to higher doses were reduced (Figs. 6 and 7). Similar inhibition was observed with the thromboxane synthetase inhibitors imidazole and carboxyheptylimidazole. As shown in Figs. 6 and 7, FPL 55712 antagonized the residual contraction to LTs.

The responses of GPP to LTB_4, LTC_4, and LTD_4 were indistinguishable. However, when a strip of guinea pig ileum was superfused in series with GPP, LTB_4 contracted only GPP, whereas LTC_4 and LTD_4 contracted both tissues (Fig. 8).

Release of Cyclo-oxygenase Products from Trachea

As previously described, LTC_4 and LTD_4 contract GPT in doses an order of magnitude higher than those required to contract GPP (20). The trachea released significant amounts of PGE_2 (50 ng g^{-1} control; 75 ng g^{-1} + LTD_4) and PGI_2 (75 ng g^{-1} control; 80 ng g^{-1} + LTD_4) but only small amounts of $PGF_{2\alpha}$ and TXA_2 (assayed as TXB_2).

DISCUSSION

The increase in relative potency of LTC_4 during a 1-min contact with GPISM, the decrease in duration of response, and the inhibition of these changes by the γ-GT inhibitor δ-D-glutamyl(O-carboxy)phenylhydrazine strongly suggest that LTC_4 was converted to LTD_4 by γ-GT present in the smooth muscle strips.

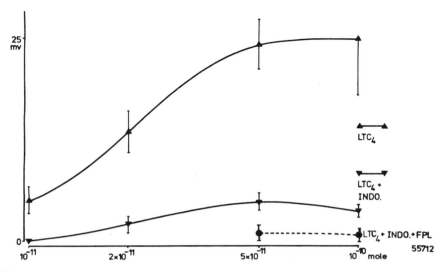

FIG. 6. Dose-response curves of guinea pig parenchymal strips after LTC₄ administration and the effects of indomethacin and FPL 55712. (▲) Contractions in response to LTC₄. (▼) Contractions in response to LTC₄ in the presence of indomethacin. (●) Contractions in response to LTC₄ in the presence of indomethacin and FPL 55712.

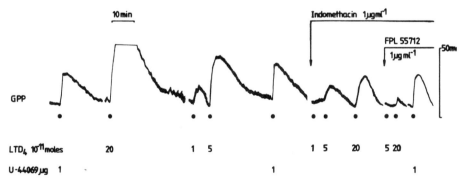

FIG. 7. The effects of indomethacin and FPL 55712 on LTD₄-induced contractions of guinea pig parenchyma (GPP). GPP was superfused with Tyrode solution 5 ml min⁻¹. In the first part of the trace, LTD₄ (1, 5, and 20 × 10⁻¹² mole) caused dose-related contractions of GPP. In the second part, after indomethacin, the response to LTD₄ (1.0 × 10⁻¹² mole) was abolished and that to 5 and 20 × 10⁻¹² mole greatly reduced. FPL 55712 abolished the residual response to LTD₄ (5 × 10⁻¹² mole) and greatly reduced that to 20 × 10⁻¹² mole. Responses to U-44069 (thromboxane analog) were not abolished by indomethacin or FPL 55712. (From ref. 19.)

After LTC₄ passage through the pulmonary circulation of guinea pig lung, whether it was given by bolus injection or infusion, the responses of GPISM were also increased. Under the same conditions, however, responses to LTD₄ were either slightly reduced or unchanged. Similar increases in biological activity of LTC₄ occurred in rat isolated perfused lungs when LTC₄ was given by infusion. The

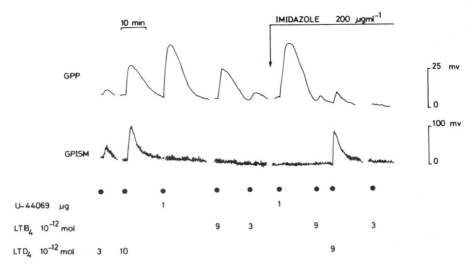

FIG. 8. Differential responses to LTB₄ and LTD₄. Strips of GPP and GPISM were superfused in series with Tyrode solution (5 ml min⁻¹). In the left-hand panel LTD₄ 3 and 10 × 10⁻¹² mole caused GPP and GPISM to contract. U-44069 1 µg ml⁻¹ (2.86 × 10⁻⁹ mole) (thromboxane analog) and LTB₄ (3 and 9 × 10⁻¹² mole) caused only GPP to contract. In the right-hand panel, in the presence of imidazole 200 µg ml⁻¹ (2.94 × 10⁻³ M) the response to U-44069 was unchanged, the contractions of GPP in response to LTB₄ and LTD₄ were reduced, but the contraction of GPISM in response to LTD₄ was not inhibited.

apparent increased activity of LTC_4 was unlikely to be due to conversion to LTD_4 by γ-GT as it was not blocked by δ-D-glutamyl(O-carboxy)phenylhydrazine. Neither was it due to synergism with cyclo-oxygenase products as the lungs (guinea pig) were treated with indomethacin to prevent LT-induced release of TXA_2 and prostaglandins (19). The activity of LTC_4 was also increased by passage through the coronary circulation of guinea pig hearts, in which no release of TXA_2 was detected. Again, this was unlikely to be due to conversion of LTC_4 to LTD_4 as LTC_4 is a more active constrictor of the coronary circulation than LTD_4.

The sensitization of tissues to other agonists by SRS-A has been recognized for many years (3). We demonstrated the same phenomenon caused by LTC_4. The increased activity of LTC_4 after passage through the coronary or pulmonary circulations may be due to the same mechanism(s), but doses of LTC_4 or LTD_4 given DIR were not potentiated. The reason for sensitization to other agonists and increased activity of LTC_4 is not fully understood, and the range of agonists which can exhibit potentiation of biological activity under these conditions remains to be investigated.

Leukotrienes C_4 and D_4 were potent constrictors of the coronary circulation of guinea pig perfused working hearts as well as of rabbit and rat hearts, although they had no direct action on isolated arterial blood vessels. As in the microvasculature of guinea pig skin (18), LTC_4 was the most potent vasoconstrictor. Both LTs caused a reduction in rate and force of contraction of the heart. As neither

LTC$_4$ (2×10^{-10} to 2×10^{-7} M) nor LTD$_4$ (2.6×10^{-10} to 3.2×10^{-7} M) had any direct action on spontaneously beating paired atria or electrically driven ventricular strips (11), the changes in rate and force were likely to be secondary to the coronary vasoconstriction.

There appears to be a difference in the mechanism by which coronary flow is reduced by LTC$_4$ and LTD$_4$ as there is a large component of reduction in flow in response to LTD$_4$ which is resistant to FPL 55712 or indomethacin. Although FPL 55712 (2 μg/ml) blocked the effects of both LTs on contractility and heart rate, the reduction in coronary flow in response to LTD$_4$ was reduced by a much smaller degree than that in response to LTC$_4$. Indomethacin (5 μg ml^{-1}) inhibited the effects on heart rate and contractility to both LTs but blocked only the initial component of reduction in flow in response to LTD$_4$ and reduced the overall reduction in response to LTC$_4$, the maximum effect occurring after 2 min. These results could account for the findings of Levi et al. (12) and Anhut et al. (1) that indomethacin either blocked or delayed cardiac anaphylaxis. The action of indomethacin suggests that cyclo-oxygenase products might have a variable role in the cardiac actions of LTs but there was no evidence of TXA$_2$ or PG release when the effluent from the heart was superfused over assay tissues.

Although LTC$_4$ and LTD$_4$ caused vasoconstriction in guinea pig skin and coronary circulation, each caused hypotension *in vivo*. The initial hypotensive effect, however, could be due to the actions of LTC$_4$ and LTD$_4$ on cardiac output. The hypertension was probably a reflex from bronchoconstriction and the prolonged hypotension caused by the release of some cyclo-oxygenase product, possibly PGI$_2$, as it was shortened by indomethacin administration (20).

LTC$_4$ and LTD$_4$ increased resistance and reduced compliance in monkeys (2) and guinea pigs (5) *in vivo*. If LTC$_4$ and LTD$_4$ cause vasoconstriction in the small blood vessels of the lung, this would contribute to the reduction in compliance. Even low doses of LTC$_4$ and LTD$_4$ released TXA$_2$ and PG-like substances from guinea pig lung *in vitro* (19), which probably explains why the bronchoconstrictor action of LTC$_4$ and LTD$_4$ in this species was blocked by indomethacin (20).

Guinea pig parenchyma was contracted by LTB$_4$, LTC$_4$, LTD$_4$, and LTE$_4$. Leukotriene E$_4$, which is probably a metabolite of LTC$_4$ and LTD$_4$ was less potent than the other LTs and caused a longer-lasting response.

As the contractions of GPP in response to LTB$_4$, LTC$_4$ and LTD$_4$ were either completely blocked or reduced by indomethacin or imidazole, it appears that a large component of their actions in guinea pig lung were due to release of TXA$_2$ and prostaglandins. The direct action of LTC$_4$ and LTD$_4$ which was blocked by FPL 55712 was only a small proportion of their total constrictor action. Rabbit, rat, and human parenchymal strips were much less sensitive to LTC$_4$ and LTD$_4$ than GPP, and the contractions were not blocked by indomethacin but were blocked by FPL 55712. This probably indicates that the responses of rat, rabbit, and human lung were due to the direct actions of LTC$_4$ and LTD$_4$. Histological examination of GPP showed the presence of smooth muscle and blood vessels so that constriction of the vessels by LTs might have contributed to the contractions (19).

Gryglewski et al. (9) showed that in guinea pig lung the major release of TXA$_2$ occurred from parenchymal tissues, and the low levels of PGE$_2$ and PGI$_2$ which were generated from GPT are in agreement with this. The cyclo-oxygenase products [TXA$_2$ and PG-like substances generated by LTs in GPP (small airways)] augmented the responses of LTB$_4$, LTC$_4$, and LTD$_4$ on this tissue. In contrast, only low levels of PGE$_2$ and PGI$_2$ which tend to relax GPT (large airways) were released from this tissue by LTD$_4$. These findings probably explain why GPP is more sensitive than GPT to contractions by LTC$_4$ and LTD$_4$.

The potent biological actions of the leukotrienes suggest that they have a pathological role in various conditions. They might contribute to inflammation, to respiratory disease where constriction of the airways and edema of the bronchial mucosa occurs, and to conditions involving vasoconstriction, e.g., angina, cardiac ischemia, or perhaps migraine.

ACKNOWLEDGMENTS

We thank the Medical Research Council, the Asthma Research Council, and the Lilian Roxon Memorial Trust (Australia) for financial support. We also thank Dr. J. Rokach, Merck Frosst Laboratories, for the gift of synthetic LTC$_4$, LTD$_4$, LTE$_4$, and δ-D-glutamyl(O-carboxy)phenylhydrazine; Dr. J. E. Pike, The Upjohn Company, for U-44069; Dr. A. W. Ford-Hutchinson for purified LTB$_4$; and Dr. A. M. White, Ciba/Geigy, for carboxyheptylimidazole.

REFERENCES

1. Anhut, H., Bernauer, S., and Peskar, B. A. (1977): Radioimmunological determination of thromboxane release in cardiac anaphylaxis. *Eur. J. Pharmacol.*, 44:85–88.
2. Bach, M. K., Brashler, J. R., Johnson, H. G., and McNee, M. L. (1981): Pharmacology of the SRSs: some aspects of their selective actions and modulation of their synthesis. In: *SRS-A and Leukotrienes*, edited by P. J. Piper, pp. 161–179. Wiley, New York.
3. Chakravarty, N. (1960): The occurrence of a lipid soluble smooth muscle stimulating principle ('SRS') in anaphylactic reaction. *Acta Physiol. Scand.*, 48:167–177.
4. Dahlén, S.-E., Hedqvist, P., Hammerström, S., and Samuelsson, B. (1980): Leukotrienes are potent constrictors of human bronchi. *Nature*, 228:484–486.
5. Drazen, J. M., Austen, K. F., Lewis, R. A., Clark, D. A., Goto, G., Marfat, A., and Corey, E. J. (1980): Comparative airway and vascular activities of leukotrienes C-1 and D in vivo and in vitro. *Proc. Natl. Acad. Sci. USA*, 77:4354–4358.
6. Emmerson, J., and McKay, D. (1979): The zig-zag tracheal strip. *J. Pharmacol.*, 31:798–803.
7. Ford-Hutchinson, A. W., Bray, M. A., Doig, M. V., Shipley, M. E., and Smith, M. J. H. (1980): Leukotriene B: a potent chemokinetic and aggregating substance released from polymorphonuclear leukocytes. *Nature*, 286:264–265.
8. Griffith, O. W., and Meister, A. (1979): Translocation of intra-cellular glutathione to membrane-bound γ-glutamyl transpeptidase as a discrete step in the γ-glutamyl cycle: glutathionuria after inhibition of transpeptidase. *Proc. Natl. Acad. Sci. USA*, 76:268–272.
9. Gryglewski, R. J., Dembinska-Kiec, A., Grodzinska, L., and Panczenko, B. (1976): Differential generation of substances with prostaglandin-like and thromboxane-like activities by guinea-pig trachea and lung strips. In: *Lung Cells in Disease*, edited by A. Bouhuys, pp. 289–307. Elsevier/North Holland, Amsterdam.
10. Jose, P., Niederhauser, U., Piper, P. J., Robinson, C., and Smith, A. P. (1976): Degradation of PGF$_{2\alpha}$ in the human pulmonary circulation. *Thorax*, 31:713–719.
11. Letts, L. G., and Piper, P. J. (1981): The effects of leukotrienes (LT) D$_4$ and C$_4$ on the guinea-pig isolated heart. *J. Physiol. (Lond.)*, 317:94–95.

12. Levi, R., Allan, G., and Zavecz, J. H. (1976): Prostaglandins and cardiac anaphylaxis. *Life Sci.*, 18:1255–1264.
13. Lulich, K. M., Mitchell, H. W., and Sparrow, M. P. (1976): The cat lung strip as an in vitro preparation of peripheral airways: a comparison of β-adrenoceptor agonists, autacoids and anaphylactic challenge on the lung strip and trachea. *Br. J. Pharmacol.*, 58:71–79.
14. Morris, H. R., Taylor, G. W., Piper, P. J., Samhoun, M. N., and Tippins, J. R. (1980): Slow reacting substances (SRSs): the structure identification of SRSs from rat basophil leukaemia (RBL-1) cells. *Prostaglandins*, 19:185–201.
15. Morris, H. R., Taylor, G. W., Piper, P. J., and Tippins, J. R. (1980): Structure of slow reacting substance of anaphylaxis from guinea-pig lung. *Nature*, 285:105–106.
16. Murphy, R. C., Hammarström, S., and Samuelsson, B. (1979): Leukotriene C: a slow-reacting substance from murine mastocytoma cells. *Proc. Natl. Acad. Sci. USA*, 76:4275–4279.
17. Palmer, M. A., Piper, P. J., and Vane, J. R. (1973): Release of rabbit aorta contracting substance (RCS) and prostaglandins induced by chemical or mechanical stimulation of guinea-pig lung. *Br. J. Pharmacol.*, 49:226–242.
18. Peck, J. J., Piper, P. J., and Williams, T. J. (1981): The effects of leukotrienes C₄ and D₄ on the microvasculature of guinea-pig skin. *Prostaglandins*, 21:315–321.
19. Piper, P. J., and Samhoun, M. N. (1981): The mechanism of action of leukotrienes C₄ and D₄ in guinea-pig isolated perfused lung and parenchymal strips of guinea-pig, rabbit and rat. *Prostaglandins*, 24:793–803.
20. Piper, P. J., Samhoun, M. N., Tippins, J. R., Williams, T. J., Palmer, M. A., and Peck, M. J. (1981): Pharmacological studies on pure SRS-A, and synthetic leukotrienes C₄ and D₄. In: *SRS-A and Leukotrienes*, edited by P. J. Piper, pp. 81–99. John Wiley, New York.
21. Piper, P. J., and Vane, J. R. (1969): Release of additional factors in anaphylaxis and its antagonism by anti-inflammatory drugs. *Nature*, 223:29–35.
22. Piper, P. J., Vane, J. R., and Wyllie, J. H. (1970): Inactivation of prostaglandins by the lungs. *Nature*, 255:600–604.
23. Samuelsson B., Borgeat, P., Hammarström, S., and Murphy, R. C. (1979): Introduction of a nomenclature: leukotrienes. *Prostaglandins*, 17:785–787.
24. Sirois, P., Roy, S., and Borgeat, P. (1981): The action of leukotriene B₄ (LTB₄) on the lung. *Prostaglandins Med.*, 5:429–444.
25. Ueno, A., Tanaka, K., Katori, M., Hayashi, M., and Arai, Y. (1981): Species differences in increased vascular permeability by synthetic leukotrienes C₄ and D₄. *Prostaglandins*, 21:637–647.

Leukotrienes and Other Lipoxygenase Products,
edited by B. Samuelsson and R. Paoletti.
Raven Press, New York © 1982.

Interaction of Leukotrienes With Cyclo-oxygenase Products in Guinea Pig Isolated Trachea

Priscilla J. Piper and J. R. Tippins

*Department of Pharmacology, Institute of Basic Medical Sciences,
Royal College of Surgeons of England, Lincoln's Inn Fields, London WC2A 3PN,
England*

Various groups (4,5,10) have suggested that guinea pig tracheal (GPT) muscle tone is modulated by the release of prostaglandins (PG), particularly PGE_2 and $PGF_{2\alpha}$. Cyclo-oxygenase inhibitors (e.g., indomethacin and aspirin) have been shown to potentiate the response to a number of agonists (e.g., acetylcholine, histamine, 5-hydroxytryptamine). Using bioassay and thin-layer chromatography, these same groups identified PGE_2 and $PGF_{2\alpha}$ released by the trachea.

We have shown (9) that leukotriene (LT) D_4 is a major arachidonic acid (AA) metabolite in sensitized guinea pig lungs and the LTC_4 and LTD_4 have potent bronchoconstrictor activity (11). This response of the guinea pig trachea is potentiated by indomethacin, and we therefore attempted to determine the mechanism of this potentiation.

METHODS

The action of a number of enzyme inhibitors (i.e., indomethacin, carboxyheptylimidazole, and tranylcypromine on the response of the isolated trachea was determined by immersing the trachea, cut according to the method of Emmerson and McKay (3), in oxygenated Tyrode solution in a 15-ml organ bath. A cumulative dose-response curve to LTD_4 (10^{-10} to 3×10^{-8} M) was prepared. The inhibitor was added to the bath 15 min before the dose-response curve was repeated.

The release of cyclo-oxygenase products by LTD_4 or anaphylaxis was determined by superfusing a tracheal strip (3) above a bank of tissues which responded to the various AA metabolites, i.e., rabbit aorta [PGG_2, thromboxane A_2 (TXA_2)], rabbit celiac artery (TXA_2), bovine coronary artery (PGI_2), rat stomach strip, and rat colon (PGE_2 and $PGF_{2\alpha}$).

The release of cyclo-oxygenase products by LTD_4 or anaphylaxis was also determined by radioimmunoassay. Tracheal rings were incubated in 0.5 ml Tyrode solution in the presence of LTD_4 or egg albumin, with or without the inhibitors. The cyclo-oxygenase products were then extracted into diethyl ether at pH 3 and assayed for PGE_2, $PGF_{2\alpha}$, 6-keto-$PGF_{1\alpha}$, and TXB_2 (7).

RESULTS

LTC_4 and LTD_4 (10^{-10} to 3×10^{-8} M) produced a dose-related contraction of the isolated trachea. The response to LTD_4 was potentiated by 1 μg ml^{-1} indomethacin (maximum 145 $\pm$ 37%), 1 μg ml^{-1} carboxyheptylimidazole (maximum 44 $\pm$ 9%), and 3.3 μg ml^{-1} tranylcypromine (maximum 78 $\pm$ 10%) (mean $\pm$ SEM, $p < 0.001$, Student's t-test, $N \geq 5$).

Trachea from sensitized animals, challenged with egg albumin, or trachea from normal animals, contracted with LTD_4, released a PGE-like material which caused a prolonged relaxation of the rabbit celiac artery and a contraction of the rat stomach strip. The sensitized tracheal strip also released a material which contracted the rabbit aorta and rat stomach strip. PGI_2 release was not detected.

Normal trachea released PGE_2, $PGF_{2\alpha}$, PGI_2, and TXA_2 as determined by radioimmunoassay. PGE_2 (50 ng g^{-1}) and PGI_2 (75 ng g^{-1}) were released from control tracheas. With LTD_4 these amounts were increased to 75 ng g^{-1} and 80 ng g^{-1}, respectively. Less than 30 ng g^{-1} of $PGF_{2\alpha}$ and TXA_2 were released, and these amounts were not altered greatly by the addition of LTD_4. Indomethacin inhibited the release of PGE_2 and PGI_2, whereas the release of $PGF_{2\alpha}$ and TXA_2 was not altered significantly from the controls. Carboxyheptylimidazole inhibited the release of TXA_2 at the concentration which potentiated the response to LTD_4 and increased the release of the PGs. Tranylcypromine, at the concentration which potentiated the response to LTD_4, potentiated the release of PGI_2. At 100 μg ml^{-1}, release of PGI_2 was inhibited, but there was no effect on the response of the GPT to LTD_4. Release of PGE_2 was also increased in the presence of tranylcypromine (3.3 μg ml^{-1}). Similar results were obtained using sensitized tracheas challenged with 10 μg egg albumin.

DISCUSSION

The release of PGE_2 and $PGF_{2\alpha}$ has been reported to be important in the modulation of tracheal muscle tone (4,5,10). However, the release of other AA metabolites with potent activity on bronchial tone (PGG_2, PGH_2, PGI_2, and TXA_2) has not been fully investigated, although it has been suggested that these metabolites may play an important role (8). In our work, indomethacin potentiated the response to LTD_4 in a manner similar to the potentiation demonstrated by other groups (e.g., ref. 10) except that there was little evidence for inhibition of the response to low agonist concentrations. Carboxyheptylimidazole, which is a specific thromboxane synthetase inhibitor, blocked the synthesis of TXA_2 and potentiated the release of the other mediators. This same concentration (1 μg ml^{-1}) also potentiated the response of the GPT to LTD_4. Tranylcypromine, a prostacyclin synthetase inhibitor, at 3.3 μg ml^{-1} potentiated the response to LTD_4. However, this concentration also potentiated the release of PGI_2 and was therefore not acting by inhibiting PGI_2 synthesis.

TXA_2 is a potent bronchoconstrictor (6) and has been shown to mediate the response of guinea pig lung parenchymal strips elicited by the leukotrienes (12). In the trachea, TXA_2 does not appear to have this same role.

PGI_2 has been shown to be able to reverse the bronchoconstriction induced by other agonists (1), and we have also been able to relax the trachea, maximally contracted by LTD_4, with PGI_2 as well as PGE_2. However, the use of tranylcypromine to block the synthesis of PGI_2 has not clarified the role of PGI_2. This is probably due to the fact that tranylcypromine has a wide spectrum of activity which may have affected the response to the leukotriene.

Our results suggest that the potentiation of the response of GPT to LTD_4 by indomethacin was due mainly to the inhibition of PGE_2 and PGI_2 synthesis. There was no evidence of an alteration in cyclo-oxygenase metabolism in tracheas from sensitized animals as has been postulated in guinea pig lung (2). This may have been due, however, to the limitations of our method of assaying cyclo-oxygenase metabolites.

$PGF_{2\alpha}$ and TXA_2 appear to play very little part in modulating the response of the trachea to LTD_4. Although the exact relationship between tracheal stimulation by LTD_4 and the release of cyclo-oxygenase products has not been determined, it seems possible that PGE_2 and PGI_2 release may represent a physiological antagonism to the potent constrictor effect of LTD_4 during anaphylaxis.

REFERENCES

1. Bianco, S., Robuschi, M., Ceserani, R., Gandolfi, C., and Kamburoff, P. L. (1978): Prevention of a specifically induced bronchoconstriction by PGI_2 and 20-methyl-PGI_2 in asthmatic patients. *Pharmacol. Res. Commun.*, 10:657–674.
2. Boot, J. R., Cockerill, A. F., Dawson, W., Mallen, D. N. B., and Osborn, D. J. (1978): Modification of prostacyclin and thromboxane release by immunological challenges from guinea-pig lung. *Int. Arch. Allergy Appl. Immunol.*, 57:159–164.
3. Emmerson, J., and McKay, D. (1979): The zig-zag tracheal strip. *J. Pharm. Pharmacol.*, 31:798.
4. Farmer, J. B., Farrar, D. G., and Wilson, J. (1974): Antagonism of tone and prostaglandin-mediated responses in a tracheal preparation by indomethacin and SC-19220. *Br. J. Pharmacol.*, 52:559–565.
5. Grodzinska, L., Panczenko, B., and Gryglewski, R. J. (1975): Generation of prostaglandin-E like material by the guinea-pig trachea contracted by histamine. *J. Pharm. Pharmacol.*, 27:88–91.
6. Hamberg, M., Hedqvist, P., Strandberg, K., Svensson, J., and Samuelsson, B. (1975): Prostaglandin endoperoxides. IV. Effects on smooth muscle. *Life Sci.*, 16:451–462.
7. Jose, P., Niederhauser, U., Piper, P. J., Robinson, C., and Smith, A. P. (1976): Degradation of prostaglandin $F_{2\alpha}$ in the human pulmonary circulation. *Thorax*, 31:713–719.
8. Moncada, S., and Vane, J. R. (1979): Pharmacology and endogenous roles of prostaglandin endoperoxides, thromboxane A_2 and prostacyclin. *Pharmacol. Rev.*, 30:293–331.
9. Morris, H. R., Taylor, G. W., Piper, P. J., and Tippins, J. R. (1980): Structure of slow-reacting substance of anaphylaxis from guinea-pig lung. *Nature*, 285:104–106.
10. Orehek, J., Douglas, J. S., Lewis, A. J., and Bouhuys, A. (1973): Prostaglandin regulation of airway smooth muscle tone. *Nature*, 245:84–85.
11. Piper, P. J., Samhoun, M. N., Tippins J. R., Williams, T. J., Palmer, M. A., and Peck, M. J. (1980): Pharmacological studies on pure SRS-A, SRS and synthetic leukotriene C_4 and D_4. In: *SRS-A and Leukotrienes*, edited by P. J. Piper, pp. 81–99. Wiley, New York.
12. Piper, P. J., and Samhoun, M. N. (1981): The mechanism of action of leukotrienes C_4 and D_4 in guinea-pig isolated perfused lung and parenchymal strips of guinea pig, rabbit and rat. *Prostaglandins*, 24:793–803.

Leukotrienes and Other Lipoxygenase Products,
edited by B. Samuelsson and R. Paoletti.
Raven Press, New York © 1982.

Pulmonary and Vascular Actions of Leukotrienes

Per Hedqvist, Sven-Erik Dahlén, and *Jakob Björk

*Department of Physiology, Karolinska Institute, S-104 01 Stockholm, Sweden; and
Department of Experimental Medicine, Pharmacia AB, S-751 04 Uppsala, Sweden

For more than 40 years after the original discovery, slow-reacting substance of anaphylaxis (SRS-A), presumed mediator of important events in hypersensitivity reactions such as asthma, remained an intriguing entity of unknown composition and chemical structure (24). However, very recently SRS-A was found to belong to a newly discovered family of bioactive substances, the leukotrienes (LTs), derived from arachidonic acid and closely related polyunsaturated fatty acids (23,24). This was first demonstrated using an SRS generated from murine mastocytoma cells (18), and it is now evident that both immunologically and nonimmunologically generated SRS-A can be identified as a mixture of closely related leukotrienes, principally the cysteinyl-containing leukotrienes of the C, D, and E series (15,24). Furthermore, LTC_4 and LTD_4 rapidly proved to induce cutaneous edema in the guinea pig (9,14) and to possess smooth-muscle-stimulating properties almost identical to what previously was known to be characteristic for SRS-A (19); in low concentrations it contracts isolated human and guinea pig bronchi, virgin guinea pig uterus, guinea pig ileum, but not estrous rat uterus or rabbit bronchi or cecum (14).

Here we wish to report on the profile of biological actions exerted by various leukotrienes, with special reference to their effects in the pulmonary and cardiovascular systems. Our data indicate that the cysteinyl-containing leukotrienes (LTC, LTD, LTE) are bronchoconstrictors of remarkable potency, and that together with another leukotriene, LTB, they induce microvascular alterations—macromolecular leakage and leukocyte adhesion to the endothelium in postcapillary venules—that closely resemble the early vascular events in acute inflammatory reactions.

CARDIOVASCULAR EFFECTS OF LEUKOTRIENES

The cardiovascular effects of LTC_4 were recently documented in guinea pigs and monkeys *(Macaca irus)* (4,25). LTC_4, when injected intravenously into guinea pigs, caused an initial transient hypertension (30 to 60 sec), presumably a manifestation of its vasoconstrictor properties (3), followed by a hypotensive period of longer duration (5 to 15 min). The cause of the hypotension is not known, although peripheral plasma exudation and decreased venous return might contribute to the effect.

The cardiovascular alterations caused by right atrial injection of LTC_4 in the monkey were similar to those seen in the guinea pig: an initial brisk systemic hypertension followed by a biphasic long-lasting reduction of mean arterial pressure (25) (Fig. 1). Cardiac output decreased after LTC_4, apparently in parallel with the long-lasting hypotension. Whether the reduced cardiac output reflects myocardial depression, decreased venous return, or a combination of these factors is not known. LTC_4 also caused an initial and short-lived (20 to 30 sec) rise in pulmonary arterial pressure that appeared before any visible change in transpulmonary pressure, indicating that LTC_4 also may have vasoconstrictor properties in the pulmonary vascular bed. Thereafter, the pulmonary arterial pressure usually decreased, apparently in parallel with the reduction of systemic arterial pressure. It may be noted that there was a transient but marked decrease in circulating leukocytes but not in platelets after injection of LTC_4. There was a concomitant slight hemoconcentration, as indicated by a rise in hematocrit. The reason for the decrease in leukocytes is not evident, but it is conceivable that LTC_4 in relatively high concentrations may induce leukocyte trapping in various microvascular beds, similar to what has been documented for LTB_4 (3).

PULMONARY EFFECTS OF LEUKOTRIENES IN VIVO

Intravenous injections of either LTC_4 or histamine increased the tracheal insufflation pressure in anesthetized and artificially ventilated guinea pigs (4,9,14) (Fig. 2a). However, LTC_4 differed from histamine in two important respects; it was at least 100 times as potent, and it caused a long-lasting increase in insufflation

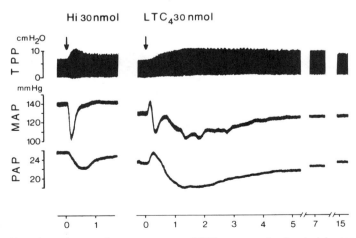

FIG. 1. Alterations of transpulmonary pressure (TPP), mean systemic arterial pressure (MAP), and pulmonary arterial pressure (PAP) induced by right atrial injection of histamine (Hi) and LTC_4 in the artificially ventilated monkey (Macaca irus).

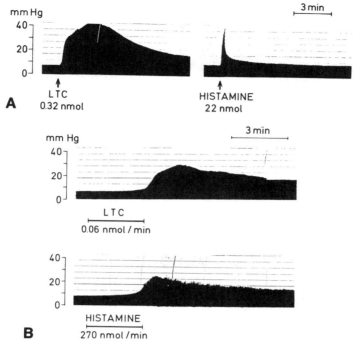

FIG. 2. Increase in tracheal insufflation pressure evoked by LTC_4 and histamine, when given intravenously (**A**) or as aerosol (**B**), in the artificially ventilated guinea pig.

pressure, in marked contrast to the transient effect of histamine. When given as an aerosol, LTC_4 was even more potent relative to histamine, the molar difference for equal responses being more than 1,000-fold (Fig. 2B). Independently of whether administered intravenously or as aerosols, LTD_4 and LTE_4 elicited airway alterations that were identical to those of LTC_4, as regards the pattern of response and the effective dose range (4).

We recently studied the bronchoconstrictor activity of LTC_4 in artificially ventilated monkeys (25). LTC_4 and histamine, when injected into the right atrium, were virtually equipotent in increasing the transpulmonary pressure, although the time course of their actions differed in the same way as in the guinea pig (Fig. 1). The histamine effect peaked after 10 to 15 sec, and it was terminated within 1 min, whereas LTC_4 elicited a maximal increase in transpulmonary pressure only after 2 min, and the response remained at this level for an additional several minutes before it slowly subsided. On the other hand, when given as an aerosol, LTC_4 was at least 100 times as potent as histamine (Fig. 3). Histamine (1,000 to 5,000 nmole) elicited a transient, readily reversible increase in transpulmonary pressure, whereas LTC_4 (20 nmole) caused severe bronchoconstriction, as evidenced by a marked rise in transpulmonary pressure and a simultaneous drastic fall in arterial PO_2. The bron-

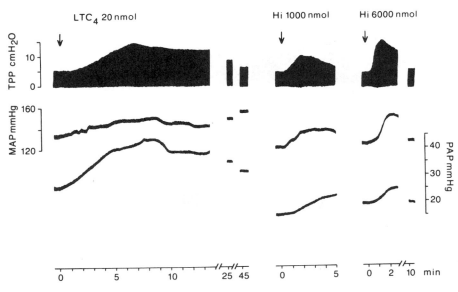

FIG. 3. Effects of LTC$_4$ and histamine aerosols (2 min) on transpulmonary pressure (TPP), mean systemic arterial pressure (MAP), and pulmonary arterial pressure (PAP) in the artificially ventilated monkey.

choconstriction and the decrease in PO$_2$ were very long-lasting, and they normalized only after 45 to 50 min in spite of the animals being subjected to intermittent positive end-expiratory pressure in order to combat formation of atelectases. Interestingly, the increase in transpulmonary pressure induced by LTC$_4$ was accompanied by a marked decrease in dynamic compliance, whereas airway resistance was little affected. This observation seems to indicate that cysteinyl-containing leukotrienes are particularly inclined to affect the peripheral airways, as previously suggested for crude SRS-A (8).

PULMONARY EFFECTS OF LEUKOTRIENES IN VITRO

The interpretation of apparent airway effects of leukotrienes *in vivo*, in particular when the drugs are administered systemically, may be complicated by simultaneous circulatory and metabolic alterations and by reflex activation of autonomic nerves. In order to overcome these possibly erroneous factors, we investigated the effects of leukotrienes on isolated airway smooth muscle preparations (5–7,14). Guinea pig tracheal spirals and parenchymal strips, representing central and peripheral airways, respectively, both responded with contraction upon administration of leukotrienes. Leukotrienes C$_4$, D$_4$, and E$_4$ were far more potent than histamine, and they evoked contractions with a characteristically slower onset and longer duration. The parenchymal strip was particularly sensitive to the leukotrienes, the ED$_{50}$ values for LTC$_4$ and histamine being 0.6 nM and 6 μM, respectively. Whereas LTC$_4$ was thus approximately 10,000 times more potent than histamine in this preparation, it was only 100 times as potent as histamine on tracheal spirals, the ED_{50} values for the two substances being 50 nM and 5 μM, respectively. The preferential and remarkable contractile activity of LTC_4 in parenchymal strips seems to provide further weight to the concept that cysteinyl-containing leukotrienes have a predominantly peripheral site of action in the airways.

Isolated human and monkey airway preparations also respond with contraction to administered LTC_4 (7,25). Human bronchi were found to be particularly sensitive, LTC_4 eliciting contractions in the same low dose range as in guinea pigs and being at least 1,000 times more potent than histamine. Monkey airways were less reactive to both substances, but LTC_4 elicited significant contractions in approximately 100 times lower concentrations.

These results clearly demonstrate that cysteinyl-containing leukotrienes are bronchoconstrictors of unusual potency. The finding that LTC_4 may induce severe bronchoconstriction in a subhuman primate, together with its exquisite potency in isolated human bronchi, strongly suggests that the human bronchus is also highly susceptible to these substances.

MODE OF ACTION AND STRUCTURE REQUIREMENTS FOR LEUKOTRIENES AS BRONCHOCONSTRICTORS

At present, little is known about the precise mechanism by which leukotrienes constrict airway smooth muscle. The typically slow onset seems to imply that they act by releasing other substances with smooth-muscle-stimulating capacity. Leukotrienes C_4, D_4, and E_4 undoubtedly may release prostaglandins and thromboxanes from isolated perfused guinea pig lungs (10,22), and nonsteroidal anti-inflammatory drugs (e.g., indomethacin) have been reported to block the increase in insufflation pressure caused by intravenous injection of LTC_4 or LTD_4 (22). We have also noted that indomethacin apparently diminishes the rise in insufflation pressure evoked by intravenous LTC_4 in guinea pigs (4). However, the effect was not specific for leukotrienes; serotonin and prostaglandin $F_{2\alpha}$ were also affected, and it could be related, at least in part, to the extensive tachyphylaxis of leukotrienes (Fig. 4). On the other hand, the airways appeared to react to aerosolized leukotrienes as expected, even after indomethacin treatment. Furthermore, neither indomethacin nor blockade of receptors for histamine, serotonin, catecholamines, and acetylcholine influenced the response to LTC_4, LTD_4, and LTE_4 in isolated airways of the guinea pig (6,14).

In agreement with our previous findings with LTC_4 and LTD_4 on isolated human and guinea pig airways (7,14), cysteinyl-containing leukotrienes in general seem to produce contractions of guinea pig lung strips that parallel the response curve to histamine (Fig. 5). LTC_4, LTD_4, LTE_4, as well as LTC_3 and LTC_5 (5), derived from eicosatrienoic acid (n-9) and eicosapentaenoic acid, respectively, were all equipotent and evoked half-maximal contraction of parenchymal strips at approximately 1 nM, as compared with 5 μM for histamine. It has been suggested that the stereochemistry at the Δ^{11} double bond may be of importance for the contractile activity of cysteinyl-containing leukotrienes (15). However, when noncumulative

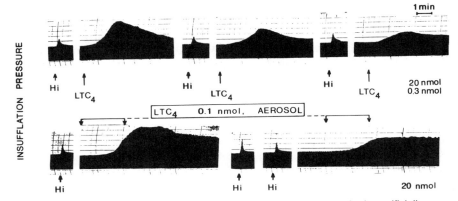

FIG. 4. Effects of LTC_4 and histamine on tracheal insufflation pressure in the artificially ventilated guinea pig. Note tachyphylactic response to repeated intravenous injection of LTC_4, followed by normal and tachyphylactic responses to aerosol administrations.

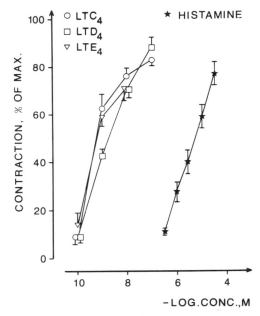

FIG. 5. Noncumulative dose-response curves for contractile activities of LTC_4, LTD_4, LTE_4, and histamine in parenchymal strips of guinea pig lungs. Mean values ± SEM (% of maximal contraction) for six to eight observations in separate tissues at each drug concentration.

dose-response curves were raised for 11-*trans*-LTC_4 and 11-*trans*-LTE_4, these two substances were found to be equipotent with their stereoisomers LTC_4 and LTE_4 (Fig. 6). On the other hand, the parent intermediate LTA_4, as well as LTB_4, were considerably less potent, although still more active than histamine. It may be concluded, therefore, that the cysteinyl substituent at $C = 6$ appears to be of major

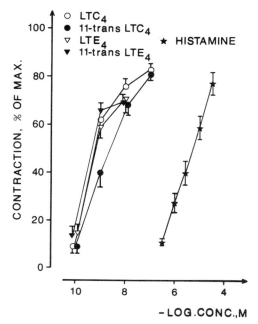

FIG. 6. Noncumulative dose-response curves for contractile activities of LTC_4, 11-*trans* LTC_4, LTE_4, 11-*trans* LTE_4, and histamine in parenchymal strips of guinea pig lungs. Mean values ± SEM (% of maximal contraction) for six to eight observations in separate tissues at each drug concentration.

importance for the high bronchoconstrictor activity of leukotrienes, rather than the number of double bonds or the stereochemistry at C11.

It has been suggested that LTD_4 is a more potent smooth muscle stimulant than are other cysteinyl-containing leukotrienes, and that effects seen with LTC_4 are due to a considerable extent to its rapid transformation into LTD_4 (22). Recent observations pertaining to the metabolism of radiolabeled LTC_3 injected into the right atrium of the monkey seemingly provide circumstantial evidence for this view (13). Thus chromatographic analysis of blood samples drawn from the aortic root revealed that 40% of the recovered radioactivity was accounted for by LTD_3 and LTE_3 15 sec after injection of LTC_3, and that LTE_3 predominated after 2 min. The outcome was quite different, however, when radiolabeled LTC_3 was added to guinea pig lung strips (6) (Fig. 7). As illustrated in Fig. 7, the contraction response to LTC_3 was half-maximal after 3 min and fully developed within 10 min, and it remained at this level throughout the experiment. High-pressure liquid chromatography (HPLC) analysis of bath fluid withdrawn at regular intervals showed that LTC_3 was intact at 3 min. After 10 min and 30 min, 25% and 70%, respectively, of the radioactivity was present as LTD_3, and LTE_3 appeared as a major component only 90 min after administration of LTC_3. Leukotrienes are not likely to be stored to any significant extent in tissues. As a consequence, it may be assumed that the metabolic pattern

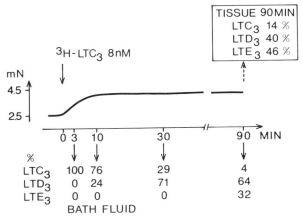

FIG. 7. Time course for contractile effect and metabolism of radiolabeled LTC₃ administered to bath fluid surrounding a parenchymal strip of guinea pig lung.

in the bath fluid reflects the conditions in the tissue at the level of the effector cells, as also indicated in Fig. 7. These results provide further support for the view that cysteinyl-containing leukotrienes are equipotent bronchoconstrictors, and therefore that the initial metabolism of LTC_4 into LTD_4 and LTE_4 does not lead to any significant change of the activity.

MICROVASCULAR EFFECTS OF LEUKOTRIENES

It is well known that SRS-A causes leakage from microvessels, leading in turn to tissue edema (19). Apparently LTC_4 and LTD_4 mimic the edema-promoting action of crude SRS-A, and they seem to be remarkably potent in this respect (9,14). When injected intradermally into guinea pigs, as little as 0.1 pmole of either compound was sufficient for inducing significant extravasation of Evan's blue. Higher doses produced dose-dependent and similar increases in spot diameter, the only difference relating to the appearance of the blue spot, being homogeneous with LTD_4 and having a central blanching with LTC_4. These observations prompted us to further study the microvascular actions of leukotrienes in the hamster cheek pouch *in vivo* (3).

Leukotrienes C_4 and D_4, when applied topically to the cheek pouch for a 3-min period in concentrations ranging between 0.3 and 20 nM, elicited a direct, dose-dependent contraction of arterioles, particularly terminal arterioles (Fig. 8). The vasoconstriction ceased rapidly after the 3-min period, and it was consistently followed by a dose-dependent and reversible leakage of macromolecules, as indicated by extravasation of fluorescent dextran (27) (Fig. 8). Notably, plasma leakage induced by these leukotrienes appeared at postcapillary venules, in accordance with previous observations that this vessel segment is the target of action for substances that cause reversible changes in vascular permeability (16). According to dose-response curves assessed noncumulatively, LTC_4, LTD_4, and LTE_4 induced a sig-

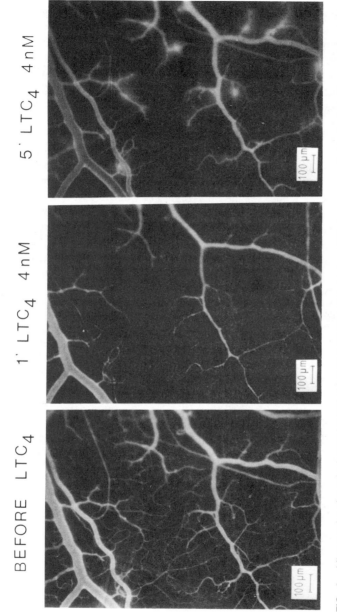

BEFORE LTC$_4$ 1' LTC$_4$ 4nM   5' LTC$_4$ 4nM

100 µm   100 µm   100 µm

FIG. 8. Micrographs of terminal vascular bed of hamster cheek pouch *in vivo*. LTC$_4$ 4 nM was applied topically to the pouch for 3 min. Note arteriolar constriction, followed by leakage of fluorescein-labeled dextran from postcapillary venules.

nificant increase of vascular permeability at much lower concentrations than did histamine (Fig. 9). LTC_4 was most potent in this respect, being approximately 5,000 times more potent than histamine. LTA_4 and LTB_4, on the other hand, were at least 100 times less active than the cysteinyl-containing leukotrienes, although they were still more potent than histamine (1).

The vascular effects of leukotrienes were further compared with those of two other vasoactive substances, angiotensin II and histamine. Angiotensin was only slightly more potent than LTC_4, LTD_4, and LTE_4 in causing arteriolar constriction, and it did not elicit macromolecular leakage. Histamine, on the other hand, dose-dependently increased postcapillary leakage but did not result in vasoconstriction. This seems to indicate that vasoconstriction and plasma leakage induced by leukotrienes need not be causally related. In line with this contention, low concentrations of LTC_4 (1 nM or less), superfusing the cheek pouch for 20 min induced maximal leakage while causing only slight and pulsatory vasoconstriction (1). It is conceivable that administration of a vasodilator together with leukotrienes might potentiate the increase in plasma leakage caused by a submaximal dose of leukotrienes, as has been reported in the guinea pig for PGE_2 and LTD_4 (21) and in the guinea pig, rabbit, and rat for PGE_2 and LTB_4 (2,28). However, macromolecular leakage induced by LTC_4 is not affected by indomethacin or mepyramine, indicating that leukotrienes per se are highly effective (1).

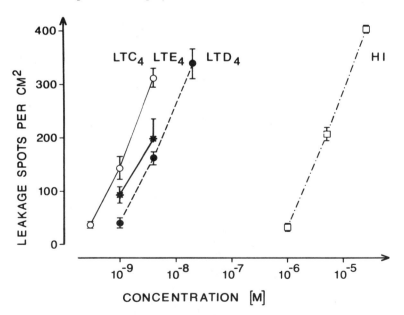

FIG. 9. Noncumulative dose-response curves for macromolecular leakage induced by LTC_4, LTD_4, LTE_4, and histamine in hamster cheek pouches *in vivo*. Mean values ± SEM for four to six observations in separate animals at each concentration.

LTB_4, when administered to the hamster cheek pouch in the same dose-range as LTC_4, did not elicit vasoconstriction and had no effect on plasma leakage. On the other hand, LTB_4 caused a dramatic increase in leukocyte adhesion to the endothelium in small venules (Fig. 10). Upon addition of LTB_4 to the medium superfusing the cheek pouch (final concentration 500 pM), marginating leukocytes immediately started to roll slower, and although blood flow was unchanged there was soon a pronounced increase in the number of leukocytes adhering to the endothelium (Figs. 10 and 11). This effect was dose-dependent, reached its maximum after 6 to 8 min, and remained at this level until LTB_4 was withdrawn, whereupon it gradually subsided to reach the control value within 5 to 15 min. Even during a short-lasting LTB_4 superfusion (6 to 10 min), the number of interstitial white cells seemed to increase (compare Fig. 10, left and right). This is consistent with the chemotactic stimulant property of LTB_4, documented *in vitro* (11,12,17,20) and by white cell accumulation in the guinea pig peritoneal cavity 5 hr after intraperitoneal injection (26).

The observations on the hamster cheek pouch imply that leukotrienes, in minute concentrations, have important and distinct microcirculatory actions. LTB_4 appears to be concerned primarily with leukocyte behavior by stimulating leukocyte adhesion to the endothelium in postcapillary venules, followed by migration into the surrounding interstitial space. Cysteinyl-containing leukotrienes do not have this property, at least not in low concentrations, but they are of the utmost potency in promoting plasma leakage, another important vascular event in inflammation.

CONCLUSIONS

Although only recently discovered, leukotrienes have rapidly proven to possess a great number of biological actions. LTC_4 and its metabolites LTD_4 and LTE_4 are prominent smooth muscle stimulants with a hitherto unsurpassed potency as bronchoconstrictors and inducers of plasma leakage from postcapillary venules. The remarkable potency of LTC_4 in pulmonary and vascular smooth muscle implies that introducing a glutathionyl side chain at C-6 markedly increases the spasmogenic properties inherent in LTA_4. Furthermore, the agonist potency is not changed by successive reduction of the peptide chain, yielding LTD_4 and LTE_4. This indicates that the cysteinyl substituent induces stereospecific alterations that are responsible for the high agonist potency. LTB_4, on the other hand, formed by double hydroxylation of LTA_4, acquires a different profile of biological activity. It is considerably less spasmogenic than are the cysteinyl-containing leukotrienes, but it seems to have a specific ability to stimulate leukocytes. It is thus likely that the steric configuration of LTB_4 and cysteinyl-containing leukotrienes make them activate different receptive structures and elicit distinct biochemical transductions, but much work remains before the identity of such mechanisms can be established.

The findings reported in this chapter further substantiate that cysteinyl-containing leukotrienes have actions that warrant a mediator role in airway and skin manifes-

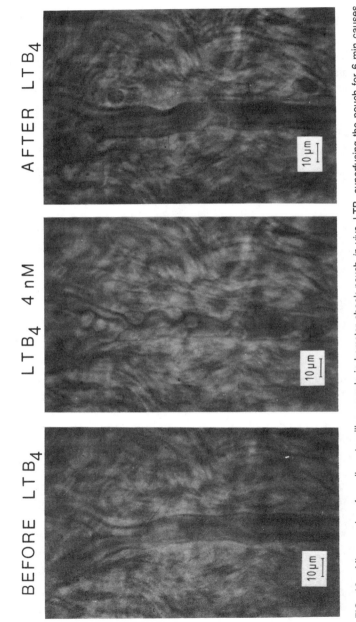

FIG. 10. Micrographs of small postcapillary venule in hamster cheek pouch *in vivo*. LTB$_4$ superfusing the pouch for 6 min causes massive adhesion of leukocytes to the vessel wall. Note leukocytes outside vessel after LTB$_4$.

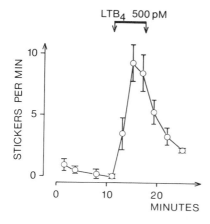

FIG. 11. Hamster cheek pouch *in vivo*. Number of leukocytes adhering (stickers) to the endothelium of small postcapillary venule ($\phi \simeq 15$ μM) before, during, and after superfusion of the pouch with LTB_4 500 pM for 6 min. Mean values $\pm$ SEM, $N = 6$.

tations of immunological origin, but they also raise the possibility that these leukotrienes may be involved in inflammatory reactions of more general character. The ability of LTB_4 to cause leukocyte adhesion to the endothelium in postcapillary venules *in vivo* and to stimulate leukocyte chemotaxis *in vitro* points to leukotrienes possibly affecting another event of great importance in the host's defense against various noxious stimuli. Therefore it may be more than chance that the combined microvascular effects of LTB_4 and cysteinyl-containing leukotrienes are strikingly reminiscent of the early phase of acute inflammation.

ACKNOWLEDGMENTS

The research reported herein was supported by grants from the Swedish Medical Research Council (project 04X-4342), the Swedish Society for Medical Research (O. Lundbergs Minnesfond), Karolinska Institute, and the Swedish Society Against Heart and Chest Diseases.

REFERENCES

1. Björk, J., Dahlén, S.-E., Hedqvist, P., and Arfors, K.-E. (1981): Effect of leukotrienes on vascular permeability and leukocyte adhesion. In: *The Inflammatory Process*, edited by P. Venge and A. Lindborn, pp. 103–112. Almgvist and Wiksell, Stockholm.
2. Bray, M. A., Cunningham, F. M., Ford-Hutchinson, A. W., and Smith, M. J. H. (1981): Leukotriene B_4: a mediator of vascular permeability. *Br. J. Pharmacology*, 72:483–486.
3. Dahlén, S.-E., Björk, J., Hedqvist, P., Arfors, K.-E., Hammarström, S., Lindgren, J.-Å,and Samuelsson, B. (1981): Leukotrienes promote plasma leakage and leukocyte adhesion in postcapillary venules: in vivo effects with relevance to the acute inflammatory response. *Proc. Natl. Acad. Sci. USA*, 78:3887–3891.
4. Dahlén, S.-E., and Hedqvist, P. (1982): Pulmonary and cardiovascular actions of leukotrienes in guinea pig in vivo. *(to be published)*.
5. Dahlén, S.-E., Hedqvist, P., and Hammarström, S. (1982): Effects of some naturally occurring cysteinyl-substituted leukotrienes in the guinea pig parenchymal strip. *Acta Physiol. Scand. (in press)*.
6. Dahlén, S.-E., Hedqvist, P., and Hammarström, S. (1982): On the mode of action of leukotriene-induced contraction of airway smooth muscle. *(to be published)*.

7. Dahlén, S.-E., Hedqvist, P., Hammarström, S., and Samuelsson, B. (1980): Leukotrienes are potent constrictors of human bronchi. *Nature*, 288:484–486.
8. Drazen, J. M., and Austen, F. K. (1974): Effects of intravenous administration of slow reacting substance of anaphylaxis, histamine, bradykinin and prostaglandin $F_{2\alpha}$ on pulmonary mechanics in the guinea pig. *J. Clin. Invest.*, 53:1679–1685.
9. Drazen, J. M., Austen, F. K., Lewis, R. A., Clark, D. A., Goto, G., Marfat, A., and Corey, E. J. (1980): Comparative airway and vascular activities of leukotrienes C-1 and D in vivo and in vitro. *Proc. Natl. Acad. Sci. USA*, 77:4354–4358.
10. Folco, G., Hansson, G., and Granström, E. (1981): Leukotriene C_4 stimulates TXA_2 formation in isolated sensitized guinea pig lungs. *Biochem. Pharmacol. (in press)*.
11. Ford-Hutchinson, A. W., Bray, M. A., Doig, M. V., Shipley, M. E., and Smith, M. J. H. (1980): Leukotriene B, a potent chemokinetic and aggregating substance released from polymorphonuclear leukocytes. *Nature*, 286:264–265.
12. Goetzl, E. J., and Pickett, W. C. (1980): The human PMN leukocyte chemotactic activity of complex hydroxy-eicosatetraenoic acids (HETEs). *J. Immunol.*, 125:1789–1791.
13. Hammarström, S., Bernström,K., Örning, L., Dahlén, S.-E., Hedqvist, P., Smedegård, G., and Revenäs, B. (1981): Rapid in vivo metabolism of leukotriene C_3 in the monkey, Macaca irus. *Biochem. Biophys. Res. Commun.*, 101:1109–1115.
14. Hedqvist, P., Dahlén, S.-E., Gustafsson, L., Hammarström, S., and Samuelsson, B. (1980): Biological profile of leukotrienes C_4 and D_4. *Acta Physiol. Scand.*, 110:331–333.
15. Lewis, R. A., Drazen, J. M., Austen, K. F., Clark, D. A., and Corey, E. J. (1980): Identification of the C(6)-S-conjugate of leukotriene A with cysteinyl as a naturally occurring slow reacting substance of anaphylaxis (SRS-A): importance of the 11-cis-geometry for biological activity. *Biochem. Biophys. Res. Commun.*, 96:271–277.
16. Majno, G., Palade, G. E., and Schoefl, G. I. (1961): Studies on inflammation. II. The site of action of histamine and serotonin along the vascular tree: a topographic study. *J. Biophys. Biochem. Cytol.*, 11:607–626.
17. Malmsten, C. L., Palmblad, J., Udén, A.-M., Rådmark, O., Engstedt, L., and Samuelsson, B. (1980): Leukotriene B_4: a highly potent and stereospecific factor stimulating migration of polymorphonuclear leukocytes. *Acta Physiol. Scand.*, 110:449–451.
18. Murphy, R. C., Hammarström, S., and Samuelsson, B. (1979): Leukotriene C: a slow reacting substance from murine mastocytoma cells. *Proc. Natl. Acad. Sci. USA*, 76:4275–4279.
19. Orange, R. P., and Austen, F. K. (1969): Slow reacting substance of anaphylaxis. *Adv. Immunol.*, 10:105–144.
20. Palmer, R. M. J., Stephney, R. J., Higgs, G. A., and Eakins, K.-E. (1980): Chemokinetic activity of arachidonic acid lipoxygenase products on leukocytes of different species. *Prostaglandins*, 20:411–418.
21. Peck, M. J., Piper, P. J., and Williams, T. J. (1981): The effect of leukotrienes C_4 and D_4 on the microvasculature of guinea pig skin. *Prostaglandins*, 21:315–321.
22. Piper, P. J., Samhoun, M. N., Tippins, J. R., Williams, T. J., Palmer, M. A., and Peck, M. J. (1980): Pharmacological studies on pure SRS-A, and synthetic leukotrienes C_4 and D_4. In: *SRS-A and Leukotrienes*, edited by P. J. Piper, pp. 81–99. Wiley, New York.
23. Samuelsson, B., Borgeat, P., Hammarström, S., and Murphy, R. C. (1980): Leukotrienes: a new group of biologically active compounds. *Adv. Prostaglandin Thromboxane Res.*, 6:1–18.
24. Samuelsson, B., Hammarström, S., Murphy, R. C., and Borgeat, P. (1980): Leukotrienes and slow reacting substance of anaphylaxis (SRS-A). *Allergy*, 35:375–381.
25. Smedegård, G., Revenäs, B., Hedqvist, P., Dahlén, S.-E., Hammarström, S., and Samuelsson, B. (1982): Leukotriene C_4 affects pulmonary and cardiovascular dynamics in monkey. *Nature (in press)*.
26. Smith, M. J. H., Ford-Hutchinson, A. W., and Bray, M. A. (1980): Leukotriene B: a potent mediator of inflammation. *J. Pharm. Pharmacol.*, 32:517–518.
27. Svensjö, S., Arfors, K.-E., Arturson, G., and Rutili, G. (1978): The hamster cheek pouch preparation as a model for studies of macromolecular permeability of the microvasculature. *Ups. J. Med. Sci.*, 83:71–79.
28. Wedmore, C. V., and Williams, T. J. (1981): Control of vascular permeability by polymorphonuclear leukocytes in inflammation. *Nature*, 289:646–650.

Leukotrienes and Other Lipoxygenase Products,
edited by B. Samuelsson and R. Paoletti.
Raven Press, New York © 1982.

Cardiovascular, Respiratory, and Hematologic Effects of Leukotriene D_4 in Primates

L. Casey, *J. Clarke, J. Fletcher, and **P. Ramwell

*Casualty Care Research Program Center and *Hyperbaric Medicine Program Center,
Naval Medical Research Institute, Bethesda, Maryland 20014; and
**Georgetown University Medical Center, Washington, D.C. 20007*

Slow-reacting substance of anaphylaxis (SRS-A) is thought to be an important mediator of asthma and other immediate hypersensitivity reactions. Recently it was determined that SRS-A consists of leukotrienes (LT) C_4 and D_4, which are formed from arachidonic acid by the lipoxygenase pathway (2). Intravenous administration of either crude SRS-A, LTC_4, or LTD_4 in the guinea pig produces a decrease in respiratory compliance and a modest increase in resistance, thus suggesting that their site of action is on the peripheral rather than the central airways (5,6).

LTC_4 and LTD_4 have been recently synthesized and are available in limited quantities (15). We have evaluated the respiratory, cardiovascular, and hematologic effects of LTD_4 in subhuman primates.

METHODS

Five monkeys *(Macaca mulatta)* weighing 4 to 8 kg and five baboons *(Papio anubis)* weighing 30 to 40 kg were anesthetized with ketamine (15 mg/kg) and intubated with the largest possible endotracheal tube. A catheter was placed in the femoral artery for monitoring arterial blood pressure, and in the monkeys a double-lumen thermodilation catheter (Edwards Laboratory) was inserted via the femoral vein into the pulmonary artery. All pressures were measured with transducers (Hewlett-Packard 267AC or 267BC) and recorded on an eight-channel recorder (Sanborn model 958-100).

The primate was placed in a special restraining chair designed to maintain an upright posture. A continuous infusion of ketamine was administered to maintain a very light level of sedation. An esophageal balloon catheter was placed in the distal esophagus and inflated with 0.25 ml air. The transpulmonary pressure (P_{tp}) was measured by a Validyn differential pressure transducer bridged between the tracheal tube and the esophageal catheter. Airflow ($\overset{*}{V}$) was measured with a Fleisch no. 1 pneumotachograph heated above body temperature and coupled to a Validyn differential pressure transducer. $\overset{*}{V}$ was integrated to give volume (V). Resistance

(R_l) and compliance (C_{dyn}) calculations were performed by an analog computer (Buxco Eelctronics, Inc.) which generates a signal proportional to the R_l and C_{dyn}. A Bird micronebulizer was connected to the endotracheal tube, and the leukotriene solution was aerosolized during tidal ventilation.

Cardiac output was measured by the thermal dilution method and calculated on a PDP-12 computer (Digital Equipment Corp.). Arterial blood gases were measured on a blood gas analyzer (Instrumentations Laboratories, model 313). Platelet and white blood cell counts were determined by an electronic particle counter (Coulter, model ZBI). Lactic acid was measured spectrophotometrically using Calbiochem-Behring rapid lactate reagents. Tromethamine PGF$_{2\alpha}$ was obtained from the Upjohn Company.

A sterile solution of LTD$_4$ (1 mg/ml) was stored in methanol at $-70°C$. The activity of the LTD$_4$ was determined by ultraviolet (UV) absorption at 280 nm and by an increase in insufflation pressure after intravenous administration to an anesthetized guinea pig (9,11). The activity was checked before and after completion of these studies.

RESULTS

During preliminary studies we found no significant change in dynamic compliance or resistance in rhesus monkeys after repeated doses of LTD$_4$ over the range of 0.1 to 30 µg/kg. Thereafter all studies were done using 20 µg/kg in the five rhesus monkeys. A bolus injection of LTD$_4$ 20 µg/kg produced no significant change in respiratory mechanics (Fig. 1). Although there were minor changes in some of the parameters, they are no different from the transient changes seen during baseline observations. In contrast, the intravenous administration of an equivalent dose of tromethamine PGF$_{2\alpha}$ (19 µg/kg) produced severe changes in respiratory mechanics. In fact, the flow rate fell so low that the respiratory analyzer was unable to calculate the resistance or compliance and thus required 30 sec to reset, after which it demonstrated a 300% increase in resistance and a 50% decrease in compliance.

Table 1 contains the cardiovascular, respiratory, and hematologic parameters, comparing baseline values to those 5 min after injection of the leukotriene. After the bolus injection of LTD$_4$ (20 µg/kg), all animals showed a decrease in mean arterial pressure ($p < 0.005$). This decrease occurred rapidly after the intravenous injection, although in two of the animals there was a slight increase in blood pressure (systolic pressure increased by 10 mm Hg) prior to a 25 mm Hg decrease. The decreased mean arterial pressure generally returned to baseline within 10 to 15 min. There were no significant changes in heart rate, cardiac output, central venous pressure, pulmonary artery pressure, or pulmonary capillary wedge pressures.

At the end of 5 min, there was alveolar hyperventilation, as indicated by the decrease in PaCO$_2$. This was the result of both an 18% increase in respiratory rate

---

FIG. 1. Effects of PGF$_{2\alpha}$ and LTD$_4$ on pulmonary mechanics in rhesus monkeys. PGF$_{2\alpha}$ and LTD$_4$ were given intravenously into the superior vena cava.

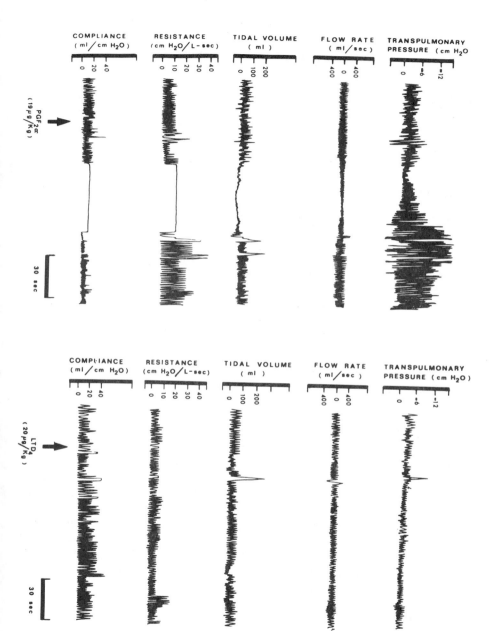

TABLE 1. *Effect of intravenous LTD_4 (20 $\mu g/kg$) in rhesus monkeys*

Parameter[a]	Baseline	At 5 min post LT	Significance
MAP	111 ± 4	89 ± 6	$p < 0.001$
PAP	14.6 ± 1.4	14.9 ± 1.4	N.S.
CVP	2.9 ± 1.0	2.4 ± 0.5	N.S.
PCWP	4.1 ± 0.5	± 0.6	N.S.
HR	192 ± 4	194 ± 4	N.S.
CO (l/min)	1.04 ± 0.17	1.13 ± 0.22	N.S.
RR	45 ± 9	53 ± 8	N.S.
V_t (ml)	58 ± 5	85 ± 13	$p < 0.05$
P_{pl} (cm H_2O)	1.8 ± 0.2	3.0 ± 0.6	$p < 0.05$
R_1 (cm $H_2O/l/sec$)	14 ± 0.2	16 ± 0.2	N.S.
C_{dyn} (ml/cm $H_2O/l/sec$)	32 ± 2	29 ± 2	N.S.
pH	7.39 ± 0.01	7.39 ± 0.02	N.S.
PCO_2	37.4 ± 2.1	31.4 ± 2.5	$p < 0.005$
PO_2	82.9 ± 5.3	87.3 ± 8.6	N.S.
HCO_3	22 ± 1.1	18.5 ± 1.4	$p < 0.005$
Lactic acid (mg/dl)	18.1 ± 4.6	17.7 ± 4.7	N.S.
WBC	$19,800 \pm 3,000$	$11,600 \pm 2,500$	$p < 0.005$
Platelets	$249,000 \pm 477,000$	$219,000 \pm 39,700$	N.S.

[a]MAP–mean arterial pressure, PAP–mean pulmonary artery pressure, CVP–central venous pressure, PCWP–pulmonary capillary wedge pressure, HR–heart rate, CO–cardiac output, RR–respiratory rate, V_t–tidal volume, P_{pl}–pleural pressure, C_{dyn}–dynamic compliance, R_1–resistance.

and a 46% increase in tidal volume. There was still no change in either C_{dyn} or R_1. The small but consistent decrease in plasma bicarbonate was not the result of increased lactic acid.

Twenty $\mu g/kg$ of LTD_4 produced a 58% decrease in the white blood cell count but did not significantly affect the platelet counts.

We also evaluated the respiratory and hematologic effects of LTD_4 in five baboons (Fig. 2). The intravenous administration of LTD_4 (1 $\mu g/kg$) caused 40% and 35% decreases in C_{dyn}, respectively, in two animals, with 450% and 85% increases in R_1. There were no significant changes in C_{dyn} or R_1 in the third animal, even after 3 $\mu g/kg$. Two animals received a 5-min aerosolization of 1 ml saline solution of LTD_4 (40 $\mu g/ml$) (Fig. 3). There were 38% and 27% reductions in C_{dyn}, respectively, without significant changes in R_1.

Intravenous administration of LTD_4 (1 $\mu g/kg$) caused acute neutropenia which then returned to baseline within 30 min; 3 $\mu g/kg$, on the other hand, caused rebound granulocytosis. There were no effects on lymphocyte or platelet counts (Figs. 4 and 5).

All baboons showed a transient decrease in mean arterial pressure by about 25 mm Hg that returned to baseline within 10 to 15 min. The initial hypertensive phase was never observed. Electrocardiograms were monitored in two of the animals. There was an increased heart rate and deepening of the inverted T-waves in leads

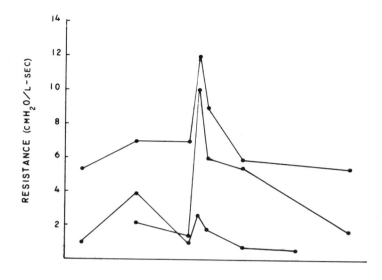

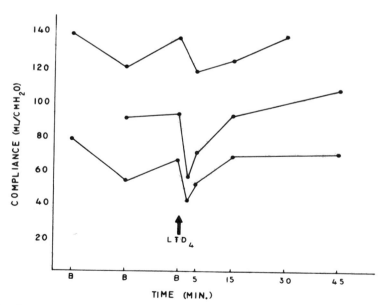

FIG. 2. Effect of intravenous LTD$_4$ (1 µg/kg) on baboon pulmonary mechanics. Each point represents the mean of at least five breaths. B represents baseline measurements made at 15-min intervals.

LTD$_4$ EFFECTS IN PRIMATES

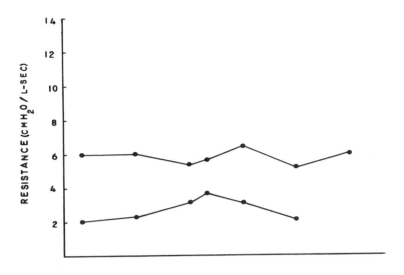

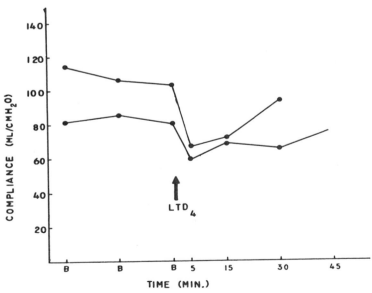

FIG. 3. Effect of aerosolized LTD$_4$ (1 ml of a 40 μg/ml solution) on baboon pulmonary mechanics. Each point represents the mean of at least five breaths. B represents baseline measurements made at 15-min intervals.

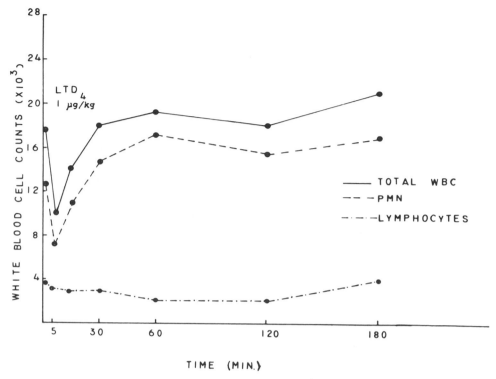

FIG. 4. Effect of intravenous LTD$_4$ (l μg/kg) on the white blood cell (WBC) count of a baboon. PMN = polymorphonuclear neutrophils.

II, III, and AVF, but no significant ST-T wave changes to suggest acute myocardial ischemia or infarction.

CONCLUSIONS

From our evaluation of five rhesus monkeys and five baboons, we conclude that primates (unlike guinea pigs) manifest only mild changes in respiratory mechanics after either intravenous or aerosolization of active synthetic leukotrienes (Figs. 2 and 3).

LTB$_4$ *in vitro* is a very potent chemoattractant, and *in vivo* in rabbits it caused a transient neutropenia (3). LTD$_4$ *in vitro* has no effect on white cell migration, but we demonstrated that intravenous LTD$_4$ produces a neutropenic response similar to that of LTB$_4$, although with higher doses of LTD$_4$ the initial neutropenia is followed by marked granulocytosis (Figs. 4 and 5). There were no changes in platelet or lymphocyte counts.

In contrast to the widely held belief that human lung is very sensitive to SRS-A, recent studies have found that human parenchymal lung strips are much more sensitive to PGF$_{2\alpha}$ than to SRS-A (8). This study is consistent with the observation

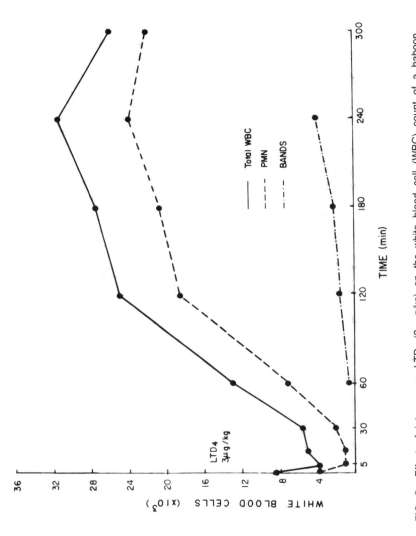

FIG. 5. Effect of intravenous LTD$_4$ (3 μg/kg) on the white blood cell (WBC) count of a baboon. PMN = polymorphonuclear neutrophils.

that aerosol challenge of normal humans with either SRS-A or $PGF_{2\alpha}$ produces little response, whereas asthmatics develop bronchoconstriction (10,12). A previous report of increased resistance and decreased compliance in monkeys after aerosolized SRS-A utilized a highly selected population of monkeys with bronchial hyperreactivity to *Ascaris* (14). The combination of the above studies with our findings suggests that hypersensitivity reactions cannot be explained solely by increased formation of leukotrienes.

Crude SRS-A as well as synthetic LTC_4 and LTD_4 can stimulate prostaglandin and thromboxane formation, as well as potentiate the effects of histamine (1,4,7). Guinea pig bronchial smooth muscle contraction can be inhibited by nonsteroidal anti-inflammatory drugs (13). Thus it appears that, at least in some species, many of the leukotriene effects are mediated via stimulation of prostaglandin or thromboxane production.

Apparently, during the process of sensitization mechanisms other than just increased leukotriene formation must occur. Some possible explanations could be (a) formation of leukotriene receptors; (b) generation of additional mediators; (c) sensitization of cholinergic nerves; (d) alterations of the β-adrenergic system; or (e) a decrease in a plasma inactivator of leukotrienes.

ACKNOWLEDGMENT

Synthetic LTD_4 was kindly provided by Dr. J. Rokach, Merck Frosst Laboratories, Quebec.

REFERENCES

1. Adkinson, N., Newball, H., Findlay, S., Adams, G., and Licktenstein, L. (1979): Origin of PGF2α production following anaphylactic challenge of human lung. *Monogr. Allergy*, 14:122–125.
2. Borgeat, P., and Samuelsson, B. (1979): Arachidonic acid metabolism in polymorphonuclear leukocytes: effect of ionophore A23187. *Proc. Natl. Acad. Sci. USA*, 76:2148–2152.
3. Bray, M., Cunningham, F., Davidson, E., Ford-Hutchinson, A., and Smith, M. (1981): Differing biological potencies of isomers of leukotriene B4. *Br. J. Pharmacol.*, 73:210P.
4. Brocklehurst, W. (1958): Ph.D. thesis, University of London.
5. Drazen, J., and Austen, K. (1974): Effects of intravenous administration of slow-reacting substance of anaphylaxis histamine, bradykinin, and prostaglandin F2α on pulmonary mechanics in the guinea pig. *J. Clin. Invest.*, 53:1679–1685.
6. Drazen, J., Lewis, R., Wasserman, S., Orange, R., and Austen, K. (1979): Differential effects of a partially purified preparation of slow-reacting substance of anaphylaxis on guinea pig tracheal spirals and parenchymal strips. *J. Clin. Invest.*, 63:1–5.
7. Dinaz, M., Engineer, D., Morris, H., Piper, P., and Sirois, P. (1978): The release of prostaglandins and thromboxanes from guinea-pig lung by slow-reacting substance of anaphylaxis, and its inhibition. *Br. J. Pharmacol.*, 64:211–218.
8. Ghelani, A., Halroyde, M., and Sheard, P. (1980): Response of human isolated bronchial and lung parenchymal strips to SRS-A and other mediators of asthmatic bronchospasm. *Br. J. Pharmacol.*, 71:107–112.
9. Hedqvist, P., Dahlén, S., Gustafason, L., Hammarström, S., and Samuelsson, B. (1980): Biologic profile of leukotriene C4 and D4. *Acta Physiol. Scand.*, 110:331–333.
10. Herxheimer, H., and Stresemann, E. (1963): The effect of slow reacting substance (SRS-A) in guinea pigs and in asthmatic patients. *J. Physiol. (Lond.)*, 165:78P.
11. Holme, G., Brunet, G., Piechuta, H., Masson, P., Girard, Y., and Rokach, J. (1980): The activity of synthetic leukotriene C-1 on guinea pig trachea and ileum. *Prostaglandins*, 20:717–728.

12. Mathe, A. (1976): Studies on actions of prostaglandins in the lung. *Acta Physiol. Scand.[Suppl.]*, 441.
13. O'Donnell, M., Falcone, K., and Kiell, R. (1981): Pharmacologic characterization of slow-reacting substance of anaphylaxis (SRS-A) induced bronchoconstriction in guinea pigs. *Fed. Proc.*, 40:721.
14. Patterson, R., Orange, R., and Harris, K. (1978): A study of the effect of slow-reacting substances of anaphylaxis on the rhesus monkey airway. *J. Allergy Clin. Immunol.*, 62:371–377.
15. Rokach, J., Girard, Y., Guinden, Y., Atkinson, J., Lawe, M., Masson, P., and Holme, G. (1980): The synthesis of a leukotriene with SRS-like activity. *Tetrahedron Lett.*, 21:1485–1488.

Leukotrienes and Other Lipoxygenase Products,
edited by B. Samuelsson and R. Paoletti.
Raven Press, New York © 1982.

Leukotriene D_4: A Potent Vasoconstrictor of the Pulmonary and Systemic Circulations in the Newborn Lamb

K. Yokochi, P. M. Olley, E. Sideris, F. Hamilton, D. Huhtanen, and F. Coceani

Research Institute and Department of Pediatrics, The Hospital for Sick Children, Toronto, Canada M5G 1X8

Arachidonic acid may be transformed during lipoxygenase-catalyzed reactions to several biologically active compounds. Among these, the leukotrienes (LT) C_4 and D_4 have been identified as the active constituents of slow-reacting substance of anaphylaxis (SRS-A) (1). Lipoxygenases occur in lung tissue of guinea pigs (2) and humans (5). Leukotrienes may be mediators of bronchoconstriction in asthma or of hypersensitivity reactions, and they could be released into the circulation in other conditions of pulmonary stress. Their vascular actions may therefore be important in the pathophysiology of several disease states including those which afflict the premature infant. We have therefore studied the actions of synthetic LTD_4 on the pulmonary and systemic circulations of the lamb.

METHODS

Mixed-breed lambs aged 4 to 7 days underwent left thoracotomy during which C and C electromagnetic flow probes were placed on the right and left pulmonary arteries, using surgical techniques described elsewhere (4). The animals were allowed to recover from surgery and were then used to study LTD_4. On the day of the experiment, which was never less than 5 days after surgery, two cardiac catheters were introduced, one with its tip in the aortic root and the other with its tip in either the left or the right pulmonary artery (PA) just beyond the flow probe. The PA catheter was used to make selective injections into one lung. Catheter insertion was performed under brief ketamine anesthesia, and the animal was allowed to recover completely before being studied.

Synthetic LTD_4 was received as an aqueous solution of 1 mg/ml and was diluted with appropriate volumes of tris buffer (150 mM, pH 7.4) on the day of the experiment. Bolus injections of 1 ml in doses ranging from 0.01 to 5.0 µg/kg were given either into the catheterized lung or into the aortic root. The order and site of injection was randomly varied from experiment to experiment.

Mean blood flow was recorded continuously from each lung along with mean aortic and pulmonary artery pressures. Cardiac output was obtained by summing the two flows, and pulmonary and systemic resistances were calculated by dividing pressure by flow and were expressed as "units" (mm Hg/L/min). Six experiments were carried out in four lambs, and LTD_4 was tested in a further three lambs pretreated with indomethacin 3 mg/kg/day for 3 days.

RESULTS

PA injections caused a dose-dependent increase in total pulmonary vascular resistance (TPVR) and reduced the proportion of pulmonary blood flow to the injected lung (Fig. 1). A dose of 1.0 μg/kg into the PA increased TPVR from 21.9 ± 3.45 to 74.4 ± 13.7 units, this change being highly significant ($p < 0.001$), whereas the proportion of total pulmonary blood flow going to the injected lung fell from 43 ± 9% to 18 ± 6% ($p < 0.001$). PA injections of 1.0 μg/kg caused an initial fall in systemic vascular resistance (SVR) from a baseline value of 114.9 ± 33.7 to 75.5 ± 29.2 units at 10 sec postinjection, after which SVR rose to 250 ± 48 units. These changes were significant at the $p < 0.01$ and <0.001 levels, respectively (Fig. 2).

LTD_4 injections of 1.0 μg/kg into the aortic root increased SVR from 127 ± 47 to 256 ± 110 units ($p < 0.05$) without an initial fall (Fig. 3). The threshold for

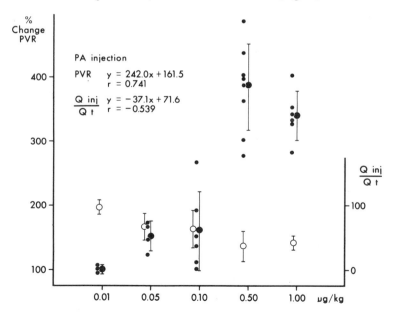

FIG. 1. Change in total pulmonary vascular resistance (PVR, *solid circles*) and in the ratio of flow in the injected lung to total pulmonary blood flow (*open circles*) plotted against dose of LTD_4. Values in all three figures are the mean ± 1 SD.

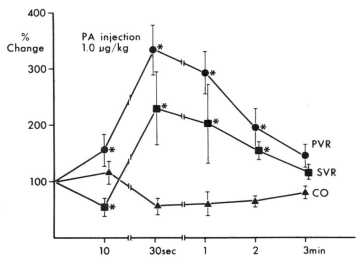

FIG. 2. Time course of changes in pulmonary vascular resistance (PVR, *circles*), systemic vascular resistance (SVR, *squares*), and cardiac output (CO, *triangles*).

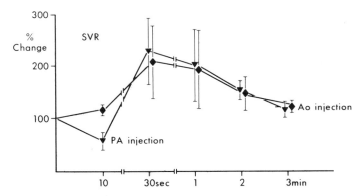

FIG. 3. Effect on systemic vascular resistance (SVR) of LTD₄ 1.0 μg/kg given into the pulmonary artery (*triangles*) and the aortic root (*diamonds*).

all these effects was between 0.01 and 0.5 μg/kg, and the effects persisted for 2 to 3 min.

Pretreatment with indomethacin altered the baseline values, and comparisons are therefore made as percentage changes. PA injections of 1.0 μg/kg increased TPVR to 154 ± 44% of baseline in pretreated animals compared with a 328 ± 48% increase in untreated animals; the difference is significant ($p < 0.001$).

Indomethacin pretreatment also abolished the initial fall in SVR observed after PA injections in untreated animals. The rise in SVR was not significantly altered: 163 ± 9% compared with 228 ± 62%.

CONCLUSION

LTD_4 given as a bolus injection is a more potent pulmonary and systemic vasoconstrictor than prostaglandin (PG) $F_{2\alpha}$ or D_2 in the same experimental model (3). It has a direct action on the pulmonary circulation as demonstrated by the redistribution of pulmonary blood flow away from the injected lung to the noninjected lung. The systemic constrictor effects are similar for pulmonary artery and aortic injections, suggesting that there is little or no inactivation of this leukotriene in the pulmonary circulation. The initial fall in systemic arterial pressure and resistance, which is seen with pulmonary artery injections but not with aortic root injections, suggests that LTD_4 induces the release of a systemic vasodilator by the lungs. This initial vasodilation is abolished by pretreatment with the cyclo-oxygenase inhibitor indomethacin, implying that the vasodilator released may be PGI_2 or PGE_2. Pretreatment with indomethacin increases resting tone in the pulmonary and systemic vascular beds, which may explain the reduced sensitivity to the constrictor action of LTD_4 on both circulations in pretreated animals. Alternatively, the vasoconstrictor action of LTD_4 in the lamb may be partially mediated by release of thromboxane A_2 or $PGF_{2\alpha}$. This possibility is supported by the finding that LTC_4 and LTD_4 release thromboxane and prostaglandins from isolated guinea pig lungs (Chapters 14 and 15, *this volume*).

These findings support the view that the leukotrienes which make up SRS-A have the potential to cause significant hemodynamic effects if released into the circulation. It remains to be shown if such release occurs in human disease.

ACKNOWLEDGMENT

LTD_4 was kindly provided by Dr. J. Rokach of Merck-Frosst Laboratories, Canada.

REFERENCES

1. Hedqvist, P., Dahlen, S. -E., Gustaffson, L., Hammarström, S., and Samuelsson, B. (1980): Biological profile of leukotrienes C_4 and D_4. *Acta Physiol. Sand.*, 110:331–333.
2. Hamberg, M. (1976): On the formation of thromboxane B_2 and 12L-hydroxy-5,8,10,14-eicosatetraenoic acid (12 ho-20:4²): In tissues from the guinea pig. *Biochim. Biophys. Acta.*, 431:651–654.
3. Lock, J. E., Olley, P. M., and Coceani, F. (1980): Direct pulmonary vascular responses to prostaglandins in the conscious newborn lamb. *Am. J. Physiol.*, 238:H631–H638.
4. Lock, J. E., Hamilton, F., Luide, H., Coceani, F., and Olley, P. M. (1980): Direct pulmonary vascular responses in the conscious newborn lamb. *J. Appl. Physiol.*, 48:188–196.
5. Walker, J. L. (1980): Interrelationships of SRS-A production and arachidonic acid 12-lipoxygenase. In: *Advances in Prostaglandin and Thromboxane Research, Vol. 6*, edited by B. Samuelsson, P. W. Ramwell, and R. Paoletti, pp. 115–119. Raven Press, New York.

Leukotrienes and Other Lipoxygenase Products,
edited by B. Samuelsson and R. Paoletti.
Raven Press, New York © 1982.

SRS-A, Leukotrienes, and Immediate Hypersensitivity Reactions of the Heart

Roberto Levi, James A. Burke, and *Elias J. Corey

*Department of Pharmacology, Cornell University Medical College, New York, New York 10021; and *Department of Chemistry, Harvard University, Cambridge, Massachusetts 02138*

Anaphylaxis is a generalized form of immediate-type hypersensitivity affecting several organs simultaneously. Fatalities result from severe respiratory obstruction, irreversible cardiovascular collapse, or both (6). Acute circulatory failure may be due to a fall in peripheral resistance and blood volume depletion. On the other hand, ventricular arrhythmias and cardiac arrest can be the primary cause of the cardiovascular collapse (7,16). Although cardiac arrhythmias might be expected to result from the asphyxiating effects of bronchospasm and laryngeal edema, a wealth of clinical observations suggests that cardiac dysrhythmia and pump failure can occur without being preceded by respiratory distress (as reviewed in ref. 16). Proof that the heart is a primary target organ in immediate hypersensitivity reactions has been found in laboratory studies conducted in rodents as well as in subhuman primates (11,13,15,32,33).

"Cardiac anaphylaxis," a term coined by Feigen 20 years ago (14), refers to an experimental model of immediate hypersensitivity in which the heart is the target tissue. This model has been used extensively in our laboratory to determine the release and investigate the role of various active substances liberated from the heart during antigen-antibody reactions. The model uses guinea pigs which are sensitized actively or passively. The heart is excised and placed in a Langendorff apparatus where it beats spontaneously while being perfused at constant pressure. The heart is then challenged with an intracoronary bolus injection of the specific antigen (11,15,27). Characteristically, the antigenic challenge of the sensitized heart causes tachycardia, a brief and short-lasting increase in ventricular contraction followed by prolonged contractile failure, severe dysrhythmias, and a marked decline in coronary flow (11,27) (Fig. 1). The picture is that of an acute and dramatic derangement in cardiac function incompatible with the maintenance of adequate cardiac output.

Analysis of the coronary venous effluent by biologic, fluorometric, and radioimmunoassay reveals that the anaphylactic heart releases a variety of vasoactive substances. These include histamine, prostaglandin $F_{2\alpha}$ ($PGF_{2\alpha}$), PGE_2, PGD_2, the metabolites of thromboxane A_2 (TXB_2) and prostacyclin (6-keto-$PGF_{1\alpha}$), and, in-

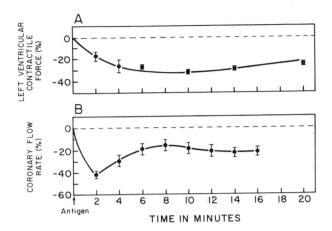

FIG. 1. Anaphylaxis in the isolated guinea pig heart. Time courses of the changes in left ventricular contractile force (**A**) and coronary flow rate (**B**) following antigenic challenge (points are means ± SEM; $N = 9$ to 10). The hearts, obtained from guinea pigs passively sensitized with homologous anti-DNP-bovine γ-globulin, were challenged with DNP-bovine serum albumin.

TABLE 1. *Release of mediators during cardiac anaphylaxis in vitro*

Mediator	Amount released	N
Histamine	2.34 ± 0.79 μg/g	7
SRS-A	525 ± 92 units/g	7
$PGF_{2\alpha}$	12.18 ± 0.14 ng/g	9
TXB_2	109.1 ± 11.3 ng/g	4
6-Keto-$PGF_{1\alpha}$	95 ± 16.4 ng/g	4

Values (means ± SEM) refer to total amounts released into the coronary venous effluent within 14 min following antigenic challenge.
Data are from refs. 1, 3, and 19.

deed, slow-reacting substance of anaphylaxis (SRS-A) (Table 1). Histamine release is rapid and relatively short-lived; it precedes that of various metabolites of arachidonic acid, whose release is delayed but more prolonged (1,3,4,11,15,19) (Fig. 2).

Histamine receptor antagonists inhibit both the tachycardia and arrhythmias which characterize anaphylaxis (21). A combination of H_1 and H_2 histamine receptor antagonists is needed, since both H_1 and H_2 receptors mediate the chronotropic and arrhythmogenic effects of histamine (17,24). The intracardiac administration of histamine enhances the automaticity of sinoatrial and idioventricular pacemaker cells by acting at H_2 receptors (22,25,26,31), whereas histamine delays impulse conduction by stimulating H_1 receptors located at the atrioventricular node (23,24). These findings suggest that tachycardia and arrhythmias, whether of conduction or

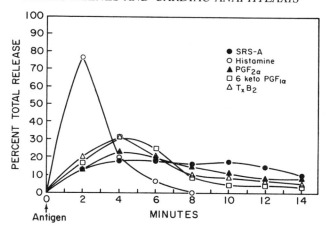

FIG. 2. Anaphylaxis in the isolated guinea pig heart. Time course of mediator release following antigenic challenge. Points are mean percentage of the total quantity of mediator, or metabolite, released in each experiment. $N = 4$ to 7.

automaticity, are generated by the release of endogenous cardiac histamine as a result of the antigen-antibody reaction.

Pharmacologic inhibition of the cyclo-oxygenase pathway in the arachidonic acid cascade enhances the release of histamine from the anaphylactic heart; at the same time, tachycardia and arrhythmia are even more severe than expected on the basis of the enhanced histamine release (18). This suggests that during cardiac anaphylaxis prostaglandins modulate histamine release and the effects of released histamine. Indeed, the intracardiac administration of PGE_2 at concentrations comparable to those released from the anaphylactic heart attenuates the chronotropic effect of exogenous histamine. However, PGE_2 enhances the immunologic release of endogenous cardiac histamine (1). Thus PGE_2 released in small amounts during cardiac anaphylaxis is likely to attenuate the response of the heart to released histamine, without reducing the amounts of histamine released. On the other hand, cyclo-oxygenase products other than PGE_2, possibly PGD_2, must control and curb the quantity of histamine released from the heart during anaphylaxis (1).

One of the striking features of cardiac anaphylaxis is the precipitous fall in coronary flow, the magnitude of which is directly proportional to the severity of the arrhythmia (15). Yet in the presence of indomethacin, despite the fact that anaphylactic arrhythmias are more severe, there is less of a fall in coronary flow (1,3,18). Antagonism of prostaglandin effects at their receptor sites with the compound N 0164 also reduces the anaphylactic decline in coronary flow (1,2), as does the specific inhibition of thromboxane synthetase by 1-(2-isopropyl phenyl)imidazole (3). Thus, it appears that vasoconstricting prostaglandins, particularly TXA_2, may be responsible for the marked decline in coronary flow during cardiac anaphylaxis (3).

As mentioned above, pharmacologic inhibition of cyclo-oxygenase potentiates anaphylactic histamine release from the heart, and the resulting tachycardia is even greater than expected considering the larger quantities of histamine released. Although the easiest explanation of this finding is that indomethacin prevents the generation of prostaglandins that curb histamine release (e.g., PGD_2) and its effects (e.g., PGE_2) (1,18), with indomethacin more SRS-A is released from the anaphylactic heart (29). Thus we hypothesized that SRS-A might potentiate the chronotropic effect of histamine. A precedent for this could be found in the literature, namely, that SRS-A potentiates the histamine-induced contraction of the guinea pig ileum (12). Indeed, we found that histamine-free SRS-A from either rat peritoneal mast cells or guinea pig lung potentiates and prolongs the chronotropic and arrhythmogenic effects of histamine in the guinea pig heart (8,19,20). The specific anti-SRS compound FPL 55712 (5) antagonizes this potentiation (19). Moreover, we found that FPL 55712 diminishes the severity of anaphylactic cardiac dysfunction: It antagonizes the contractile failure and reduces the tachycardia, while abbreviating the duration of both; it also partially attenuates the fall in coronary flow (8,19). This indicated that SRS-A, aside from potentiating and prolonging the cardiac effects of histamine, must be primarily involved in the protracted contractile failure and, at least partially, in the decreased coronary perfusion associated with cardiac anaphylaxis. Subsequent studies with crude SRS-A obtained from guinea pig lungs immunologically challenged in the presence of indomethacin revealed that the administration of SRS-A causes dose-dependent decreases in left ventricular contractile force and coronary flow rate in the isolated guinea pig heart. Both dose-response curves are shifted to the right in a parallel fashion by the compound FPL 55712 (8) (Fig. 3). With 200 units of SRS-A (1 unit SRS-A = 5 ng histamine in terms of magnitude of contraction of the guinea pig ileum) the contractile force of the left ventricle and the coronary flow decrease by as much as 50%. Since the anaphylactic heart releases approximately 500 units of SRS-A (Table 1), the contribution of SRS-A to anaphylactic cardiac dysfunction can be readily appreciated. Chemically pure SRS obtained from peritoneal macrophages (30) also causes the ventricular myocardium to fail and the coronary flow to decline (10). As with crude SRS-A, the effects of chemically pure SRS on contraction and coronary flow are long-lasting and are antagonized by FPL 55712. Thus, it appears that SRS has negative inotropic effects, decreases coronary flow, and potentiates the tachyarrhythmic effects of histamine.

We sought a definitive confirmation of the role of SRS in immediate-type hypersensitivity reactions of the heart by attempting to reproduce the functional changes caused by natural SRS, whether crude or chemically pure, with the administration of leukotrienes C_4 and D_4 (LTC_4 and LTD_4), the newly recognized components of SRS-A (28). We have found that LTC_4 and LTD_4 are powerful depressants of myocardial contractile force; in fact, marked negative inotropic effects are observed with doses of LTD_4 and LTC_4 in the picogram and nanogram range, respectively (9). Thus, although qualitatively similar to LTC_4, LTD_4 appears to be much more potent than LTC_4 in decreasing the force of contraction of the heart. Concomitant with, but independent of, their negative inotropic effect, leukotrienes cause a large decrease in coronary flow (9) (Table 2). Both effects are long-lasting and antag-

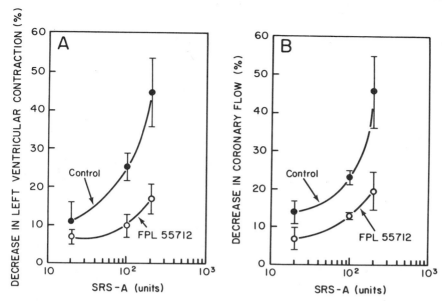

FIG. 3. Effects of partially purified SRS-A on the contractile force (**A**) and coronary flow (**B**) of the isolated guinea pig heart and their modification by compound 55712. SRS-A was obtained from guinea pig lungs immunologically challenged in the presence of indomethacin, followed by ethanol extraction, desalting, and histamine separation by Amberlite XAD-7 column chromatography. The concentration of FPL 55712 was 100 ng/ml. Each point is the mean (± SEM) of three observations (8).

TABLE 2. *Effects of leukotrienes on cardiac contraction and coronary flow[a]*

Dose (ng)	Myocardial contractile force (% change from control)	Coronary flow (% change from control)
LTC_4		
0.3	-10 ± 2[b]	-5 ± 2[b]
25	-53 ± 4	-58 ± 2
LTD_4		
0.01	-5 ± 1	-5 ± 1
1	-40 ± 7	-32 ± 3

[a]Leukotrienes were administered by intracoronary bolus injection in isolated guinea pig hearts.
[b]Values are means of peak changes (± SEM; N = 3 to 5).

onized by FPL 55712 but not by indomethacin. This suggests that depression of contractility and coronary flow by leukotrienes is not secondary to the release of prostaglandins. In addition, we have found that, similar to SRS-A, LTD_4 potentiates histamine-induced tachycardia. The concentration at which LTD_4 potentiates histamine-induced tachycardia (i.e., 2 ng/ml) is equivalent, as measured by bioassay in the guinea pig ileum, to 20 units SRS-A/ml. It is at this concentration that crude

SRS-A potentiates the chronotropic effect of histamine in the guinea pig heart. Since LTC_4 at 2 ng/ml is ineffective in this respect, it is likely that LTD_4, but not LTC_4, is responsible for the potentiation of histamine effects seen with crude SRS-A.

Thus, we found that FPL 55712 inhibits and abbreviates the contractile failure of cardiac anaphylaxis while reducing the decrease in coronary flow, and that negative inotropism and decreased coronary flow result from the intracardiac administration of SRS-A. We have also established that the cardiac effects of SRS-A are reproduced by the administration of pure synthetic leukotrienes and abolished by FPL 55712. All of this evidence indicates that SRS-A can reproduce a substantial part of the phenomenology associated with immediate hypersensitivity reactions of the heart and that SRS-A may mediate cardiac contractile failure and the decrease in coronary flow rate associated with cardiac anaphylaxis.

CONCLUSIONS

Histamine appears to be a major mediator of anaphylactic cardiac dysfunction, causing intense tachycardia and severe arrhythmias. It is clear, however, that histamine cannot account for all aspects of cardiac dysfunction in anaphylaxis and that a number of other mediators contribute to the overall derangement in cardiac function (Fig. 4). Thus the prolonged contractile failure and the precipitous decline of coronary flow can be ascribed to the effects of lipoxygenase and cyclo-oxygenase products of the arachidonic acid cascade.

Following an antigen-antibody reaction, histamine is promptly released, and tachycardia and arrhythmias ensue. After a short delay, SRS-A and prostaglandins are generated, causing a marked decrease in coronary flow and a protracted negative inotropic effect. Although at high concentrations $PGF_{2\alpha}$, PGD_2, and most likely TXA_2 are capable of decreasing the force of myocardial contraction (2), the amounts generated in the heart during anaphylaxis are probably not high enough to induce such a response. On the other hand, SRS-A, which is composed of LTC_4 and LTD_4 (28), has powerful negative inotropic effects and indeed is released in massive

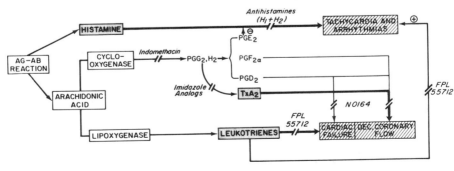

FIG. 4. Cardiac anaphylaxis. Proposed scheme for the actions and interactions among the various mediators of immediate-type hypersensitivity reactions of the heart.

concentrations from the anaphylactic heart. Thus contractile failure is probably caused primarily by SRS-A, possibly with a minor contribution by PGD_2, $PGF_{2\alpha}$, and TXA_2. In contrast with contractile failure, anaphylactic coronary constriction appears to be mediated primarily by SRS-A and TXA_2, as evidenced by the reproducibility of the effects of SRS-A and TXA_2 on coronary flow with synthetic analogs or their abolition with receptor antagonists and synthesis inhibitors. Again, a minor functional role could be played here by $PGF_{2\alpha}$ and PGD_2. Moreover, the metabolites of arachidonic acid also appear to interact with histamine by modulating its release and its effects. SRS-A potentiates and prolongs the tachyarrhythmic effects of histamine, whereas PGE_2 attenuates these effects. In addition, PGE_2 and PGD_2 could increase and decrease, respectively, the quantity of histamine released from the anaphylactic heart. Thus an intricate interrelationship exists between the various mediators of immediate hypersensitivity. During the course of anaphylactic reactions the heart becomes the target of mediators released intracardially, as well as of mediators reaching the left side of the heart from the lung. Thus the heart is subjected to multiple assaults which can culminate in fatal cardiac dysfunction.

ACKNOWLEDGMENTS

This work was supported by USPHS-NIH grants GM 20091 and 07547, HL 18828, RGP50GM26145-03, and by a grant from the National Science Foundation. Compound FPL 55712 was donated by P. Sheard of Fisons Ltd., England. Dr. Arleen B. Rifkind was most helpful in critically reviewing the manuscript. Ms. Jacqueline F. Hirsch provided excellent technical and secretarial assistance.

REFERENCES

1. Allan, G., and Levi, R. (1980): Pharmacological studies on the role of prostaglandins in cardiac hypersensitivity reactions. In: *Prostaglandins in Cardiovascular and Renal Function*, edited by A. Scriabine, A. M. Lefer, and F. A. Keuhl, Jr., pp. 223–237. Spectrum Publications, New York.
2. Allan, G., and Levi, R. (1980): The cardiac effects of prostaglandins and their modification by the prostaglandin antagonist N-0164. *J. Pharmacol. Exp. Ther.*, 214:45–49.
3. Allan, G., and Levi, R. (1981): Thromboxane and prostacyclin release during cardiac immediate hypersensitivity reactions in vitro. *J. Pharmacol. Exp. Ther.*, 271:157–161.
4. Anhut, H., Peskar, B. A., and Bernauer, W. (1978): Release of 15-keto-13,14-dihydro-thromboxane B_2 and prostaglandin D_2 during anaphylaxis as measured by radioimmunoassay. *Naunyn Schmiedebergs Arch. Pharmacol.*, 305:247–252.
5. Augstein, J., Farmer, J. B., Lee, T. B., Sheard, P., and Tattersall, M. L. (1973): Selective inhibitor of slow reacting substance of anaphylaxis. *Nature*, 245:215–217.
6. Austen, K. (1978): The anaphylactic syndrome. In: *Immunological Diseases*, 3rd ed., Vol. II, edited by M. Samter, pp. 885–899. Little Brown, Boston.
7. Baraka, A., and Sfeir, S. (1980): Anaphylactic cardiac arrest in a parturient. *JAMA*, 243:1745–1746.
8. Burke, J. A., and Levi, R. (1980): Slow-reacting substance of anaphylaxis (SRS-A): direct and indirect cardiac effects. *Fed. Proc.*, 39:389.
9. Burke, J. A., Levi, R., and Corey, E. J. (1981): Cardiovascular effects of pure synthetic leukotrienes C and D. *Fed. Proc.*, 40:1015.
10. Burke, J. A., Levi, R., Rouzer, C. A., Scott, W. A., and Corey, E. J. (1981): Cardiovascular effects of "slow reacting substances" derived from anaphylactic guinea pig lung and resident mouse peritoneal macrophages: comparison with effects of pure synthetic leukotrienes C and D. *Adv. Prostaglandin Thromboxane Res.*, Vol. 10 (in press).

11. Capurro, N., and Levi, R. (1975): The heart as a target organ in systemic allergic reactions: comparison of cardiac anaphylaxis in vivo and in vitro. Circ. Res., 36:520–528.
12. Chakravarty, N. (1959): A method for the assay of 'slow reacting substance.' Acta Physiol. Scand., 46:298–313.
13. Feigen, G. A., and Prager, D. J. (1969): Experimental cardiac anaphylaxis: physiologic, pharmacologic and biochemical aspects of immune reactions in the isolated heart. Am. J. Cardiol., 24:474–491.
14. Feigen, G. A., Vurek, G. G., Irvin, W. S., and Peterson, J. K. (1961): Quantitative adsorption of antibody by the isolated heart and the intensity of cardiac anaphylaxis. Circ. Res., 2:177–183.
15. Levi, R. (1972): Effects of exogenous and immunologically released histamine on the isolated heart: a quantitative comparison. J. Pharmacol. Exp. Ther., 182:227–238.
16. Levi, R., and Allan, G. (1980): Histamine-mediated cardiac effects. In: Drug-Induced Heart Disease, edited by M. Bristow, pp. 377–395. Elsevier-North Holland, Amsterdam.
17. Levi, R., Allan, G., and Zavecz, J. H. (1976): Cardiac histamine receptors. Fed. Proc., 35:1942–1947.
18. Levi, R., Allan, G., and Zavecz, J. H. (1976): Prostaglandins and cardiac anaphylaxis. Life Sci., 18:1255–1264.
19. Levi, R., and Burke, J. (1980): Cardiac anaphylaxis: SRS-A potentiates and extends the effects of released histamine. Eur. J. Pharmacol., 62:41–49.
20. Levi, B., Burke, J., and Holland, B. (1979): Slow reacting substance of anaphylaxis (SRS-A): a possible role in cardiac hypersensitivity reactions. Fed. Proc., 38:261.
21. Levi, R., and Capurro, N. (1973): Histamine H₂-receptor antagonism and cardiac anaphylaxis. In: International Symposium on Histamine H₂-Receptor Antagonists, edited by C. J. Wood and M. A. Simkins, pp. 175–186. Smith, Kline & French, London.
22. Levi, R., Capurro, N., and Lee, C. H. (1975): Pharmacological characterization of cardiac histamine receptors: sensitivity to H₁- and H₂-receptor agonists and antagonists. Eur. J. Pharmacol., 30:328–335.
23. Levi, R., Ganellin, C. R., Allan, G., and Willens, H. J. (1975): Selective impairment of atrioventricular conduction by 2-(2-pyridyl)ethylamine and 2-(2-thiazolyl)ethylamine, two histamine H₁-receptor agonists. Eur. J. Pharmacol., 34:237–240.
24. Levi, R., Owen, D. A. A., and Trzeciakowski, J. (1982): Actions of histamine on the heart and vasculature. In: Pharmacology of Histamine Receptors, edited by R. Ganellin and M. Parsons. J. Wright & Sons, London (in press).
25. Levi, R., and Pappano, A. J. (1978): Modification of the effects of histamine and norepinephrine on the sinoatrial node pacemaker by potassium and calcium. J. Pharmacol. Exp. Ther., 204:625–633.
26. Levi, R., and Zavecz, J. H. (1979): Acceleration of idioventricular rhythms by histamine in guinea pig heart. Circ. Res., 44:847–855.
27. Levi, R., Zavecz, J. H., and Ovary, Z. (1978): IgE-mediated cardiac hypersensitivity reactions. Int. Arch. Allergy Appl. Immunol., 57:529–534.
28. Lewis, R. A., Austen, K. F., Drazen, J. M., Clark, D. A., Marfat, A., and Corey, E. J. (1980): Slow reacting substances of anaphylaxis: identification of leukotrienes C-1 and D from human and rat sources. Proc. Natl. Acad. Sci. USA, 77:3710–3714.
29. Liebig, R., Bernauer, W., and Peskar, B. A. (1975): Prostaglandin, slow-reacting substance, and histamine release from anaphylactic guinea-pig hearts, and its pharmacological modification. Naunyn Schmiedebergs Arch. Pharmacol., 289: 65–76.
30. Rouzer, C. A., Scott, W. A., Cohn, Z. A., Blackburn, P., and Manning, J. M. (1980): Mouse peritoneal macrophages release leukotriene C in response to a phagocytic stimulus. Proc. Natl. Acad. Sci. USA, 77:4928–4932.
31. Trzeciakowski, J., and Levi, R. (1980): The cardiac pharmacology of tiotidine (ICI 125,211): a new histamine H₂-receptor antagonist. J. Pharmacol. Exp. Ther., 214:629–634.
32. Weichman, B. M., Hostelley, L. S., Bostick, S. P., Levi, R., and Chakrin, L. W. (1981): Immunologic histamine release in vitro from the heart and lung of the cynomolgus monkey. Int. Arch. Allergy Appl. Immunol., 64:456–463.
33. Zavecz, J. H., and Levi, R. (1977): Separation of primary and secondary cardiovascular events in systemic anaphylaxis. Circ. Res., 40:15–19.

Leukotrienes and Other Lipoxygenase Products,
edited by B. Samuelsson and R. Paoletti.
Raven Press, New York © 1982.

Comparative Biological Activity of Natural and Synthetic Leukotrienes

S. R. Baker, J. R. Boot, W. Dawson, W. B. Jamieson,
D. J. Osborne, and W. J. F. Sweatman

*Lilly Research Centre Limited, Erl Wood Manor, Windlesham,
Surrey, GU20 6PH, England*

5(S)-Hydroxy-6(R)-S-cysteinylglycine-7,9-*trans*,11,14-*cis*-eicosatetraenoic acid (leukotriene D_4, LTD_4) has three optically active centers and four double bonds giving 128 possible isomers. We have prepared eight of these isomers, including the natural product, and have compared some aspects of the *in vitro* and *in vivo* pharmacology of each.

METHODS

Synthesis of LTD_4 and Its Isomers

Synthesis of the isomers has been described elsewhere (2).

Biological Preparations

The compounds were dissolved in Tyrode solution and concentrations measured by ultraviolet (UV) spectroscopy assuming $\lambda_{max}\epsilon = 40,000$. Guinea pig ileum preparations, parenchymal strips, and tracheal chains were prepared as previously described (2,5). The parenchymal strips were suspended in Krebs Hensleit solution at 37°C, and contractions were recorded isotonically (4). Each preparation was allowed to equilibrate for at least 1 hr before commencing an experiment. The biological activities of the compounds were assayed in a dose-respondent fashion in at least three experiments for each of the smooth muscle preparations.

Pulmonary resistance and compliance were measured in the anesthetized guinea pig using a method based on that of Amdur and Mead (1).

RESULTS AND DISCUSSION

The eight isomers of LTD_4 are divided into four classes, each class having different stereochemistry about the 9 and 11 double bonds (Fig. 1). Each isomer was obtained in both the natural 5S,6R and unnatural 5R,6S forms (Fig. 2). The two isomers of 9-*cis*-LTD_4 readily undergo a 1,7-hydride shift reaction, giving rise

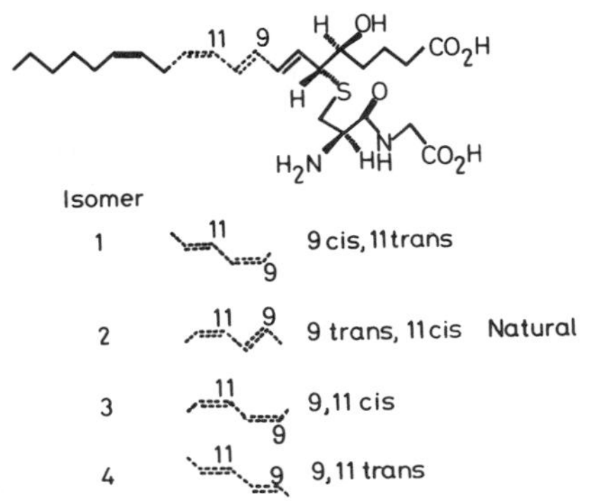

Isomer

1	9 cis, 11 trans
2	9 trans, 11 cis Natural
3	9,11 cis
4	9,11 trans

FIG. 1.   Geometric isomers of LTD$_4$.

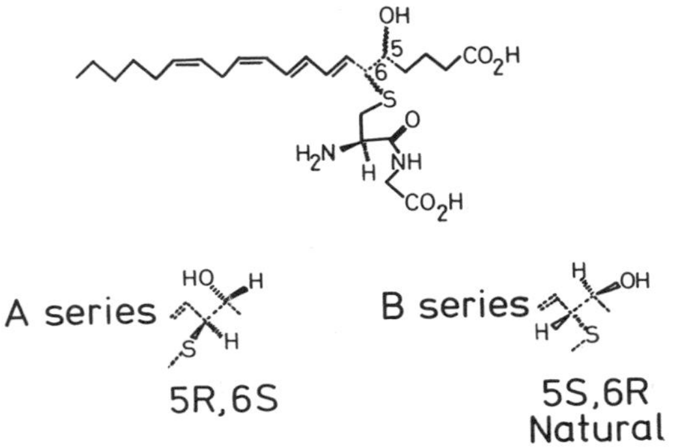

A series   5R,6S

B series   5S,6R   Natural

FIG. 2.   Optical isomers of LTD$_4$.

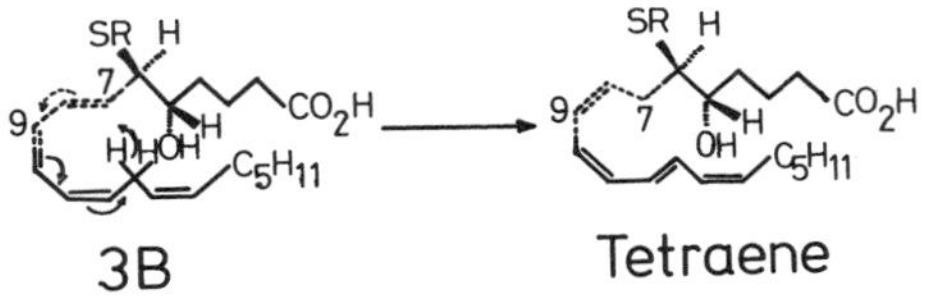

3B   Tetraene

FIG. 3.   1,7-Hydride shift reaction of 9-cis-LTD$_4$.

to the 8,10-cis,12-trans,14-cis-eicosatetraenoic acid analogs (Fig. 3). After high-pressure liquid chromatography (HPLC), 10 eicosatetraenoic acid derivatives were available for pharmacological investigation.

In Vitro Pharmacology

The comparative spasmogenic activity of the 10 isomers were investigated using isolated guinea pig ileum, parenchymal strips, and tracheal chains. Figure 4 shows a typical experiment on guinea pig ileum, each dose-response curve being derived from four measurements. It was found that on all three tissues the four natural 5S,6R isomers were at least 100 times more active than their corresponding 5R,6S

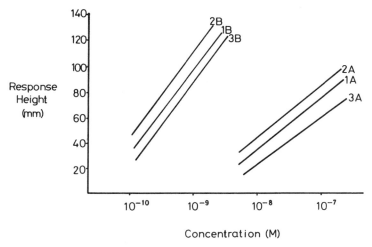

FIG. 4. A typical series of *in vitro* dose-response curves to the isomers of LTD_4 on guinea pig ileum.

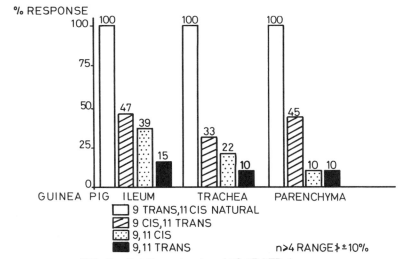

FIG. 5. Relative potencies of 5S 6R LTD_4 isomers.

analogs. Figures 5 and 6 show the effect of changing the 9 and 11 double bond stereochemistry on the spasmogenicity of the 5S,6R and 5R,6S isomers. The rank order of potency of the isomers on all three tissues was 9-*trans*, 11-*cis* > 9-*cis*, 11-*trans* > 9,11-*cis* > 9,11-*trans*. These data are interesting when compared with those obtained using the natural substances (6).

A similar series of experiments were carried out with LTC_4 and its corresponding isomers. The results obtained from this study indicated a similar order of activity, but the LTC_4 isomers were 5 to 10 times less active than their LTD_4 analogs.

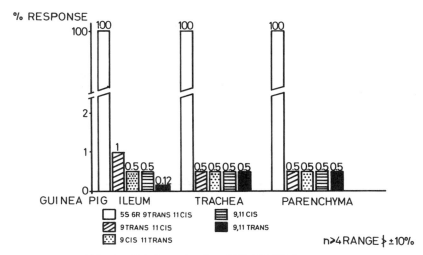

FIG. 6. Relative potencies of 5R 6S LTD₄ isomers.

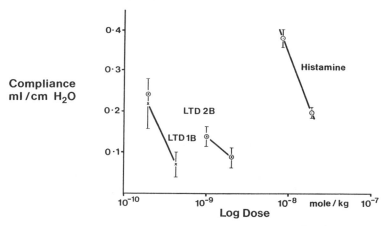

FIG. 7. Effect of LTD₄ 2B, 1B and histamine on anesthetized guinea pig pulmonary compliance.

The two 8,10-*cis*,12-*trans*,14-*cis*-LTD₄ analogs proved to be at least 500 times less active than their parent leukotrienes.

These *in vitro* results enable us to draw some conclusions about the structural requirements of the leukotriene receptor. The presence of both a double bond at C7 and the natural 5S,6R stereochemistry appears to be of prime importance in receptor binding, whereas the stereochemistry of the 9 and 11 double bonds and the exact nature of the peptide side chain seem to be less important.

There is the possibility that the limited biological activity of the 5R,6S isomer could be due to approximately 1% contamination with the more active 5S,6R isomer, but this is unlikely as all the isomers were at least 99.5% optically pure as measured

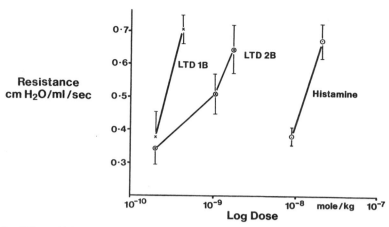

FIG. 8. Effect of LTD$_4$ 2B, 1B and histamine on anesthetized guinea pig pulmonary resistance.

by HPLC and in some cases > 99.8% pure. In addition, it is significant that the dose-response curves of the two series are not parallel (Fig. 4).

In Vivo Pharmacology

The effect of LTD$_4$ and two synthetic isomers 5S,6R 9-*cis*,11-*trans*-LTD$_4$ and 5R,6S LTD$_4$ were compared on the respiratory system of the anesthetized guinea pig. LTD$_4$ and 5S,6R 9-*cis*,11-*trans*-LTD$_4$ were potent bronchoconstrictors (2 × 10^{-9} mole/kg i.v.), producing a dose-related fall in compliance and a rise in resistance, whereas histamine was approximately 100 times less active. These preliminary *in vivo* results also indicate that isomers of LTD$_4$ with the natural 5S,6R stereochemistry were approximately 100 times more active than the corresponding 5R,6S configuration (Figs. 7 and 8).

The relative potencies obtained suggested that 5S,6R 9-*cis*,11-*trans*-LTD$_4$ was more potent *in vivo* than LTD$_4$ itself. This may be due to metabolic stability, as it has been shown that 5S,6R 9-*cis*,11-*trans*-LTD$_4$ is resistant to biological inactivation by soybean lipoxygenase (3).

REFERENCES

1. Amdur, M. O., and Mead, J. (1958): Mechanics of respiration in unaesthetized guinea pigs. *Am. J. Physiol.*, 192:364–368.
2. Baker, S. R., Boot, J. R., Jamieson, W. B., Osborne, D. J., and Sweatman, W. J. F. S. (1981): The comparative *in vitro* pharmacology of LTD$_4$ and its isomers. (*In press.*)
3. Baker, S. R., Boot, J. R., and Osborne, D. J. (1981): A comparative study of the physical properties and enzymatic reactions of SRS-A and LTD$_4$ isomers. (*In press.*)
4. Constantine, J. W. (1965): The spirally cut tracheal strip preparation. *J. Pharm. Pharmacol.*, 17:384–385.
5. Drazen, J. M., and Schneider, M. W. (1978): Comparative responses of tracheal spirals and parenchymal strips to histamine and carbachol. *J. Clin. Invest.*, 61:1441–1447.
6. Lewis, R. A., Austen, K. F., Drazen, J. M., Clark, D. A., Margat, A., and Corey, E. J. (1980): Slow reacting substances of anaphylaxis: identification of leukotrienes C-1 and D from human and rat sources. *Proc. Natl. Acad. Sci. USA*, 77:3710–3714.

Leukotrienes and Other Lipoxygenase Products,
edited by B. Samuelsson and R. Paoletti.
Raven Press, New York © 1982.

Antagonists of SRS-A and Leukotrienes

P. Sheard, M. C. Holroyde, A. M. Ghelani, J. R. Bantick,
and T. B. Lee

*Fisons Limited, Pharmaceutical Division, Research and Development Laboratories,
Loughborough, Leicestershire LE11 0QY, England*

In 1973 the Fisons compound FPL 55712 (Fig. 1) was reported to be a potent, dose-dependent, selective, receptor antagonist of slow-reacting substance of anaphylaxis (SRS-A) (5). This was the first such compound to be described.

Subsequently the compound has been evaluated in other laboratories, reviewed by Sheard (12), as an antagonist of SRS-A from various sources. The essentially competitive nature of the antagonism was indicated in several of these studies by (a) parallel shifts of SRS-A dose-response curves in the presence of increasing concentrations of FPL 55712 without a reduction in the maximum response, and (b) the demonstration that the effect of the antagonist could be overcome by increasing the concentration of SRS-A. However, the slope of the regression line calculated from a Schild plot to determine pA_2 in our early studies was 0.57 compared to a theoretical value of 1.0 for true competitive antagonism (5). We had assumed that one possible reason for this was the impure nature of the SRS-A used in our studies at that time. Thus it was of interest to investigate the antagonism by FPL 55712 of the synthetic leukotrienes (LT) C_4 and D_4. Results are shown in Table 1.

The slopes of the Schild plot regression lines obtained with both of the pure leukotrienes were much closer to unity than those obtained with unpurified SRS-A, and we consider that the antagonism demonstrated by FPL 55712 against LTC_4 and LTD_4 is one of true competition.

By way of comparison, studies in Merck Frosst Laboratories using guinea pig ileum have yielded pA_2 values of 6.4 ± 0.1 and 7.1 ± 0.1 for FPL 55712 against LTC_4 and LTD_4, respectively (7,11; Holme, *personal communication*). FPL 55712 has also recently been reported to be equally potent as an antagonist of synthetic

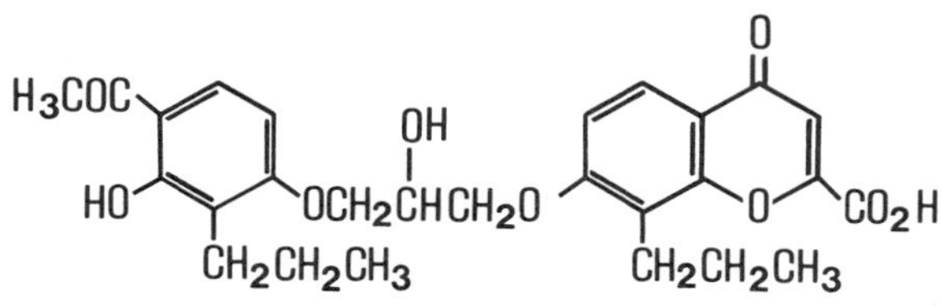

FIG. 1. Chemical structure of FPL 55712.

229

TABLE 1. *Antagonism of leukotrienes C_4 and D_4 by FPL 55712[a] on guinea pig ileum*

	pA_2[b]: Method of		Slope of Schild plot
Agonist	Lockett and Bartlet (10)	Arunlakshana and Schild (4)	
LTC_4	7.4 ± 0.3 (3)[c]	7.1 (2)	0.88 (2)
LTD_4	7.63 (2)	7.59 ± 0.1 (6)	0.84 ± 0.05 (6)

NB: Throughout this paper concentrations or doses quoted for FPL compounds used in our laboratories refer to the free acid form.
[a]Tested as the lysine salt.
[b]Mean ± SEM (where applicable).
[c]Figures in parentheses denote number of experiments.

LTE_4 on the guinea pig ileum with an IC_{50} of 3.5×10^{-8} M, i.e., 0.018 μg ml^{-1} (14).

Since the first publication on FPL 55712, the compound has been utilized in numerous laboratories throughout the world as a research tool to confirm the presence of SRS-A in various biological fluids. It has also been used to assess the possible role of SRS-A in the mediation of certain antigen-induced reactions in various animal species. Of particular interest in this respect are those studies which involved the use of human tissue *in vitro* or which were carried out *in vivo* in human patients and are therefore particularly relevant to the consideration of SRS-A as a mediator of such reactions in man.

Adams and Lichtenstein (1,2) conducted an elegant series of experiments on antigen-induced contraction of passively sensitized preparations of human bronchial strips. They showed that the histamine H_1 antagonist diphenhydramine (10^{-5} M, 2.5 μg ml^{-1}), administered 5 min before antigen, delayed the onset and slowed the rate of contraction but had no effect on the amplitude or duration of the response (2). However, the combined use of diphenhydramine and FPL 55712 (2×10^{-5} M, 10 μg ml^{-1}) produced a delay in the onset of the response and also decreased the amplitude and duration of the contractile response (1). In addition, when the compounds were administered after the response to antigen had developed (i.e., 30 min after antigen), diphenhydramine (3×10^{-4} M, 75 μg ml^{-1}) had no effect, but FPL 55712 (5×10^{-5} M, 25 μg ml^{-1}) completely reversed the antigen-induced contraction over a period of 16 min (2). The concentrations of FPL 55712 used in these experiments were considerably higher than those required to inhibit SRS-A-induced contractions of human bronchus (Table 2). Adams and Lichtenstein have postulated that this is due to the much higher tissue concentrations of SRS-A obtained following antigen challenge than are achieved by the addition of exogenous SRS-A, as is also the case with histamine. These experiments provide evidence for the involvement of both histamine and SRS-A in antigen-induced contraction of sensitized human bronchus, with histamine mediating the early phase and SRS-A the later, prolonged phase of the reaction.

In other studies carried out by Wanner and his colleagues, SRS-A has been implicated in a further consequence of antigen-antibody reactions in human airways,

TABLE 2. *Comparative activities of FPL 55712, FPL 57231, and FPL 59257*

Tissue and agonist	IC_{50} (μg ml^{-1})		
	FPL 55712[a]	FPL 57231[b]	FPL 59257[a]
Guinea pig ileum			
SRS-A[gp]	0.019 ± 0.003 (28)[c]	0.03 ± 0.013 (4)	0.2[d] (2)
LTD$_4$	0.032 ± 0.004 (9)	0.033 ± 0.012 (3)	0.26 ± 0.005[d] (3)
Histamine	21.4 ± 5.9 (5)	43.7 ± 6.3 (3)	> 100 (3)
5-Hydroxytrypt- amine	28.5 ± 12.3 (5)	10.8 ± 2.8 (3)	> 100 (3)
PGF$_{2\alpha}$	4.9 ± 2.6 (5)	12.3 ± 0.9 (5)	> 100 (3)
Guinea pig lung strip			
SRS-A[gp]	0.087 ± 0.019 (7)	0.09 (1)	0.92 (1)
LTD$_4$	0.78 ± 0.11 (4)	0.65 ± 0.18 (3)	1.51 ± 0.42 (4)
Human lung strip			
SRS-A[gp]	0.33 ± 0.05 (5)		2.12 (2)
LTD$_4$	0.58 (1)	3.0 ± 1.1 (3)	
Human bronchial strip			
SRS-A[gp]	0.22 ± 0.03 (3)	0.63 ± 0.44 (3)	1.57 (2)

Results are given as the mean $\pm$ SEM.
[a]Tested as the lysine salt.
[b]Tested as the sodium salt.
[c]Figures in parentheses denote number of experiments.
[d]At time of maximum inhibition, i.e., on response immediately following that in presence of compound.

i.e., the reduction in tracheal mucous velocity (TMV). In initial studies in dogs with hypersensitivity to *Ascaris suum*, Wanner et al. (13) showed that neither histamine nor acetylcholine could be considered as mediators responsible for the antigen-induced fall in TMV, as inhalation of both agents stimulated mucous transport. However, inhalation of an aerosol from a 1% solution of FPL 55712 before and during antigen challenge in these dogs prevented the normal antigen-induced depression of TMV and a marked increase in TMV resulted. In subsequent studies in ragweed-sensitive patients (3), in whom the control response to antigen was again a fall in TMV, the inhalation of aerosol from a 1% solution of FPL 55712 before and after (Fig. 2), or only after, antigen inhalation produced an increase in TMV. Given alone, FPL 55712 had no effect on TMV. These results suggest that the SRS-A released following antigen challenge impairs mucous transport in the upper airways in man. Following blockade of this effect by FPL 55712, the stimulating effect of other mediators results in an increase in TMV.

Thus, human studies with FPL 55712 have led to the association of SRS-A with two pathophysiological aspects of allergic obstructive lung disease: bronchoconstriction and impaired sputum clearance. The further use of SRS-A/leukotriene antagonists and the natural or synthetic leukotrienes themselves will lead to a clearer understanding of the role of SRS-A in various inflammatory disease states.

FPL 55712 is one of a series of chromone-2-carboxylic acids which has also given rise to compounds with potent antiallergic (sodium cromoglycate-like) activity. FPL 55712 itself has some antiallergic activity with an ED_{50} in rat passive

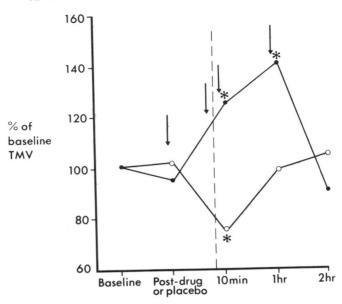

FIG. 2. Effect of antigen challenge *(at vertical interrupted line)* on tracheal mucous velocity (TMV) in six asymptomatic subjects with ragweed asthma after pre- and postantigen treatment *(arrows)* with placebo (○) or 1% FPL 55712 (●). * $p < 0.05$ compared to baseline. (From Ahmed et al. (3) with permission.)

cutaneous anaphylaxis (PCA) of 5.6 mg kg^{-1} i.v. (12), and an ED$_{50}$ as an inhibitor of histamine release in rat passive peritoneal anaphylaxis of 3.2×10^{-8} mole (16 μg) rat^{-1} i.p. (6). Furthermore, when inhaled for 60 sec as an aqueous aerosol from a 0.5% solution 10 min before antigen, FPL 55712 inhibited the antigen-induced fall in FEV$_1$ in an asthmatic volunteer by approximately 60%.

During further studies of analogs of FPL 55712 we were interested to find two propionic acid analogs which, although retaining selective anti-SRS-A activity, had little if any antiallergic activity and thus had a relatively cleaner spectrum of activity as antagonists of SRS-A. These two compounds, FPL 57231 and FPL 59257 (Fig. 3), were only slightly active in rat PCA at the high dose of 20 mg kg^{-1} i.v. (13% and 18% inhibition, respectively). In addition, when inhaled by the same asthmatic volunteer under identical conditions to the FPL 55712 inhalation, FPL 57231 did not inhibit the antigen-induced fall in FEV$_1$. (FPL 59257 has not yet been tested against antigen challenge in man.) The use of compounds such as FPL 57231 and FPL 59257 should be of benefit in those experimental situations where the use of an antagonist of SRS-A which also has some antiallergic properties might cloud the interpretation of the results obtained. Therapeutically, however, it seems desirable to have a compound which possesses both properties.

The comparative activities of FPL 55712, 57231, and 59257 are shown in Table 2. Note that in 28 recent experiments the mean IC$_{50}$ for FPL 55712 as an antagonist

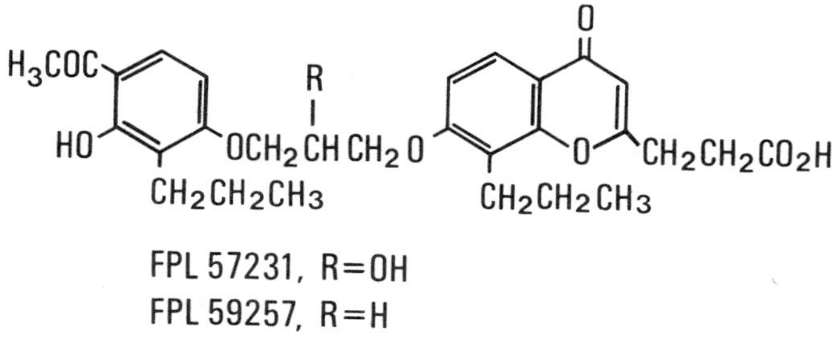

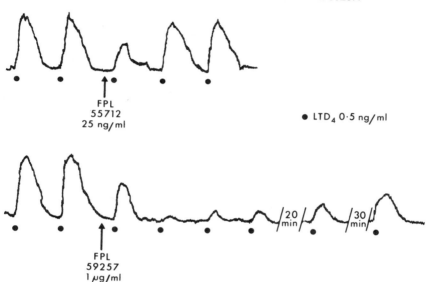

FPL 57231, R=OH
FPL 59257, R=H

FIG. 3. Chemical structures of FPL 57231 and FPL 59257.

FPL
55712
25 ng/ml

● LTD$_4$ 0·5 ng/ml

FPL
59257
1 μg/ml

FIG. 4. Antagonism of LTD$_4$-induced contractions of guinea pig ileum by FPL 55712 and FPL 59257.

of unpurified SRS-A^{gp} on the guinea pig ileum was 0.019 ± 0.003 μg/ml compared with earlier reported values of 0.005 ± 0.001 μg/ml (5,12). This confirms a slight shift in potency which had been noted in our laboratories during the intervening years between these studies. FPL 57231 was generally similar in potency, whereas FPL 59257 was less potent than FPL 55712. However, both of the later compounds retained the selectivity of FPL 55712. Most interestingly, FPL 59257 had a longer duration of action; whereas the inhibitory effect of FPL 55712 (and of FPL 57231) on the guinea pig ileum could be removed readily by only one or two washes, that of FPL 59257 was much more persistent. In fact, after exposing the tissue to concentrations of FPL 59257 greater than 1 μg/ml, the response to LTD (or SRS-A) frequently could not be fully re-established (Fig. 4). It was also notable that the maximum inhibitory effect of FPL 59257 was observed against the LTD (or SRS-

A) response which immediately followed the response obtained in the presence of the compound (Fig. 4). Thus FPL 59257 appears to equilibrate only slowly with the leukotriene/SRS-A receptor and then to dissociate slowly from it.

The activities of FPL 55712, 57231, and 59257 have also been investigated in an *in vivo* model of SRS-A-induced bronchoconstriction. Consistent bronchoconstrictor responses were produced at 40-min intervals in anesthetized guinea pigs by the forced inhalation of aerosols of SRS-A[gp] [partially purified by Amberlite chromatography (9)]. In our hands this technique produces more consistent responses than the repetitive intravenous administration of SRS-A.

The guinea pigs were artificially respired to a pressure of 8 to 10 cm water and bronchoconstriction measured as an increase in tracheal overflow pressure, expressed as a percentage of the maximum possible bronchoconstriction (obtained by clamping off the trachea). The compounds to be tested were administered intravenously immediately before exposure to nebulized SRS-A, and the antagonism was expressed as a percent reduction of the previous control response to SRS-A. The ED_{50} (dose required to inhibit the response to SRS-A by 50%) was then calculated. Results are shown in Table 3. An ED_{50} was readily obtained with FPL 55712 and FPL 57231, as the effects of these compounds are short-lasting and control responses to SRS-A are quickly re-established.

In the case of FPL 59257, however, an exact ED_{50} could not be determined due to persistent attenuation of control responses to SRS-A following a single administration of this compound. Although FPL 59257 was less potent than FPL 55712 on ileum and airway tissues *in vitro* (Table 2), the compounds were similar in potency *in vivo* in the guinea pig (Table 3) and in man (8).

Plasma clearance of FPL 55712 is rapid. The drug is efficiently extracted by the liver and excreted unchanged in the bile. Little renal excretion occurs. FPL 59257 is also rapidly cleared from the plasma with a half-life of circa 20 min following a dose of 10 mg/kg i.v. in the rat (cf. FPL 55712 plasma $t_{1/2}$ circa 10 min). Thus the long duration of activity of FPL 59257 *in vivo* appears to be due to the compound

TABLE 3. *Antagonism of bronchoconstriction induced by inhalation of nebulized SRS-A in the anesthetized guinea pig*

Compound	ED_{50} (mg kg^{-1} i.v.)	N
FPL 55712[a]	12.8 ± 2.5	5
FPL 57231[b]	14.9 ± 3.4	4
FPL 59257[a]	—[c]	

Results are given as the mean ± SEM.
[a]Tested as the lysine salt.
[b]Tested as the sodium salt.
[c]ED_{50} not obtained (see text). In individual animals, 5 mg kg^{-1} gave 51% inhibition, 10 mg kg^{-1} gave 52% and 53% inhibition, and 15 mg kg^{-1} gave 77% inhibition.

becoming tightly bound to, and slowly dissociated from, the airway tissue or the leukotriene/SRS-A receptor, as also suggested by the *in vitro* studies.

Persistent activity such as has been demonstrated with FPL 59257 should enhance the potential therapeutic value of an SRS-A antagonist.

ACKNOWLEDGMENTS

We thank Mr. J. Fuher, Mr. P. A. West, and Mrs. M. E. Mather for their help in various aspects of this work. The synthetic LTC_4 and LTD_4 were obtained from Prof. E. J. Corey.

REFERENCES

1. Adams, G. K., and Lichtenstein, L. M. (1977): Antagonism of antigen-induced contraction of guinea-pig and human airways. *Nature*, 270:255–257.
2. Adams, G. K., and Lichtenstein, L. M. (1979): In vitro studies of antigen-induced broncho-spasm: effect of antihistamine and SRS-A antagonist on response of sensitised guinea-pig and human airways to antigen. *J. Immunol.*, 122:555–562.
3. Ahmed, T., Greenblatt, D. W., Birch, B., Marchetti, B., and Wanner, A. (1981): Abnormal mucociliary transport in allergic patients with antigen-induced bronchospasm: role of SRS-A. *Am. Rev. Resp. Dis.*, 124: .
4. Arunlakshana, O., and Schild, H. O. (1959): Some quantitative uses of drug antagonists. *Br. J. Pharmacol.*, 14:48–58.
5. Augstein, J., Farmer, J. B., Lee, T. B., Sheard, P., and Tattersall, M. L. (1973): Selective in-hibition of slow reacting substance of anaphylaxis. *Nature [New Biol.]*, 245:215–216.
6. Buckle, D. R., Outred, D. J., Ross, J. W., Smith, H., Smith, R. J., Spicer, B. A., and Gasson, B. C. (1979): Aryloxyalkyloxy- and aralkyloxy-4-hydroxy-3-nitrocoumarins which inhibit hista-mine release in the rat and also antagonise the effects of a slow reacting substance of anaphylaxis. *J. Med. Chem.*, 22:158–168.
7. Holme, G., Brunet, G., Piechuta, H., Masson, P., Girard., Y., and Rokach, J. (1980): The activity of synthetic leukotriene C-1 on guinea-pig trachea and ileum. *Prostaglandins*, 20:717–728.
8. Holroyde, M. C., Altounyan, R. E. C., Cole, M., Dixon, M., and Elliott, E. V. (1981): Selective inhibition of bronchoconstriction induced by leukotrienes C and D in man. This volume.
9. Lee, T. B., Fuher, J., Holroyde, M. C., Mann, J., and Bantick, J. R. (1979): An improved technique for the partial purification of SRS-A. *J. Pharm. Pharmacol.*, 31:866–867.
10. Lockett, M. F., and Bartlet, A. L. (1956): A method for the determination of pA_2 at two minutes. *J. Pharm. Pharmacol.*, 8:18–26.
11. Rokach, J., Girard, Y., Guindon, Y., Atkinson, J. G., Larue, M., Young, R. N., Masson, P., Hamel, R., Piechuta, H., and Holme, G. (1981): The synthesis of leukotrienes. In: *SRS-A and Leukotrienes*, edited by P. J. Piper, pp. 65–72. Wiley, Chichester.
12. Sheard, P. (1981): Effects of anti-allergic compounds on SRS-A and leukotrienes. In: *SRS-A and Leukotrienes*, edited by P. J. Piper, pp. 209–218. Wiley, Chichester.
13. Wanner, A., Zarzecki, S., Hirsch, J., and Epstein, S. (1975): Tracheal mucous transport in experimental canine asthma. *J. Appl. Physiol.*, 39:950–957.
14. Welton, A. F., Crowley, H. J., Miller, D. A., and Yaremko, B. (1981): Biological activities of a chemically synthesized form of leukotriene E_4. *Prostaglandins*, 21:287–296.

Leukotrienes and Other Lipoxygenase Products,
edited by B. Samuelsson and R. Paoletti.
Raven Press, New York © 1982.

Selective Inhibition of Bronchoconstriction Induced by Leukotrienes C and D in Man

M. C. Holroyde, R. E. C. Altounyan, M. Cole, M. Dixon, and E. V. Elliott

Fisons Limited, Pharmaceutical Division, Research and Development Laboratories, Loughborough, Leicestershire, LE11 0QY, England

Over a number of years the belief has grown that slow reacting substance of anaphylaxis (SRS-A) may be an important mediator of allergen-induced bronchoconstriction in man. The evidence for this is relatively circumstantial but may be summarized as follows: (a) SRS-A is released from sensitized human lung following exposure to antigen (14). (b) This release is inhibited by sodium cromoglycate, a drug used in the prophylactic treatment of asthma in man (13). (c) SRS-A produces a long-lasting contraction of human airway preparations *in vitro* (2,6) but of few other smooth muscle preparations. (d) SRS-A has been detected in the sputum of asthmatic patients (15). (e) Antihistamines are relatively ineffective in the treatment of asthmatic bronchoconstriction in man, suggesting that other mediators may be involved.

Possibly the most direct attempt to examine the effects of SRS-A in man was that of Herxheimer and Stresemann (7) who administered a partially purified preparation of guinea pig SRS-A by aerosol to volunteers, producing reductions in vital capacity in nine asthmatics but no response in four normal subjects. In a subsequent study by the same authors (8) the results were considerably more variable. Unfortunately, due to the relatively impure nature of the SRS-A used in these studies, the hyperreactive airways of the asthmatic patient may have been stimulated by substances other than SRS-A.

Recent studies have indicated that the term SRS-A really refers to a biological activity which can be ascribed to a mixture of leukotrienes (LT), with LTD being predominant (9–11).

The availability of small quantities of synthetic LTC_4 and LTD_4 has allowed us to carry out a preliminary study of the pulmonary effects of these substances in two normal volunteers, and to assess the activity of two antagonists of SRS-A developed in our laboratories: FPL 55712 (1) and FPL 59257 (Chapter 22, *this volume*).

Both subjects (M. C., female, age 50; E. E., male, age 36) were smokers but were nonatopic. Both were healthy and were not taking any medication at the time of the study. Each gave fully informed consent before participation in the study.

Aerosols were generated from freshly prepared solutions of LTC and LTD in saline by using a "Hi-Flow" nebulizer (O.E.M. Medical Inc., Richmond, Va.) driven by compressed air at a flow rate of 4 liters min^{-1}. The aerosol was inhaled via a face mask for 90 sec, during which time approximately 1 ml of solution was nebulized. We have observed that leukotrienes lose biological activity during nebulization, possibly due to oxidation. Consequently, any solution remaining in the nebulizer after use (approximately 1 ml) was discarded.

At intervals, pulmonary function was monitored by measuring FEV_1 (forced expiratory volume in 1 sec) with a Vitalograph dry spirometer. As changes in FEV_1 were very small, partial expiratory flow-volume curves were used to derive a more sensitive measurement of bronchoconstriction (12). Each subject wore a noseclip and breathed through a pneumotachograph (Fleisch No. 4) connected to a capacitance manometer whose output was amplified and integrated to provide air flow and volume signals. After breathing out to residual volume (RV), each subject slowly inspired 2 liters and then performed a maximal expiratory maneuver. The air flow and volume signals were fed directly to a minicomputer which calculated flow rates at RV + 1.5, RV + 1.0, and RV + 0.5 liters, respectively.

At least three flow-volume and three FEV_1 determinations were done before inhalation of an aerosol, and to minimize the effects of a maximal inspiration on a subsequent flow-volume curve, the FEV_1 and flow-volume maneuvers were separated by at least 2 min. All experiments were carried out at the same time of day, and only one concentration of LTC or LTD was administered to each subject on any one day.

LTC and LTD were approximately equipotent and equieffective in producing bronchoconstriction, although subject E. E. was generally more reactive than M. C. (Fig. 1). Reductions in FEV_1 were small, varying from 3 to 6% in subject M. C., and 6 to 10% in E. E. In subject M. C. (but not E. E.) the response to LTD was

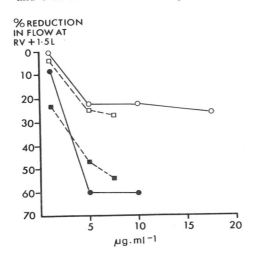

FIG. 1. Peak bronchoconstrictor responses to inhalation of various concentrations of LTC (●, ○) and LTD (■, □) in subjects E. E. (*solid symbols*) and M. C. (*open symbols*).

somewhat more rapid than to LTC. Both leukotrienes elicited bouts of coughing in both subjects, particularly at the end of each expiratory maneuver, with LTD being subjectively more irritant. Control aerosols of saline or appropriate concentrations of ethanol in saline produced no response.

On another occasion, histamine was administered to subject E. E. under identical conditions to produce a similar degree of bronchoconstriction. An aerosol of 1 mg ml^{-1} histamine acid phosphate inhaled for 90 sec reduced flow at RV + 1.5 liters by 66% (Fig. 2), but FEV_1 fell by only 2.5%. Lung function returned to normal within 10 min, whereas following inhalation of LTC or LTD impairment of flow was still detectable more than an hour later. In this experiment, LTC was approximately 200 times more potent than histamine on a molar basis. Unlike the LTs, histamine did not induce coughing.

In a further series of experiments a single concentration of LTC (10 μg ml^{-1}) was inhaled by both subjects after prior inhalation of an aerosol of sodium cromoglycate (10 mg ml^{-1}), FPL 55712 (3 mg ml^{-1}), or FPL 59257 (3 mg ml^{-1}). The drug aerosols were administered for 5 min starting 15 min before inhalation of LTC.

Sodium cromoglycate had no effect on the response to LTC, indicating that the response does not include a component which is dependent on the degranulation of mast cells. However, both FPL 55712 and FPL 59257 completely abolished the

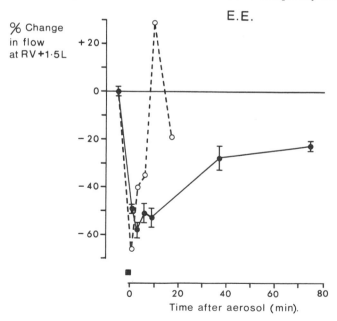

FIG. 2. Bronchoconstrictor response of subject E. E. to inhalation of aerosol of 10 μg ml^{-1} LTC (●) or 1 mg ml^{-1} histamine acid phosphate (○). Results with LTC are mean ± SEM of three experiments. Aerosols were inhaled during the period marked ■.

cough response and substantially inhibited the bronchoconstrictor response. FPL 59257 was somewhat more effective than FPL 55712 (Fig. 3), which may be a reflection of the former's longer duration of activity, as seen in animal models (Chapter 22, *this volume*).

DISCUSSION

This preliminary study has demonstrated the capacity of LTC and LTD to produce a moderate degree of bronchoconstriction in normal volunteers. The LTs were very potent and produced a long-lasting response in comparison with histamine. The magnitude of the response was similar to that produced in normal people by breathing

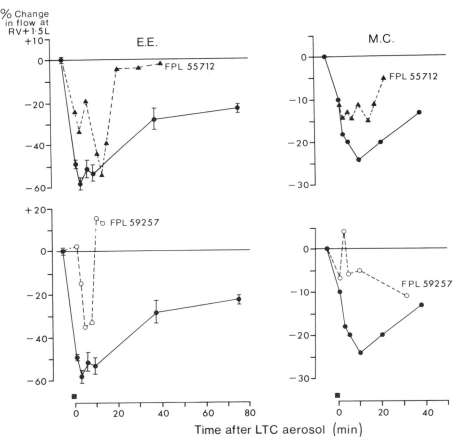

FIG. 3. Inhibition of LTC-induced bronchoconstriction in man. Responses produced in both subjects by LTC administered alone (●) and after prior inhalation of 3 mg ml⁻¹ FPL 55712 (▲) or 3 mg ml⁻¹ FPL 59257 (○). Both FPL compounds were administered as lysine salts. Concentrations are expressed as the free acid. Drug aerosols were administered 15 min before LTC, which was inhaled during the period marked ■. Responses to LTC alone are mean ± SEM of three experiments (subject E. E.) or mean of two (M. C.).

cold air (5), a procedure known to produce severe bronchoconstriction in asthmatics. By analogy, LTs might be expected to produce a severe and long-lasting bronchoconstriction in asthmatics, although the mechanisms involved in LT-induced and cold air-induced bronchoconstriction are probably quite different.

Using the data obtained so far, it is not possible to identify precisely the site of action of LTC and LTD within the lungs. The bronchoconstrictor response was manifested as a pronounced impairment of flow at low lung volumes but with little change in FEV_1. Such a response would be interpreted by some as being indicative of an effect primarily in peripheral airways (12). Indeed, it has been suggested that in guinea pig lung LTC and LTD act predominantly on peripheral airways (4). Final confirmation of their site of action in man must await more definitive tests, such as density-dependent flow-volume determinations (12).

The cough response produced by both leukotrienes was unusual in that coughing occurred not during inhalation of the leukotriene aerosols but only at the end of each forced expiratory maneuver. This suggests that LTC and LTD are not merely irritants but may sensitize sensory nerve endings so that a subsequent physical stimulus (as during the expiratory maneuver) results in cough. Certainly the cough was not simply a consequence of bronchoconstriction, as the response to histamine in the single subject examined was not accompanied by coughing.

If leukotrienes are able to sensitize cough receptors, it would be interesting to determine if these substances can also induce a state of nonspecific bronchial hyperreactivity, which is a central feature of asthma. Indeed, experiments in guinea pigs indicate that airway hyperreactivity can be produced by lipoxygenase products (3).

The present study cannot determine if LTC and LTD play a major role in asthma, but it has confirmed that these substances are potentially capable of playing such a role. It is hoped that the final answer will be provided by studying the activity of a selective antagonist such as FPL 59257 in asthmatic patients.

REFERENCES

1. Augstein, J., Farmer, J. B., Lee, T. B., Sheard, P., and Tattersall, M. L. (1973): Selective inhibitor of slow reacting substance of anaphylaxis. *Nature*, 245:215–217.
2. Brochlehurst, W. E. (1962): Slow reacting substance and related compounds. *Progr. Allergy*, 6:539–558.
3. Downs, D. D., Garland, L. G., and Shields, P. A. (1981): A possible role for lipoxygenase products in airway hyperreactivity in vivo. *Br. J. Pharmacol.*, 73:252P–253P.
4. Drazen, J. M., Austen, K. F., Lewis, R. A., Clark, D. A., Goto, G., Marfat, A., and Corey, E. J. (1980): Comparative airway and vascular activities of leukotrienes C-1 and D in vivo and in vitro. *Proc. Natl. Acad. Sci. USA*, 77:4354–4358.
5. Fanta, C. H., McFadden, E. R., and Ingram, R. H. (1981): Effects of cromolyn sodium on the response to respiratory heat loss in normal subjects. *Am. Rev. Resp. Dis.*, 123:161–164.
6. Ghelani, A. M., Holroyde, M. C., and Sheard, P. (1980): Response of human isolated bronchial and lung parenchymal strips to SRS-A and other mediators of asthmatic bronchospasm. *Br. J. Pharmacol.*, 71:107–112.
7. Herxheimer, H., and Stresemann, E. (1963): The effect of slow reacting substance (SRS-A) in guinea pigs and in asthmatic patients. *J. Physiol. (Lond.)*, 165:78P–79P.
8. Herxheimer, H., and Stresemann, E. (1966): Unsuccessful tests for antagonists to slow-reacting substance (SRS-A) in asthmatic patients. *J. Physiol. (Lond.)*, 184:82P–83P.

9. Lewis, R. A., Austen, K. F., Drazen, J. M., Clark, D. A., Marfat, A., and Corey, E. J. (1980): Slow reacting substances of anaphylaxis: identification of leukotrienes C-1 and D from human and rat sources. *Proc. Natl. Acad. Sci. USA*, 77:3710–3714.
10. Lewis, R. A., Drazen, J. M., Austen, K. F., Clark, D. A., and Corey, E. J. (1980): Identification of the C(6)-S-conjugate of leukotriene A with cysteine as a naturally occurring slow reacting substance of anaphylaxis (SRS-A): importance of the 11-cis geometry for biological activity. *Biochem. Biophys. Res. Commun.*, 96:271–277.
11. Morris, H. R., Taylor, G. W., Piper, P. J., and Tippins, J. R. (1980): Structure of slow reacting substance of anaphylaxis from guinea pig lung. *Nature*, 285:104–106.
12. Pride, N. B. (1979): Assessment of changes in airway calibre. I. Tests of forced expiration. *Br. J. Clinc. Pharmacol.*, 8:193–203.
13. Sheard, P., and Blair, A. M. J. N. (1970): Disodium cromoglycate: activity in three in vitro models of the immediate hypersensitivity reaction in lung. *Int. Arch. Allergy*, 38:217–224.
14. Sheard, P., Killingback, P. G., and Blair, A. M. J. N. (1967): Antigen-induced release of histamine and SRS-A from human lung passively sensitized with reaginic serum. *Nature*, 216:283–284.
15. Turnbull, L. S., Turnbull, L. W., Leitch, A. G., Crofton, J. W., and Kay, A. B. (1977): Mediators of immediate type hypersensitivity in sputum from patients with chronic bronchitis and asthma. *Lancet*, 2:526–529.

Leukotrienes and Other Lipoxygenase Products,
edited by B. Samuelsson and R. Paoletti.
Raven Press, New York © 1982.

Comparative Pharmacology and Antagonism of Synthetic Leukotrienes on Airway and Vascular Smooth Muscle

John G. Gleason, Robert D. Krell, Barry M. Weichman,
Fadia E. Ali, and Barry Berkowitz

*Research and Development Division, Smith Kline and French Laboratories,
Philadelphia, Pennsylvania 19101*

Asthma is a disease of mixed etiology, involving acute bronchospasm, airway hypersensitivity, pulmonary edema, and mucous hypersecretion induced by multiple mediators [slow-reacting substance of anaphylaxis (SRS-A), histamine, prostaglandins, bradykinin, etc.] (2–4,9). In recent years, intensive study has centered on the role of SRS-A in asthma, an interest which has been further heightened by the identification of SRS-A from various species as members of the leukotriene class of arachidonate metabolites (3,8,11–13). Using synthetically derived leukotrienes, we have explored the effects of these potent bronchoconstrictive agents on guinea pig airway and vascular smooth muscle and have characterized a potent, selective antagonist, SK&F 88046.

Leukotriene A_4 (LTA$_4$) was synthesized as illustrated in Fig. 1 (7). Although the final step is nonstereospecific, the desired 9-*trans* isomer predominated (~3:1). Conversion of LTA$_4$ to LTC$_4$ and LTD$_4$ was accomplished by methods similar to those described by Corey et al. (5) and others (14). Final purification of both LTC$_4$ and LTD$_4$ by medium-pressure C_{18} reverse-phase chromatography afforded a diastereomeric mixture [5(S),6(R) and 5(R),6(S)] of leukotrienes containing less than 5% of minor double-bond isomers (11-*trans*, 9-*cis*). The synthetic leukotrienes were characterized spectroscopically and by reactivity toward soybean lipoxygenase (Sigma).

Characterization of the biological activity of the synthetic LTC$_4$ and LTD$_4$ was performed *in vitro* on isolated guinea pig airway and vascular smooth muscle and *in vivo* using spontaneously breathing anesthetized guinea pigs according to standard procedures. Isolated lung parenchymal strips were extremely sensitive to the synthetic leukotrienes with threshold concentrations for eliciting a contraction normally observed in the 10^{-12} to 10^{-10} M range. The cumulative dose response curves of LTD$_4$ and LTC$_4$ were described by EC$_{50}$ values of $6.7 \pm 2.3 \times 10^{-10}$ and $2.0 \pm 0.5 \times 10^{-8}$ M, respectively (Fig. 2). Both leukotrienes elicited the same maximal contraction, which was normally 70 to 80% of the maximal contraction of the reference agonist, histamine. The SRS-A antagonist, FPL 55712, competitively antagonized the LTD$_4$ induced contraction, but had no significant effect on

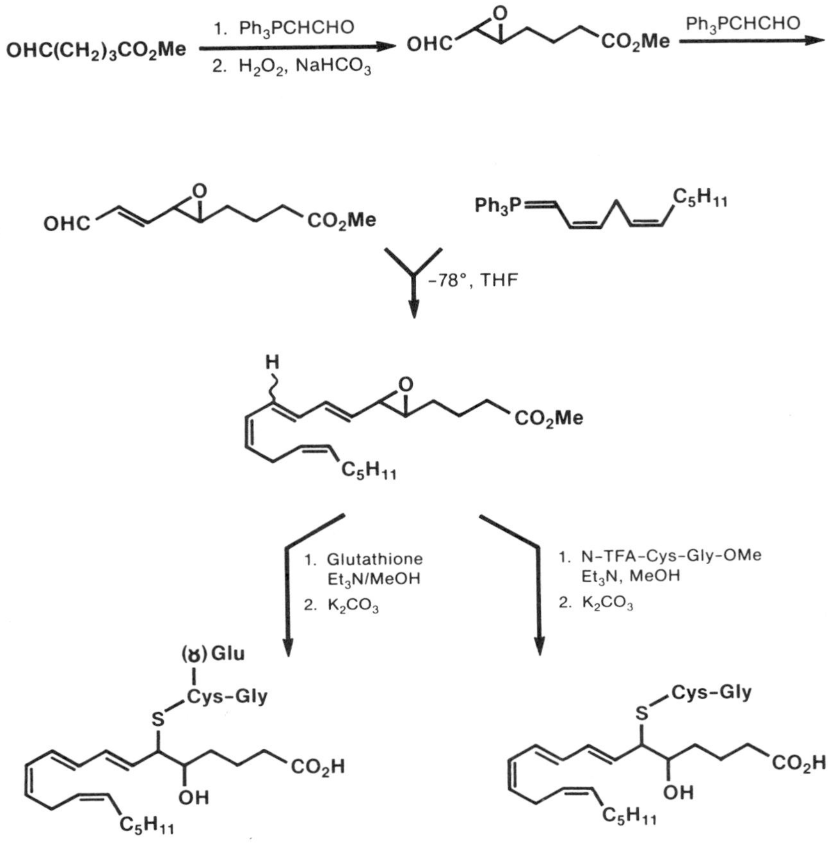

FIG. 1. Synthesis of diastereomeric LTC$_4$ and LTD$_4$.

the LTC$_4$ contraction. On trachea and bronchi, LTD$_4$ produced dose related contractions with EC$_{50}$ values of $1.7 \pm 0.3 \times 10^{-8}$ and $1.8 \pm 0.4 \times 10^{-8}$ M, respectively. However, LTC$_4$ caused a biphasic contractile response on both tissues. Whereas FPL 55712 at 1 µM was capable of shifting the LTC$_4$ dose-response curve to the right, at 10 µM, the maximal LTC$_4$-induced contraction was significantly potentiated. This potentiation by FPL 55712 was mimicked by the action of 10 µM indomethacin, an antagonist of the cyclo-oxygenase pathway. In the presence of indomethacin, 10 µM FPL 55712 was capable of antagonizing the LTC$_4$ contraction in an apparently competitive fashion. This suggested that the potentiation of the LTC$_4$ contraction by FPL 55712 may have been caused, in part, by the supplementary activity of FPL 55712 as a cyclo-oxygenase inhibitor (10).

Neither LTC$_4$ nor LTD$_4$ produced significant effects on vascular smooth muscle of guinea pig aorta. However, using the pulmonary artery, we observed that,

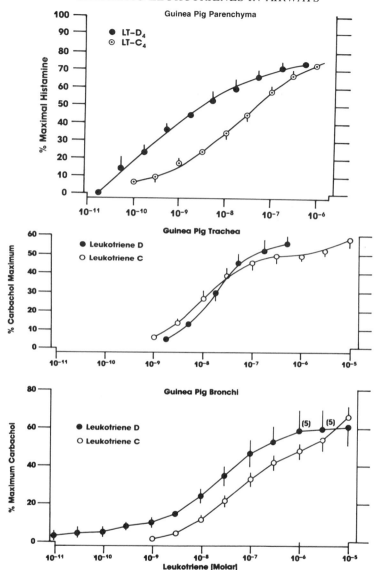

FIG. 2. Effect of LTC_4 and LTD_4 on isolated guinea pig parenchyma, trachea, and bronchi. Parenchymal strips, tracheal spiral strips, and bronchial smooth muscle rings from adult male albino Hartley strain guinea pigs (400 to 600 g) were placed in jacketed tissue baths and connected to force displacement transducers for recording isometric tension. Resting tensions were 1 g for the parenchyma and bronchi and 2 g for the trachea. The contractions are expressed relative to the maximal contraction induced by a reference agonist: histamine for the parenchyma and carbachol for the trachea and bronchi. The results are the mean ± SEM of six experiments.

although the portion of the artery proximal to the heart was unresponsive to both leukotrienes, the portion proximal to the lung produced dose-related contractions to both LTC_4 and LTD_4 (Fig. 3). Both leukotrienes were more potent than norepinephrine on this tissue, although norepinephrine (and KCl) produced similar contractions of proximal and distal sections of the artery. These results suggest a potential role for the leukotrienes on selective vascular beds.

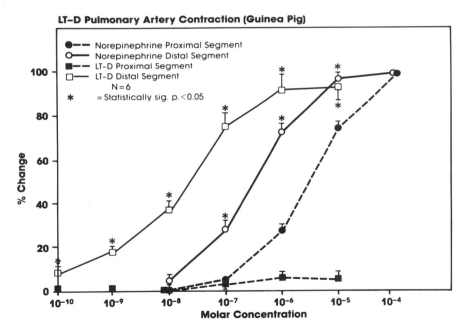

In the course of our work, we have identified a class of imidodisulfamides which appear to be selective antagonists of SRS-A (1). Representative of this class is SK&F 88046, which was equipotent with FPL 55712 against partially purified guinea pig SRS-A-induced contraction of the guinea pig ileum but appeared to be less rapidly metabolized. The effects of this antagonist on the contractile effects of LTD_4 and LTC_4 on guinea pig lung parenchymal strips are shown in Fig. 4. SK&F88046

FIG. 3. Effect of LTD_4 on guinea pig pulmonary artery. Two segments of the pulmonary artery, one proximal to the heart and the other distal, were isolated. Concentration response curves of LTD_4 and norepinephrine were run on each segment. The results are expressed as the mean ± SEM of six experiments. Results are expressed as percent maximal contraction produced by norepinephrine.

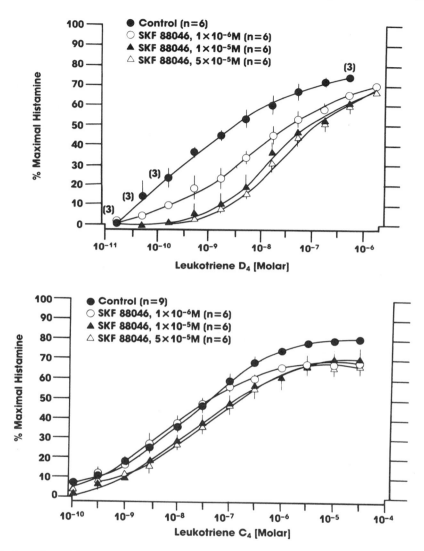

FIG. 4. Effect of SK&F 88046 on the contractile effects of LTD$_4$ and LTC$_4$ on isolated guinea pig parenchymal strips. SK&F 88046 was added to the tissue bath 30 min prior to the beginning of the cumulative concentration response curve for LTD$_4$ and LTC$_4$. All other details are identical to those of Fig. 2.

inhibited the contractile effect of LTD$_4$ in a dose-related manner at concentrations of 1 to 50 μM. However, no effect against LTC$_4$ was apparent. A similar profile of activity was observed with FPL 55712. SK&F 88046 showed no significant effect against histamine-, carbachol-, serotonin-, and KCl-induced contractions. Thus SK&F 88046 appears to be a highly specific LTD$_4$ antagonist on lung parenchyma. Moreover, the selective action of SK&F 88046 against LTD_4 and not LTC_4 suggests that the leukotrienes may be acting via two distinct receptors or sites of action.

In vivo, LTD_4 (0.3 to 10 nmole/kg) significantly increased pulmonary resistance and decreased dynamic lung compliance when injected intravenously. These effects appeared to be maximal within 30 sec of injection and were dose-related. Pretreatment of the animals with SK&F 88046 for 3 min (5 mg/kg i.v.) significantly inhibited the LTD_4-induced pulmonary changes. Representative effects of SK&F 88046 on dynamic lung compliance are shown in Fig. 5. These results demonstrate that SK&F 88046 can antagonize the action of LTD_4 *in vivo*.

CONCLUSION

These results indicate that SK&F 88046 is a potent and selective antagonist of LTD_4 both *in vitro* and *in vivo*. It is of interest that in isolated lung parenchymal strips SK&F 88046 antagonized the effects of LTD_4 but had no effect against the LTC_4-induced contraction. A similar profile of activity was also noted for FPL 55712. This observation suggests that leukotriene-mediated contraction of the pa-

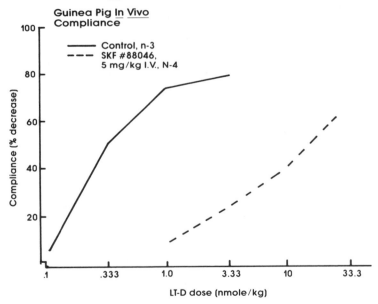

FIG. 5. Antagonism of LTD_4 by SK&F 88046 in guinea pigs *in vivo*. Spontaneously breathing male guinea pigs, anesthetized by urethane, were set up for measurement of changes in intraesophageal pressure and air flow rate from which the dynamic lung compliance was calculated. Results are expressed as percent change from predrug values. LTD_4 was injected via the jugular vein at 10 to 15-min intervals between doses. Animals receiving SK&F 88046 were pretreated (5 mg/kg i.v.) 3 min prior to the first dose of LTD_4. A similar antagonism of the LTD_4-induced increase in pulmonary resistance by SK&F 88046 was also observed.

renchymal strips may occur via two distinct sites of action. Alternatively, the possibility of allosteric interactions by these antagonists resulting in the ability of a leukotriene receptor to discriminate between LTC_4 and LTD_4 cannot be discounted. With respect to the cardiovascular activity of the leukotrienes, it is noteworthy that both LTD_4 and LTC_4 potently induced contractions of the pulmonary artery proximal to the lungs.

The exact role of the leukotrienes in the pathophysiology of asthma remains to be determined. In addition, studies are just beginning to determine if this class of compounds plays an important role in other disease states involving the cardiovascular or immunological systems. It is significant that SK&F 88046 showed potent *in vivo* antagonist activity against LTD_4. Further studies are needed to define the utility of this interesting agent.

ACKNOWLEDGMENTS

We would like to acknowledge the excellent technical and scientific contributions of Charles Kinzig, Karl Erhard, and Deborah Bryan for the preparation of the synthetic leukotrienes, and Ruth Osborn, Katherine Falcone, Margaret O'Donnell, and Bogdan Zabko-Potapovich for the biological studies described.

REFERENCES

1. Ali, F. E., Dandridge, P. A., Gleason, J. G., Krell, R. D., Kruse, C. H., Lavanchy, P. G., and Snader, K. M. (1981): *J. Med. Chem., (to be published).*
2. Austen, K. F., and Orange, R. P. (1975): Bronchial asthma: The possible role of the chemical mediators of immediate hypersensitivity in the pathogenesis of subacute chronic disease. *Am. Rev. Resp. Dis.*, 112:423–436.
3. Borgeat, P., and Sirois, P. (1981): Leukotrienes: A major step in the understanding of immediate hypersensitivity reactions. *J. Med. Chem.*, 24:121–126.
4. Chakrin, L. W., and Krell, R. D. (1980): Pathophysiology and pharmacotherapy of asthma: An overview. *J. Pharmaceut. Sci.*, 69:236–238.
5. Corey, E. J., Clark, D. A., Goto, G., Marfat, A., and Mioskowski, C. (1980): Stereospecific total synthesis of a slow-reacting substance of anaphylaxis, leukotriene C-1. *J. Am. Chem. Soc.*, 102:1436–1439.
6. Drazen, J. M., Lewis, R. A., Wasserman, S. I., Orange, R. P., and Austen, K. F. (1979): Differential effects of a partially purified preparation of SRS-A on guinea pig tracheal spirals and parenchymal strips. *J. Clin. Invest.*, 63:1–5.
7. Gleason, J. G., Bryan, D. B., and Kinzig, C. M. (1980): Convergent synthesis of leukotriene a methyl ester. *Tetrahedron Lett.*, 1129–1132.
8. Hammarstrom, S., Murphy, R. C., Samuelsson, B., Clark, D. A., Mioskowski, C., and Corey, F. J. (1979): Structure of leukotriene C—Identification of the amino acid part. *Biochem. Biophys. Res. Commun.*, 91:1266–1272.
9. Hogg, J. C., Paré, P. D., Boucher, R. C., and Michoud, M. C. (1979): The pathophysiology of asthma. *Can. Med. Assoc. J.*, 121:409–414.
10. Krell, R. D., Osborn, R., Falcone, K., and Vickey, L. (1981): *Prostaglandins (submitted for publication).*
11. Krell, R. D., O'Donnell, M., Osborn, R., Falcone, K., Vickery, L., Grous, M., Kinzig, C., Bryan, D., and Gleason, J. G. (1981): Contraction of isolated airway smooth muscle by leukotriene C_4 and its antagonism by FPL-55712. *Fed. Proc.*, 40:2586A.
12. Lewis, R. A., Austen, K. F., Drazen, J. M., Clark, D. A., Marfat, A., and Corey, E. J. (1980): Slow-reacting substance of anaphylaxis: Identification of leukotriene C-1 and D-1 from human and rat sources. *Proc. Natl. Acad. Sci. USA*, 77:3710.

13. Morris, H. R., Taylor, G. W., Piper, P. J., and Tippins, J. R. (1980): Structure of SRS-A from guinea pig lung. *Nature*, 285:104–106.
14. Rokach, J., Girard, Y., Guindon, Y., Atkinson, J. G., Larue, M., Young, R. M., Masson, P., and Holme, G. (1980): The synthesis of a leukotriene with SRS-like activity. *Tetrahedron Lett.*, 1485.

Leukotrienes and Other Lipoxygenase Products,
edited by B. Samuelsson and R. Paoletti.
Raven Press, New York © 1982.

Identification of Leukotrienes in the Sputum of Patients With Cystic Fibrosis

O. Cromwell, M. J. Walport, *G. W. Taylor, *H. R. Morris,
B. R. C. O'Driscoll, and A. B. Kay

*Department of Clinical Immunology, Cardiothoracic Institute, London, SW3 6HP
England; and *Department of Biochemistry, Imperial College, South Kensington,
London, SW7 England*

Cystic fibrosis (CF) is an hereditary disease characterized by gastrointestinal malabsorption, a high sweat sodium concentration, and severe airways obstruction. The disease is associated with the production of excessive quantities of mucus, complicated by severe bronchopulmonary infection that results in the production of copious purulent sputum. In approximately 50% of patients the airways obstruction is partially reversible by bronchodilators such as the β_2 agonists. The disease is also associated with a high incidence of atopy with postive immediate-type skin reactions to common allergens occurring in some 50% of patients, compared to 10 to 20% in the normal population (10).

Sputum from CF patients characteristically contains large numbers of neutrophils and macrophage-like cells which are potential sources of leukotrienes. In the atopic patients these mediators may be released by immunoglobulin E (IgE)-dependent mechanisms. LTC_4 and LTD_4 have been identified as the principal components of slow-reacting substance of anaphylaxis (SRS-A) (13,17,19). *In vitro*, LTC_4 and LTD_4 contract smooth muscle of the human bronchus (4, 9), lung parenchyma, pulmonary vein, and pulmonary artery (9). Other lipoxygenase products, in particular LTB_4 and to a lesser extent the monohydroxyeicosatetraenoic acids (mono-HETEs) are potent chemotractants for human granulocytes (6, 18). Neutrophils also release LTB_4 and mono-HETEs in response to a variety of stimuli including the calcium ionophore A23187 (2,6), and elevated concentrations of LTB_4 have been described in synovial fluid from rheumatoid arthritis patients (12). Furthermore, mono-HETEs promote mucus secretion *in vitro* from cultured human airways (14). These observations suggest that leukotrienes and other lipoxygenase products might be important mediators of acute inflammatory reactions, including those in the bronchi, and that in diseases such as CF it might be possible to identify them in sputum.

PATIENTS AND METHODS

Twenty-five patients with CF were studied. The mean age of the group was 23.9 years (range 15 to 42), and there were 15 females and 10 males. All patients had positive sweat tests (sodium >70 mEq.L^{-1} (7) compared to 10 to 30 mEq.L^{-1} for normals) and produced purulent sputum. They had severe airways obstruction as assessed by a forced expiratory volume in 1 sec (FEV$_1$) of less than 50% of the predicted value (except in two cases). Skin prick tests were performed with mixed pollens, cat fur, house dust, and the house dust mite (*Dermatophagoides pteronyssinus*), and *Aspergillus fumigatus*. Nine of the patients were skin-test-positive to one or more of these allergens.

Sputum samples were collected directly into 80% ethanol. After homogenization and adjustment of the final ethanol concentration to 80%, insoluble material was removed by centrifugation, and the supernatants were evaporated under vacuum. Further purification was achieved by adsorption to Amberlite XAD-8 (Rohm and Hass, Croydon, England) at pH 7.0, washing twice with deionized water, and elution with 80% ethanol and, where indicated, by Sephadex G-15 chromatography in ammonia:methanol:water (1:2:2) (15).

Extracts were reconstituted in Tyrode's solution and assayed on the isolated guinea pig ileum in the presence of atropine and mepyramine maleate (23,24). Synthetic LTD$_4$ was used as a standard. Inhibition of SRS-induced ileal contractions was achieved with the antagonist FPL 55712. Ethanol extracts containing FPL 55712-inhibitable activity were further purified by Amberlite XAD-8 chromatography and analyzed by reverse-phase high-pressure liquid chromatography (RP-HPLC) using either an isocratic system with a 5μm Nucleosil C$_{18}$ column (250 × 4.6 mm) (Jones Chromatography, Glamorgan, England) in methanol:water:acetic acid (70:30:0.1) pH 5.4 (19), or a gradient of *n*-propanol in 5% acetic acid over a μBondapak C$_{18}$ column (16). Fractions were assayed for FPL 55712-inhibitable smooth-muscle-contracting activity on guinea pig ileum, and for chemotactic activity for normal human neutrophils using a modified Boyden chamber technique (11) with 8-μm Sartorius membrane filters (V.A. Howe, London). Synthetic LTC$_4$ and LTD$_4$, and LTB$_4$ derived from human polymorphonuclear (PMN) leukocytes, were used as markers.

RESULTS

Ethanol extracts of sputum samples from 16 of 25 patients contained an activity which gave a characteristic SRS-A-like contraction of guinea pig ileum which was inhibitable, in a dose-dependent fashion, by the antagonist FPL 55712 (Fig. 1). In general, the degree of FPL 55712 inhibition of the contractions achieved by sputum extracts was less than that for comparable contractions induced by synthetic LTD$_4$. However, purification of LTD$_4$ from sputum by HPLC resulted in a preparation which was indistinguishable from synthetic LTD$_4$ in terms of its action on the ileum and inhibition by FPL 55712. There was no clear relationship between "SRS-A-like" activity in CF sputum and the atopic status.

Samples from 12 of 25 extracts were applied to columns of Sephadex G-15 which had been calibrated with synthetic LTD$_4$, phenol red, and histamine. Two peaks

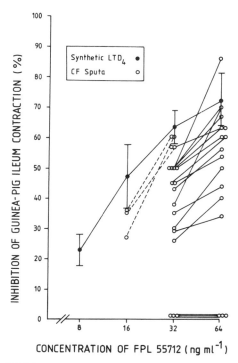

FIG. 1. Effect of FPL 55712 on contraction of the atropinized guinea pig ileum by CF sputum extracts. Synthetic LTD_4 served as a reference, and the points represent means ± SD of 6, 12, 12, and 12 estimations for FPL 55712, 8, 16, 32, and 64 ng ml^{-1}, respectively. Inhibition was with 16 and 32 ng ml^{-1} (o --- o) or 32 and 64 ng ml^{-1} (o ——— o).

of ileum-contracting activity could be identified. One was FPL 55712-inhibitable and eluted in the same volume as synthetic LTD_4. The second peak eluted slightly later but was not inhibitable by FPL 55712 (Fig. 2). The pattern of ileal contraction was slightly different from that of the first peak in that the onset was more rapid and less sustained. Of the 16 samples with FPL 55712-inhibitable activity, eight were applied to Sephadex G-15 columns; they gave two peaks of activity, the LTD_4 peak predominating. Of the nine samples with noninhibitable activity, four were applied to columns. One had FPL 55712-inhibitable and noninhibitable activities as described above, whereas three had only the FPL 55712 noninhibitable activity.

Eight samples were further purified by RP-HPLC. Each contained a peak of FPL 55712-inhibitable activity which absorbed in the ultraviolet (UV) spectrum at 280 nm and eluted with a retention time identical to that of LTD_4; two of the samples also contained activity consistent with LTC_4 (Fig. 3). A peak of neutrophil chemotactic activity was also observed and eluted with a retention time identical to that of purified LTB_4. In addition, there was a more-polar compound which gave a second peak of chemotactic activity in two of the samples (Fig. 3).

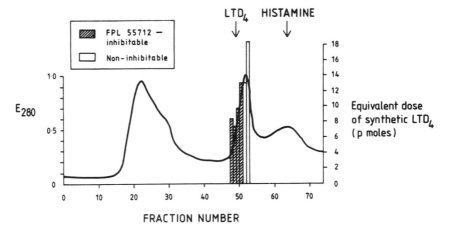

FIG. 2. Sephadex G-15 chromatography of an ethanol extract of CF sputum. Representative of nine experiments (see text).

DISCUSSION

Sputum from CF patients has been shown to contain a "slow-reacting substance" which, by HPLC, appears to be composed largely of LTD_4 (Fig. 3). Small amounts of LTC_4 were found in two samples, but LTE_4 was not detected. In addition, LTB_4 has been identified by HPLC and biologically using an assay of neutrophil chemotaxis (Fig. 3). A second peak of chemotactic activity attributable to a more-polar compound than LTB_4 was also observed. Its identity has not yet been established, but a chemotactic agent with a similar HPLC retention time has been described in rat peritoneal macrophages (5).

In 16 of the 25 extracts tested, a large proportion of the ileum-contracting activity was inhibitable by FPL 55712 at concentrations previously found to antagonize the effects of SRS-A (1). The degree of inhibition was usually less than that observed for synthetic LTD_4. This may be attributable to other agonists in the extract, such as $PGF_{2\alpha}$, or to the combination of other cyclo-oxygenase products and LTB_4, as these mixtures have previously been shown to have smooth-muscle-contracting activity (22). This might also explain the second peak of ileum-contracting activity from the Sephadex G-15 column that was not inhibitable by FPL 55712 (Fig. 2).

The isocratic system employed in HPLC (19) has the advantage over the gradient elution system (16) in that it gives better resolution of LTC_4 and LTD_4. The system was able to detect 5 ng of LTD_4 as assessed by UV absorbance, although smaller amounts were detected by bioassay of the appropriate fractions collected from the column (Fig. 3).

Although it has been possible to identify lipoxygenase products in sputum, the formidable problems associated with their quantitation have not yet been overcome, and consequently a correlation between the levels of these compounds and the severity of lung disease cannot be made at the present time.

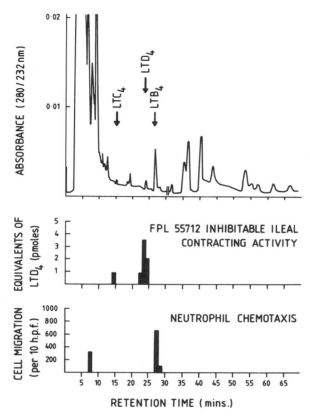

FIG. 3. Reverse-phase isocratic HPLC (see text) of an extract of CF sputum prepared by ethanol extraction and Amberlite XAD-8 chromatography. Flow rate was 1 ml min⁻¹, and 1-ml fractions were assayed for FPL 55712-inhibitable guinea pig ileum contracting activity and neutrophil chemotactic activity (see text).

Sputum is a complex mixture of various fluids (e.g., mucous and serous secretions), as well as a wide variety of cell types in various stages of disintegration. The numbers and types of cells present in sputum can often be a guide to the severity and nature of the underlying inflammatory process. Analysis of sputum mediators, as well as the presence of eosinophils, might reflect local immediate-type hypersensitivity. For instance, elevated sputum concentrations of histamine, SRS-A, and IgE have been found to be associated with chronic bronchitis and atopic asthma (23). Although atopy is a feature of CF, LTD_4 was present in many of the samples from skin-test-negative patients. This suggests that the mast cell is unlikely to be the primary source of the leukotrienes in CF sputum. It seems more likely that they are derived from inflammatory cells (e.g., PMN leukocytes and mononuclear cells) present in abundance in CF sputum.

Abnormalities in calcium metabolism are also a feature of CF, and this might be related to the release of lipoxygenase products in inflammatory cells. For instance, calcium is found in high concentrations in the glycoprotein-rich secretions of CF patients (8), and it is thought that, in general, there is increased intracellular levels of calcium in this disease (21). It is possible that this high content of cytoplasmic calcium might activate calcium-dependent phospholipase, which in turn leads to release of arachidonic acid in a fashion analogous to the action of the calcium ionophore in generating lipoxygenase products (2).

Another potential source of lipoxygenase products, in pulmonary exudates, is the alveolar macrophage, which generates cyclo-oxygenase and lipoxygenase products following phagocytosis (3,20). These cells might also be relevant to the present observation of elevated leukotriene levels in the sputum in CF.

Thus lipoxygenase products are potentially important mediators in CF and other lung diseases. Our observations might have therapeutic implications as drugs are now available which can either inhibit arachidonic acid metabolism or antagonize certain lipoxygenase products.

SUMMARY

Sputum extracts from 16 of 25 adult patients with CF produced a slow, sustained contraction of isolated guinea pig ileum which was partially inhibitable by the SRS-A antagonist FPL 55712. Following progressive purification and HPLC, the major ileum-contracting substance was identified as LTD_4. LTC_4 was also present but in smaller amounts. The sputum contained other SRS-A-like activity which was not inhibited by FPL 55712; this substance was of a smaller molecular size than LTD_4. The presence of LTB_4 was also established using HPLC and an assay of neutrophil chemotaxis. These studies raise the possibility that leukotrienes and other lipoxygenase products may contribute to the progressive lung disease characteristic of CF.

ACKNOWLEDGMENTS

LTC_4 and LTD_4 were gifts from Merck Frosst Laboratories, Quebec, Canada, and LTB_4 was a gift from Dr. A. W. Ford-Hutchinson. Fisons Laboratories, Ltd., Loughborough, kindly provided FPL 55712. The excellent technical assistance of Miss Kaye Cussen is acknowledged. The work was supported by grants from the Asthma Research Council, the Cystic Fibrosis Research Trust, and the Clinical Research Committee, Brompton Hospital, London.

REFERENCES

1. Augstein, J., Farmer, J. B., Lee, T. B., Sheard, P., and Tattersall, M. L. (1973): Selective inhibitor of slow reacting substance of anaphylaxis. *Nature [New Biol.]*, 245:215–217.
2. Borgeat, P., and Samuelsson, B. (1979): Arachidonic acid metabolism in polymorphonuclear leukocytes: effects of ionophore A23187. *Proc. Natl. Acad. Sci. USA*, 76:2148–2152.
3. Bretz, U., Dewald, B., Payne, T., and Schnyder, J. (1981): Phagocytosis stimulates the release of slow reacting substance (SRS) in cultured macrophages. *Br. J. Pharmacol. (in press)*.

4. Dahlén, S.-E., Hedqvist, P., Hammarström, S., and Samuelsson, B. (1980): Leukotrienes are potent constrictors of human bronchi. *Nature*, 288:484–486.
5. Doig, M. V. and Ford-Hutchinson, A. W. (1980): The production and characterisation of products of the lipoxygenase enzyme system released by rat peritoneal macrophages. *Prostaglandins*, 20:1007–1019.
6. Ford-Hutchinson, A. W., Bray, M. A., Doig, M. V., Shipley, M. E., and Smith, M. J. H. (1980): Leukotriene B, a potent chemokinetic and aggregating substance released from polymorphonuclear leukocytes. *Nature*, 286:264–265.
7. Gibson, L. E., and Cooke, R. E. (1959): A test for concentration of electrolytes in sweat in cystic fibrosis of the pancreas utilising pilocarpine by iontophoresis. *Paediatrics*, 29:545–549.
8. Gibson, L. E., Matthews, W. J., Minhan, P. T., and Patti, J. A. (1971): Relating mucous calcium and sweat in a new concept of cystic fibrosis. *Pediatrics*, 48:695–710.
9. Hanna, C. J., Bach, M. K., Pare, P. D., and Schellenberg, R. R. (1981): Slow reacting substances (leukotrienes) contract human airway and pulmonary vascular smooth muscle in vitro. *Nature*, 290:343–344.
10. Hodson, M. E. (1980): Immunological abnormalities in cystic fibrosis: chicken or egg? *Thorax*, 35:801–806.
11. Kay, A. B. (1970): Studies on eosinophil leucocyte migration. II. Factors specifically chemotactic for eosinophils and neutrophils generated from guinea pig serum by antigen-antibody complexes. *Clin. Exp. Immunol.*, 7:723–737.
12. Klickstein, L. B., Shapleigh, C., and Goetzl, E. J. (1980): Lipoxygenation of arachidonic acid as a source of polymorphonuclear leukocyte chemotactic factors in synovial fluid and tissue in rheumatoid arthritis and spondylo-arthritis. *J. Clin. Invest.*, 66:1166–1170.
13. Lewis, R. A., Austen, K. F., Drazen, J. M., Clark, D. A., Marfat, A., and Corey, E. J. (1980): Slow reacting substance of anaphylaxis: identification of leukotrienes C-1 and D from human and rat sources. *Proc. Natl. Acad. Sci. USA*, 77:3710–3714.
14. Marom, Z., Shelhamer, J. H., Sun, F., and Kaliner, M. (1981): The effects of arachinoids on mucus secretion from human airways. *J. Allergy Clin. Immunol.*, (*in press*).
15. Morris, H. R., Taylor, G. W., Piper, P. J., Sirois, P., and Tippins, J. R. (1978): Slow reacting substance of anaphylaxis:purification and characterisation. *FEBS Lett.*, 87:203–206.
16. Morris, H. R., Taylor, G. W., Piper, P. J., and Tippins, J. R. (1979): Slow reacting substance of anaphylaxis: studies on purification and characterisation. *Agents Actions (Suppl. 6)*, pp. 27–36.
17. Morris, H. R., Taylor, G. W., Piper, P. J., and Tippins, J. R. (1980): Structure of slow reacting substance of anaphylaxis from guinea pig lung. *Nature*, 285:104–106.
18. Nagy, L., Lee, T. H., Goetzl, E. J., Pickett, W. C., and Kay, A. B. (1981): Complement receptor enhancement and chemotaxis of human neutrophils and eosinophils by leukotrienes and other lipoxygenase products. *Clin. Exp. Immunol. (in press)*.
19. Orning, L., Hammarström, S., and Samuelsson, B. (1980): Leukotriene D: a slow reacting substance from rat basophilic leukemia cells. *Proc. Natl. Acad. Sci. USA*, 77:2014–2017.
20. Rouzer, C. A., Scott, W. A., Cohn, Z. A., Blackburn, P., and Manning, J. M. (1980): Mouse peritoneal macrophages release leukotriene C in response to phagocytic stimuli. *Proc. Natl. Acad. Sci. USA*, 77:4928–4932.
21. Shapiro, B. L., Feigal, R. J., and Lam, L. F. H. (1980): Intracellular calcium and cystic fibrosis. In: *Perspectives in Cystic Fibrosis*, edited by J. M. Sturgess, pp. 15–28. Proceedings of 8th International Cystic Fibrosis Congress, Toronto, Canada.
22. Sirois, P., Borgeat, P., Jeanson, A., Roy, S., and Girard, G. (1980): The action of leukotriene B_4 (LTB$_4$) on the lung. *Prostaglandins Med.*, 5:429–444.
23. Turnbull, L. S., Turnbull, L. W., Leitch, A. G., Crofton, J. W., and Kay, A. B. (1977): Mediators of immediate-type hypersensitivity in sputum from patients with chronic bronchitis and asthma. *Lancet*, 2:526–529.
24. Turnbull, L. W., Turnbull, L. S., Crofton, J. W., and Kay, A. B. (1979): Variations in chemical mediators of hypersensitivity in the sputum of chronic bronchitics: correlations with peak expiratory flow. *Lancet*, 2:184–186.

Leukotrienes and Other Lipoxygenase Products,
edited by B. Samuelsson and R. Paoletti.
Raven Press, New York © 1982.

Stimulus-Secretion Coupling in the Human Neutrophil: The Role of Phosphatidic Acid and Oxidized Fatty Acids in the Translocation of Calcium

Gerald Weissmann, Charles Serhan, James E. Smolen,
Helen M. Korchak, Robert Friedman, and Howard B. Kaplan

*Division of Rheumatology, Department of Medicine, New York University
School of Medicine, New York, New York 10016*

Neutrophils are phagocytic cells displaying a variety of specific membrane receptors which recognize stimuli such as the chemoattractants f-Met-Leu-Phe and the complement-derived peptide C5a, opsonized particles, immune complexes, concanavalin A, or the tumor promoter phorbol myristate acetate (PMA) (32). Following ligand-receptor interactions, the cell undergoes a sequence of reactions which has been characterized as stimulus-secretion coupling (19,30). We and other investigators have suggested that transient increments in the intracellular activity of free calcium ions may serve as a second messenger in the coupling sequence of neutrophils, as they do in several other types of secretory cell (11,12,21,27). The neutrophil responds to ligand-receptor interactions by generating superoxide anions (O_2^-), secreting lysosomal enzymes (e.g., β-glucuronidase and lysozyme), and remodeling its membrane phospholipids which are the source for the release and oxidation of membrane arachidonic acid (19, 30, 32). They also aggregate (22), a reflection of their capacity to stick to each other and to surfaces such as endothelium.

What are the major oxidation products of arachidonate formed by human neutrophils? Work from several laboratories suggests that the sum of products formed by the lipoxgenase pathway significantly exceeds that formed via the cyclo-oxygenase pathway (Table 1). The major products formed by the human neutrophil via lipoxygenase are 5-hydroxyeicosatetraenoic acid (5-HETE), which can be reincorporated in membrane lipids, and leukotriene B_4 (LTB$_4$) or 5,12-dihydroxyeicosatetraenoic acid (5,12-DHETE) (4,9,31). By the cyclo-oxygenase pathway, prostaglandin E_2 (PGE$_2$) and thromboxane B_2 (TXB$_2$) are the major products of the neutrophil (13,31). Although neutrophils respond to most lipoxygenase products and to TXB$_2$ by undergoing chemotaxis (10), the complete stimulus-secretion coupling sequence is launched only by 5,12-DHETE (10). Consequently, it has been

259

TABLE 1. *Arachidonic acid oxidation products formed by human neutrophils*

Lipoxygenase products	Cyclo-oxygenase products
5-HETE (4,9,31)[a]	PGE$_2$ (31)
5-HPETE (9)	Thromboxane B$_2$ (13,31)
11-HETE (9)	
12-HETE (31)	
15-HETE (4)	
5,12-DHETE and isomers (4,9)	

[a]Reference numbers.

postulated that products derived from arachidonate, once released extracellularly, function mainly as chemoattractants for adjacent neutrophils. They may also play yet other important modulatory roles within cells, or they may bind to receptors on neutrophils, macrophages, or lymphocytes. Thus, for example, prostaglandins of the E series, as first shown a decade ago (38), and PGI$_2$ (36) inhibit the activation of neutrophils, platelets, and lymphocytes. This chapter describes recent studies of the stimulus-secretion coupling sequence whereby these oxidation products of arachidonic acid are released from the neutrophil.

ROLES OF EXTRACELLULAR AND INTRACELLULAR CALCIUM

One of the first lines of evidence which suggested that increments in the concentration of intracellular calcium might serve as a second messenger in stimulus-secretion coupling was the observation that calcium ionophores (e.g., A23187) cause the neutrophil to generate superoxide anions, release lysosomal enzymes, and oxidize arachidonic acid (12, 27, 31). Indeed, simply adding calcium to neutrophils depleted of calcium by incubation in calcium-free media induced secretion of lysozyme (12). Such experiments have been recently supported by the finding that *any* calcium ionophore, including the water-soluble oligomeric derivative of PGB$_1$ (or PGB$_x$), also activates the neutrophil (26, 33). Thus lysosomal enzyme secretion from cytochalasin B-treated human neutrophils exposed to the ionophore A23187 or PGB$_x$ is dependent on extracellular concentrations of calcium. However, even in the absence of extracellular calcium, some lysozyme (but little β-glucuronidase) was secreted (26). This is a general secretory response of the neutrophil exposed to *any* calcium ionophore: As the level of extracellular calcium is increased, secretion of lysozyme (mainly from specific granules) and β-glucuronidase (entirely from azurophilic granules) is progressively increased (12,26).

Such experiments suggested that a rise in intracellular calcium provoked by ionophore-induced influx of extracellular calcium is a *sufficient* signal for stimulus-secretion coupling. However, responses obtained in the absence of extracellular calcium indicate that the ionophores may also be liberating calcium from the plasmalemma or cellular sites.

To examine critically whether or not extracellular calcium is necessary for the secretory responses of the neutrophil, we took the rather drastic step of exposing cells to magnesium EGTA. We studied secretion of β-glucuronidase in response to a variety of stimuli under conditions in which cells could not possibly take up exogenous calcium from the medium (29,37). For example, enzyme release in response to f-Met-Leu-Phe acting via *its* receptor or to immune complexes via a separate receptor (the Fc receptor) were only minimally inhibited. In contrast, the response to A23187 was significantly, although not completely, blocked (Table 2). These experiments indicated that influx of *extra*cellular calcium was not an absolute requirement for stimulus-secretion coupling. What, then, is the role of intracellular calcium?

We studied the effect of the chemotactic peptide f-Met-Leu-Phe and other stimuli (e.g., PMA) on the fluorescence of chlortetracycline (CTC)-loaded neutrophils. The characteristic fluorescence of CTC in cells depends, to a considerable degree, on the localization of the CTC within a lipid environment in the presence of calcium; the technique is used to monitor, indirectly, the mobilization of membrane-associated calcium. Immediately upon addition of f-Met-Leu-Phe, rapid decrements in fluorescence of CTC indicated that mobilization of the cells' membrane-associated calcium followed receptor-ligand interactions. Confirming the work of Naccache et al. (21), we have found that decreases in CTC fluorescence are regularly observed whether or not calcium is present in the extracellular medium. Indeed, decreases of CTC fluorescence were observed in response to various stimuli even when magnesium EGTA was present in the extracellular medium.

TABLE 2. *Effect of Mg-EGTA on lysosomal enzyme release and superoxide anion generation in human PMNs*

	Lysozyme release		β-Glucuronidase release		O_2^- generation	
Stimulus	Control	Plus Mg EGTA	Control	Plus Mg EGTA	Control	Plus Mg EGTA
None	8.1 ± 4.3	6.6 ± 4.5	1.9 ± 0.7	1.2 ± 0.8	2.8 ± 2.9	2.1 ± 2.7
A23187	47 ± 9	18 ± 6[a]	41 ± 12	6.5 ± 2.5[a]	23 ± 4	3.5 ± 1.0[b]
Con A	28 ± 6	20 ± 5[a]			24 ± 8	7.6 ± 4.3[b]
BSA/anti-BSA	41 ± 9	32 ± 8[a]	26 ± 12	20 ± 9[a]	24 ± 5	18 ± 7[a]
STZ	19 ± 3	14.5 ± 2.7[a]	7.8 ± 1.8	6.0 ± 1.5[b]	34 ± 6	19 ± 4[a]
PMA	19 ± 3	17 ± 4[a]	7.0 ± 5.4	4.2 ± 3.9[a]	31 ± 3	25 ± 1[a]
ZTS	27 ± 4	28 ± 5[a]	18 ± 3	20 ± 5[a]	5.8 ± 4.0	5.0 ± 4.0[a]
FMLP	45 ± 8	42 ± 8[a]	41 ± 9	36 ± 9[a]	52 ± 25	28 ± 15[a]

Data from ref. 37.
Human PMNs were preincubated with cytochalasin B (5 μg/ml), with or without Mg EGTA (5 mM), and then exposed to the indicated stimulus.
Con A = concanavalin A. BSA/anti-BSA = immune complexes of bovine serum albumin and antibody. STZ = serum-treated zymosan. PMA = phorbol myristate acetate. ZTS = zymosan-treated serum. FMLP = f-Met-Leu-Phe.
[a]$p < 0.005$ for paired samples (greater than background).
[b]$p < 0.05$ for paired samples (greater than background).

Where is the pool of membrane-associated calcium which is mobilized in the course of the neutrophil stimulation? Hoffstein has studied this question by ultrastructural, cytochemical techniques, reacting neutrophils with sodium pyroantimonate in order to demonstrate cell-associated calcium (15). The precipitates, as validated by x-ray microanalysis, were those of calcium pyroantimonate. In the phagocytosing neutrophil calcium remained visible at all areas of the plasma membrane but was lost from the internalized phagocytic vacuoles which surround opsonized, ingested particles (zymosan). Indeed, it was possible to section phagocytosing neutrophils in the process of ingesting the ligand-coated particles. Again, in areas of the plasmalemma which were not in direct contact with ingested particles, calcium pyroantimonate deposits remained undisturbed. However, in the forming phagocytic vacuole, at sites of receptor-ligand interactions, membrane calcium was regularly lost. This observation was also made when cells were exposed to stimuli which interact exclusively with the plasmalemma, such as concanavalin A-coated beads or f-Met-Leu-Phe: Calcium was again lost only from the plasmalemma.

It was possible to conclude from studies of CTC fluorescence that mobilization of membrane-associated cellular calcium is an immediate response of the cells to stimulation (21). We can now conclude from these ultrastructural, cytochemical studies that perhaps one major pool from whence this calcium derives is the plasmalemma, specifically the site of ligand-receptor interactions (15).

EFFECT OF CALCIUM ANTAGONISTS

To test the hypothesis that intracellular calcium is the second messenger in stimulus-secretion coupling, we also studied the effects of calcium antagonists. The inhibitor 8-N,N'-diethylamino octyl 3,4,5-trimethoxybenzoate hydrochloride (TMB-8) may serve as an intracellular calcium chelator, inhibiting the reversible mobilization of intracellular calcium from sarcoplasmic reticulum and platelets (6). In a reversible and noncytotoxic fashion, TMB-8 blocked lysosomal enzyme release and superoxide anion generation from neutrophils exposed to a variety of secretagogues (28,29). We also studied the effect of trifluoperazine, or stelazine (29). At low micromolar concentrations, stelazine interferes with the action of the calcium/calmodulin complex, consequently inhibiting a variety of physiologic processes in many cell types ranging from neurons to contractile cells (24). At these concentrations, stelazine blocked lysosomal enzyme secretion and superoxide anion generation from still-viable neutrophils (29,37).

Finally, we used another antagonist of calmodulin. W-7 (N-6-aminohexyl-5-chloro-1-naphthalene sulfonamide) inhibits calmodulin at a concentration of 50 to 100 μM (18,29). As shown in Fig. 1, W-7 inhibited the response of neutrophils to the chemoattractant f-Met-Leu-Phe (lysosomal enzyme release and O_2^- generation) in a dose-dependent fashion.

This series of experiments demonstrated that two chemically distinct calmodulin antagonists (W-7 and trifluoperazine) and an antagonist of the mobilization of intracellular calcium (TMB-8) inhibited neutrophil responses to two receptor-spe-

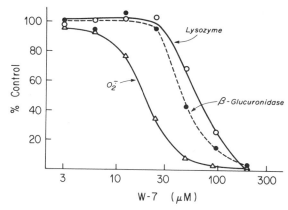

FIG. 1. Effect of W-7 on lysosomal enzyme release and superoxide anion generation by human neutrophils. Human neutrophils (5 × 10⁶/ml) were incubated for 5 min at 47°C with cytochalasin B (5 μg/ml) in the presence of increasing concentrations of W-7, a calmodulin inhibitor. Lysosomal enzyme release in response to 5 min incubation with f-Met-Leu-Phe (10^{-7} M) is expressed as the percentage of that released by cells not exposed to W-7 (36 ± 7% of lysozyme, *open circles*; and 37 ± 1% of β-glucuronidase, *closed circles*, N = 3). Superoxide anion production from f-Met-Leu-Phe-stimulated cells not exposed to W-7 was 69 ± 2 nmoles cytochrome C reduced/10⁶ cells/5 min (*open triangles*, N = 3).

cific ligands: f-Met-Leu-Phe and immune complexes. We can therefore conclude—based on cytochemical data, studies of CTC fluorescence, and data obtained with inhibitors—that it is the mobilization of intracellular calcium which is critical for stimulus-secretion coupling, although extracellular calcium, always present in physiologic fluids, may play a role in its amplification.

KINETIC ANALYSES OF NEUTROPHIL ACTIVATION:
AGGREGATION

In order to study the role of extracellular calcium, it was necessary to use additional continuous recording techniques for kinetic analysis of stimulus-secretion coupling. To this end we studied neutrophil aggregation, yet another aspect of neutrophil activation (7,22). Light and electron microscopy invariably demonstrate the aggregation of human neutrophils exposed to f-Met-Leu-Phe 30 sec after addition of the chemoattractant. Aggregation can also be monitored by means of simple nephelometry in a device used for platelet aggregometry (7,8). By this means we compared neutrophil aggregation induced by f-Met-Leu-Phe, PMA, and ionophores A23187 and PGB$_x$ (8). We found that, as f-Met-Leu-Phe causes *reversible* aggregation of the neutrophil, the aggregation curves resembled those produced by C-5-derived peptides. In contrast, two ionophores, as well as PMA, caused *irreversible* aggregation of the neutrophil and did so at concentrations at which these agents are potent secretagogues, that is, at which they provoke lysosomal enzyme release (8). We, and others, (8,22) found that the concentrations of f-Met-Leu-Phe which produce aggregation (e.g., 10^{-7} M) were higher than those at which the peptide is chemoattractant (e.g., 10^{-9} M). Since aggregation could be monitored in a continuous fashion, it became possible to analyze the kinetics of this response with respect to mobilization of *intra*cellular and uptake of *extra*cellular calcium (35). In kinetic studies, we superimposed curves of the direct recording of f-Met-Leu-Phe-induced changes in CTC fluorescence, of aggregation, and of O_2^- generation on those of lysozyme secretion of ^{45}Ca uptake obtained by rapid centrifugation techniques. The *first* (< 1 sec) response of cells to f-Met-Leu-Phe was loss of CTC fluorescence, followed by release of lysozyme (10 to 12 sec), which appeared to antecede the peak of aggregation curve (15 to 30 sec). In contrast, uptake of extracellular calcium (5 sec) significantly anteceded enzyme release and O_2^- generation. Calcium influx continued even *after* onset of lysozyme secretion and superoxide anion generation.

This experiment suggested to us that after the onset of stimulus-secretion coupling, calcium channels were open. Influx of calcium would thus augment the cell's response to a stimulus. One way to interpret these data would be to postulate that uptake of extracellular calcium follows the mobilization of cellular calcium (or, at least, CTC fluorescence decrements) which had induced new calcium channels in the plasmalemma (uptake of ^{45}Ca). One source of such calcium channels would be the generation by the cell of an endogenous calcium ionophore.

QUEST FOR AN "ENDOGENOUS" Ca IONOPHORE

We reasoned that lipid products generated at the cell membrane, perhaps those linked to the formation or oxidation of arachidonate, might act as the cell's endogenous calcium ionophore. This putative ionophore would serve in the amplification of the cell's response to a stimulus and thereby lead to extensive increments in the intracellular concentration (activity) of calcium (25).

To determine if endogenous lipids are candidates for this role, we set the following criteria for defining a calcium ionophore:

1. The agent should translocate the calcium from an aqueous phase across an intact lipid bilayer into a second aqueous compartment.

2. It should act as micromolar concentration.

3. It should be permselective with respect to the two major extracellular divalent cations, translocating calcium but not magnesium.

4. It should not lyse the bilayer.

To this end we devised a sensitive liposome assay to test the ionophoretic activity of known or putative ionophores (33). The metallochromic dye arsenazo III is readily trapped within multi- or unilamellar liposomes (34). Even when calcium is present at millimolar concentrations in the surrounding medium, arsenazo III retains its native red color (absorption maximum 550 nm), since liposomes remain impermeable to dye or calcium. When, however, lipophilic ionophores of fungal origin (such as A23187 or ionomycin) are added to or preincorporated into liposomes, calcium but not magnesium is translocated across intact bilayers to form the blue arsenazo III calcium complex (absorption maximum 656 nm). This special shift is easily monitored, and the method is sufficiently sensitive to detect nanomolar amounts of ionophores added externally to the liposome. It is also possible to generate a series of curves representing calcium translocation per micromole of membrane lipid as the molar percent preincorporated ionophore is varied. We were able to detect as little as 0.001 mole percent of ionophore A23187, e.g., 2 moles of ionophore preincorporated per 10^5 moles of membrane lipid (33). By means of this sensitive method for the detection of any putative ionophore we tested various lipids of biological origin to determine if they could translocate calcium across the intact lipid bilayers of liposomes.

A partial list of over 50 lipids tested which do *not* translocate calcium across liposomal bilayers is shown in Table 3 (25). These included phosphatidyl inositol, serine, ethanolamine and choline, sphingomyelin, and glycerol. Nor did the platelet-activating factors, acetyl glyceryl ether phosphorylcholine and various derivatives, translocate calcium. No native prostanoids of the E, I, or B series, no stable endoperoxide analog, and no hydroperoxy fatty acid, either 5 or 15, acted as a calcium ionophore in our system, except for phosphatidic acid.

PHOSPHATIDIC ACID AS A CALCIUM IONOPHORE

When a variety of secretory cells, ranging from pancreatic cells to platelets, are stimulated by ligands, a cycle is activated which, by breakdown and resynthesis of phosphatidyl inositol, transiently generates phosphatidic acid. Since elucidation of this cycle by Hokin and Hokin (16) and by Michell (20), many investigators have found that, in the course of stimulus-response coupling, such cells form diacylglycerol from phosphatidyl inositol by means of a phosphatidyl inositol-specific phospholipase C. Phosphorylation of diacylglycerol by means of a kinase and ATP generates phosphatidic acid, which later in the coupling sequence is used to regenerate phosphatidyl inositol (1–3,17,20). Indeed, some of the oxidation products

TABLE 3. *Lipids which do not translocate*
Ca^{++} across liposomal bilayers

Phospholipid and glycerol derivatives	
Phosphatidyl inositol	
Phosphatidyl serine	
Phosphatidyl ethanolamine	
Phosphatidyl choline	
Glycerol	
Platelet activating factor (AGE · phosphoryl choline and benzoyl, succinyl, maleyl derivatives)	
Prostanoids	
PGE$_1$	Endoperoxide analog I
PGE$_2$	Endoperoxide analog II
PGI$_2$	9-11 Azoprostanoid III
PGB$_1$	5-HPETE
	15-HPETE

Modified from ref. 25.

of arachidonic acid may form when arachidonic acid is generated by means of diacyl glycerol lipase (1).

We therefore studied the translocation of calcium across liposomal bilayers by known fungal ionophores as well as by phosphatidic acid. Ionomycin, A23187, and phosphatidic acid from lecithin (which contains the characteristic unsaturated fatty acids of the plasma membrane) all acted as calcium ionophores at 10 μM concentrations (25). Magnesium was not translocated, and the vesicles remained intact as judged by access to the complex of externally added EGTA. In contrast, dipalmitoyl phosphatidic acid, both acyl chains of which were saturated, was not an ionophore, nor was the probable precursor of phosphatidate, phosphatidyl inositol Fig. 2. Moreover, in data published elsewhere, we found that lecithin-derived phosphatidate was a potent ionophore when preincorporated at 1 to 5 mole percent into multilamellar vesicles.

Since phosphatidic acid, containing unsaturated fatty acids in the R2 position of the glycerol backbone, was a calcium ionophore in liposomes, we extended our studies of other possible ionophoretic lipids by analyzing a variety of native and oxidized fatty acids (25). Figure 3 shows the translocation of calcium across lipid bilayers by native and oxidized fatty acids before and after reduction by stannous chloride. Native fatty acids such as linolenic, 8,11,14-eicosatrienoic, and arachidonic acid did not translocate calcium across lipid bilayers. In contrast, when linolenic, 8,11,14-eicosatrienoic, and linoleic acids were oxidized by dimethyl sulfoxide in light, these fatty acids became potent calcium ionophores; arachidonic acid did not display this property. When the oxidized fatty acids were reduced by means of stannous chloride, their capacity to translocate calcium was lost. Moreover, treatment of known ionophores (e.g., A23187) and ionomycin with stannous chloride abolished their calcium-translocating ability.

To study the role of lipid peroxidation in this system, we next examined the relationship between reactivity of oxidized fatty acid products with thiobarbituric

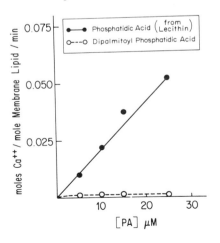

FIG. 2. Calcium uptake by multilamellar liposomes (MLV) composed of phosphatidyl choline:dicetyl phosphate:cholesterol (7:2:1) containing arsenazo III (AIII) in the presence of putative ionophores. Phosphatidic acid from ovolecithin (*closed circles*) compared with that from dipalmitoyl lecithin (*open circles*) added at increasing concentrations in DMSO. Ca translocation is expressed as micromoles of Ca translocated/micromole liposomal lipid versus standards of Ca·AIII₂ complexes. See refs. 25 and 33 for details.

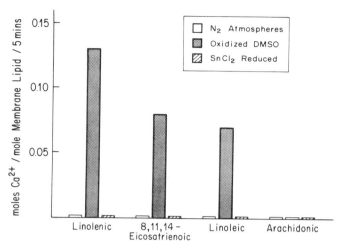

FIG. 3. Translocation of Ca across bilayers of multilamellar liposomes containing arsenazo III by unsaturated fatty acids. Fatty acids, added at 10 μM in ethanol, were exposed to a nitrogen atmosphere (*open bars*) permitted to stand in DMSO for 72 hr in sunlight (*cross-hatched bars*) or reduced by stannous chloride (*shaded bars*) at 10 mM for 4 min. Ca translocation was measured as in refs. 25 and 34.

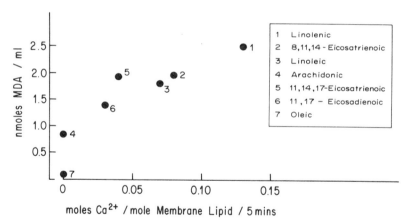

FIG. 4. Relationship between reactivity with thiobarbituric acid reactivity and Ca-translocating ability of unsaturated fatty acids. Fatty acids, oxidized for 72 hr in sunlight in DMSO, were added at 10 μM concentrations to liposomes containing arsenazo III and other aliquots reacted with thiobarbituric acid. Lipid oxidation is expressed as malondialdehyde equivalents, which was used as a standard. See refs. 25 and 33 for details.

acid (TBA) and their capacity to translocate calcium across liposomes (Fig. 4). Whereas we found a correlation of 0.9 between the extent of fatty acid oxidation as measured by TBA reactivity and the ability of a given oxidation product to translocate calcium, the structural basis for this capacity with respect to native, unoxidized fatty acid remained unclear. However, these studies (in conjunction with the stannous chloride reduction data) support the hypothesis that an appropriate configuration of oxygen atoms is required for the generation of a calcium-binding site in an oxidized fatty acid. Thus we conclude that native fatty acids are not ionophores, as previously suggested by other investigators, and that calcium translocation is not therefore a general property of amphipathic lipids (14).

PHOSPHATIDIC ACID FORMATION IN STIMULATED NEUTROPHILS

Nevertheless, phosphatidic acid remained the most viable candidate as an endogenous calcium ionophore. We therefore returned to living cells and asked the question: Is phosphatidic acid generated early enough and in sufficient quantities to account for calcium fluxes in stimulated neutrophils? More to the point: do these changes antecede such neutrophil responses as aggregation, lysosomal enzyme release, or superoxide anion generation?

Kinetic analysis was performed on neutrophil activation by 10^{-7} M f-Met-Leu-Phe (Fig. 5). At the earliest time point measurable, 5 sec, significant amounts of phosphatidic acid were formed. This phospholipid was barely detectable in unstimulated cells (not shown). Uptake of ^{45}Ca from the extracellular medium showed similar initial kinetics. The generation of phosphatidic acid and Ca influx significantly anteceded the later events of neutrophil activation (e.g., superoxide anion generation, which did not begin until 12 sec after addition of the stimulus). Moreover, the phosphatidic acid generated (e.g., 1.2 nmoles/35 million cells) at 15 sec was calculated to form at least 0.5% of the total phospholipid.

If phosphatidic acid (PA) is formed by stimulated cells at the expense of its precursor phosphatidyl inositol (PI), then decrements in the precursor should also parallel ^{45}Ca influx.

Figure 6 illustrates the time course of PI breakdown and resynthesis, together with ^{45}Ca uptake in f-Met-Leu-Phe-stimulated neutrophils. Again, at the earliest time period studied, 5 sec, PI breakdown was already profound as cells took up

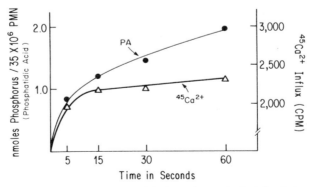

FIG. 5. Time course of phosphatidic acid (PA) formation and ^{45}Ca influx in human neutrophils after addition of 10^{-7} M f-Met-Leu-Phe. Phosphatidic acid was determined by two-dimensional thin-layer chromatography according to the method of Broekman et al. (5). ^{45}Ca uptake was determined by rapid centrifugation techniques as in ref. 19.

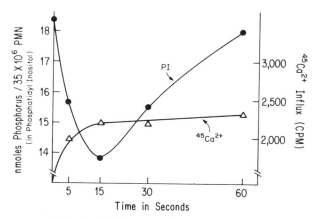

FIG. 6. Time course of phosphatidyl inositol (PI) breakdown and resynthesis, and ^{45}Ca influx, in human neutrophils after addition of 10^{-7} M f-Met-Leu-Phe. Phosphatidyl inositol was determined by two-dimensional thin-layer chromatography according to the method of Broekman et al. (5). ^{45}Ca uptake was determined by rapid centrifugation techniques as in ref. 19.

Ca from the medium. PI breakdown and calcium influx significantly anteceded other neutrophil responses such as generation of superoxide anion.

PA and PI were determined by resolution of individual lipids, utilizing two-dimensional thin-layer chromatography and phosphorus analysis of the isolated lipid spots as described by Broekman et al. (5). It should be noted that this method, in contrast to other studies which have followed the fate of radiolabeled precursors (1–3,16,17,20), measured *total* quantities of endogenous phospholipids and is independent of changes in pools or compartmentalization of labeled lipids.

We found that PI levels returned to baseline at a point when phosphatidate was still elevated (Fig. 5). Moreover, not all of the PI degraded or resynthesized could be accounted for by the phosphatidic acid formed. Thus by 15 sec 4 nmoles of PI had been lost per 35 million neutrophils at a time when only 1.2 nmoles of PA were formed. Were diacyl gylcerol formed from PI by means of the PI-specific phospholipase C—as described in platelets by Rittenhouse-Simmons (23) and Majerus' group (1)— this intermediate could be a major source of arachidonate in the neutrophil. The fate of the PI unaccounted for by transformation to PA is currently under study.

We next examined the temporal relationship between accumulation of phosphatidic acid and other parameters (simultaneously determined) of neutrophil activation. In Fig. 7 it is clear that the first measurable response of the cell to f-Met-Leu-Phe is a decrease in CTC fluorescence, which we interpret as an indirect reflection of the mobilization of membrane-bound intracellular calcium. Five seconds after addition of the stimulus—at a point where neutrophils have not yet aggregated—sufficient quantities of phosphatidic acid are already formed and calcium-45 uptake from the medium is evident. Only after a latent period of about 12 sec do the cells release lysosomal enzymes, aggregate, and generate superoxide anion. The for-

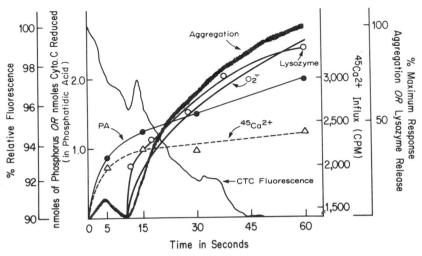

FIG. 7. Kinetic analysis of neutrophils activated by f-Met-Leu-Phe (10^{-7} M). Chlortetracycline (CTC) fluorescence was used as an indirect measure of the mobilization of intracellular calcium determined by Naccache et al. (21); aggregation (7,22) and O_2^- generation (30) were continuously recorded. Release of lysozyme, a measure of degranulation, formation of phosphatidic acid, and ^{45}Ca uptake were determined as in refs. 19, 30, and 32.

mation of phosphatidic acid and calcium influx significantly anteceded these later events.

SUMMARY

Neutrophils, stimulated via their surface receptors by ligands such as f-Met-Leu-Phe or immune complexes, release membrane calcium from intracellular stores and accumulate calcium from the extracellular medium. Both mechanisms provide for a rise in the intracellular concentration of free calcium which appears to be required for later responses of the cell: release of lysosomal enzymes and the generation of superoxide anion as neutrophils undergo an aggregation response. Exogenous ionophores, such as A23187 and PGB_x, which move calcium bidirectionally across lipid bilayers, bypass the ligand-receptor step, thereby providing the necessary rise in intracellular calcium. In the course of the remodeling of membrane lipids—within 5 sec after their exposure to a stimulus—neutrophils generate phosphatidic acid, at least in part at the expense of phosphatidyl inositol. Because phosphatidic acid is the only phospholipid whose action in model membranes mimics the action of exogenous ionophores (as it is formed early enough and in sufficient quantities to account for uptake of extracellular calcium), we propose phosphatidic acid as a promising candidate for the role of endogenous ionophore. This ionophoretic action may serve to amplify signals launched by primary changes at the plasma membrane from whence calcium is mobilized during stimulus-secretion coupling. The release of oxidation products of arachidonic acid, be it via the cyclo-oxygenase or lipoxygenase pathway, is a concomitant of neutrophil activation. Indeed, the pathways by which oxidative products of arachidonate are formed may also be influenced by stimulus-specific changes in membrane phospholipids, not the least interesting of which is the generation of phosphatidic acid.

ACKNOWLEDGMENTS

Aided by grants to Gerald Weissmann (AM-11949 and HL-19721) from the National Institutes of Health and grant AM-27223 from the National Institutes of Health to James E. Smolen. Helen M. Korchak is the recipient of a grant from the Cystic Fibrosis Foundation. Drs. Neil Pinckard and E. J. Goetzl donated the acetyl glyceryl ether phosphorylcholine, and various derivatives.

REFERENCES

1. Bell, R. L., Kennerly, D. A., Stanford, N., and Majerus, P. W. (1979): Diglyceride lipase: a pathway for arachidonate release from human platelets. *Proc. Natl. Acad. Sci. USA*, 76:3238–3241.
2. Bell, R. L., and Majerus, P. W. (1980): Thrombin-induced hydrolysis of phosphatidyl-inositol in human platelets. *J. Biol. Chem.*, 255:1790–1792.
3. Billah, M. M., Lapetina, E. G., and Cuatrecasas, P. (1980): Phospholipase A$_2$ and phospholipase C activities of platelets: differential substrate specificity, Ca^{2+} requirement, pH dependence and cellular localization. *J. Biol. Chem.*, 255:10227–10231.
4. Borgeat, P., and Samuelsson, B. (1979): Arachidonic acid metabolism in polymorphonuclear leukocytes: effects of ionophore A23187. *Proc. Natl. Acad. Sci. USA*, 76:2148–2151.
5. Broekman, M. J., Ward, J. W., and Marcus, A. J. (1980): Phospholipid metabolism in stimulated human platelets: changes in phosphatidyl inositol, phosphatidic acid, and lysophospholipids. *J. Clin. Invest.*, 66:275–283.
6. Charo, I. F., Feinman, R. D., and Detweiler, T. C. (1976): Inhibition of platelet secretion by an antagonist of intracellular calcium. *Biochem. Biophys. Res. Commun.*, 72:1462–1467.
7. Craddock, P. R., Hammerschmidt, D., White, J. G., et al. (1977): Complement (C5a)-induced granulocyte aggregation in vitro: a possible mechanism of complement-mediated leukostasis and leukopenia. *J. Clin. Invest.*, 60:260–264.
8. Friedman, R., Kaplan, H. B., Smolen, J. E., Hoffstein, S., and Weissmann, G. (1981): Leukergy rediscovered: neutrophil (PMN) aggregation as a secretory response. *Clin. Res.*, 29:532A.
9. Goetzl, E. J. (1980): Vitamin E modulates the lipoxygenation of arachidonic acid in leukocytes. *Nature*, 288:183–185.
10. Goetzl, E. J., and Pickett, W. C. (1980): The human PMN leukocyte chemotactic activity of complex hydroxy-eicosatetraenoic acids (HETEs). *J. Immunol.*, 125:1789–1791.
11. Goldstein, I. M., Hoffstein, S. T., and Weissmann, G. (1975): Influence of divalent cations upon complement-mediated enzyme release from human polymorphonuclear leukocytes. *J. Immunol.*, 115:665–780.
12. Goldstein, I. M., Horn, J. K., Kaplan, H. B., and Weissmann, G. (1974): Calcium-induced lysozyme secretion from human polymorphonuclear leukocytes. *Biochem. Biophys. Res. Commun.*, 60:807–812.
13. Goldstein, I. M., Malmsten, C. L., Kindahl, H., Kaplan, H. B., Radmark, O., Samuelsson, B., and Weissmann, G. (1978): Thromboxane generation by human peripheral blood polymorphonuclear leukocytes. *J. Exp. Med.*, 148:787–792.
14. Green, D. E., Fry, M., and Blondin, G. A. (1980): Phospholipids as the molecular instruments of ion and solute transport in biological membranes. *Proc. Natl. Acad. Sci. USA*, 77:257–261.
15. Hoffstein, S. (1979): Ultrastructural demonstration of calcium loss from local regions of the plasma membrane of surface-stimulated human granulocytes. *J. Immunol.*, 123:1395–1402.
16. Hokin, L. E., and Hokin, M. R. (1955): Effects of acetylcholine on the turnover of phosphoryl units in individual phospholipids of pancreas slices and brain cortex slices. *Biochem. Biophys. Acta*, 18:102–110.

17. Kishimoto, A., Takai, Y., Mori, T., Kikkawa, U., and Nichizuka, Y. (1980): Activation of calcium and phospholipid-dependent protein kinase by diacylglycerol, its possible relation to phosphatidyl inositol turnover. *J. Biol. Chem.*, 255:2273–2276.
18. Kobayashi, R., Tawata, M., and Hidaka, H. (1979): Ca^{2+} regulated modulator protein interaction agents: inhibition of Ca^{2+} -Mg^{2+}-ATPase of human erythrocyte ghost. *Biochem. Biophys. Res. Commun.*, 88:1037–1045.
19. Korchak, H. M., and Weissmann, G. (1980): Stimulus-response coupling in the human neutrophil: transmembrane potential and the role of extracellular Na^+. *Biochim. Biophys. Acta.*, 601:180–194.
20. Michell, R. H. (1975): Inositol phospholipids and cell surface receptor function. *Biochim. Biophys. Acta*, 415:81–147.
21. Naccache, P. H., Showell, H. J., Becker, E. L., and Sha'afi, R. I. (1979): Involvement of membrane calcium in the response of rabbit neutrophils to chemotactic factors as evidenced by the fluorescence of chlorotetracycline. *J. Cell. Biol.*, 83:179–186.
22. O'Flaherty, J. T., Showell, H. J., Ward, P. A., and Becker, E. L. (1979): A possible role of arachidonic acid in human neutrophil aggregation and degranulation. *Am. J. Pathol.*, 96:799–809.
23. Rittenhouse-Simmons, S. (1979): Production of diglyceride from phosphatidyl-inositol in activated human platelets. *J. Clin. Invest.*, 63:580–587.
24. Scharff, O. (1981): Calmodulin—and its role in cellular activation. *Cell Calcium*, 2:1–27.
25. Serhan, C., Anderson, P., Goodman, E., Dunham, P., and Weissmann, G. (1981): Phosphatidate and oxidized fatty acids are calcium ionophores: studies employing arsenazo III in liposomes. *J. Biol. Chem.*, 256:2736–2741.
26. Serhan, C., Korchak, H. M., Hoffstein, S. and Weissmann, G. (1980): PGB_x, a prostaglandin derivative mimics the action of calcium ionophore A23187 on human neutrophils. *J. Immunol.*, 125:2020–2024.
27. Simchowitz, L., Spilberg, I., and Atkinson, J. P. (1980): Superoxide generation and granule enzyme release induced by ionophore A23187: studies on the early events of neutrophil activation. *J. Lab. Clin. Med.*, 96:408–424.
28. Smith, R. J., and Iden, S. S. (1979): Phorbol myristate acetate-induced release of granule enzymes from human neutrophils: inhibition by the calcium antagonist 8-(N,N-diethylamino)-octyl,3,4,5-trimethylbenzoate hydrochloride. *Biochem. Biophys. Res. Commun.*, 91:263–271.
29. Smolen, J. E., and Weissmann, G. (1981): The roles of extracellular and intracellular calcium in lysosomal enzyme release and superoxide anion generation by human polymorphonuclear leukocytes. *Biochim. Biophys. Acta*, 677:512–520.
30. Smolen, J. E., and Weissmann, G. (1981): The secretion of lysosomal enzymes from human neutrophils: the first events in stimulus-secretion coupling. In: *Lysosomes and Lysosomal Storage Diseases*, edited by J. W. Callahan, and J. A. Lowden, pp. 31–62. Raven Press, New York.
31. Stenson, W. F., and Parker, C. W. (1979): Metabolism of arachidonic acid in ionophore-stimulated neutrophils. *J. Clin. Invest.*, 64:1457–1465.
32. Weissmann, G. (1980): Release of mediators of inflammation from stimulated neutrophils. *N. Engl. J. Med.*, 303:27–34.
33. Weissmann, G., Anderson, P., Serhan, C., Samuelsson, E., and Goodman, E. (1980): A general method, employing arsenazo III in liposomes, for the study of calcium ionophores: results with A23187 and prostaglandins. *Proc. Natl. Acad. Sci. USA*, 77:1506–1510.
34. Weissmann, G., Collins, T., Evers, A., and Dunham, P. (1976): Membrane perturbation: studies employing a calcium-sensitive dye, arsenazo III, in liposomes. *Proc. Natl. Acad. Sci. USA*, 73:510–514.
35. Weissmann, G., Serhan, C., Korchak, H., Smolen, J. E., Broekman, M. J., and Marcus, A. (1982): Neutrophils generate phosphatidic acid, an "endogenous calcium ionophore" before releasing mediators of inflammation. *Trans. Assoc. Am. Physicians* (*in press*).
36. Weissmann, G., Smolen, J. E., and Korchak, H. (1980): Prostaglandins and inflammation: receptor/cyclase coupling as an explanation of why PGE's and PGI_2 inhibit functions of inflammatory cells. In: *Advances in Prostaglandin and Thromboxane Research, Vol. 8*, edited by P. Ramwell, R. Paoletti, and B. Samuelsson, pp. 1637–1653. Raven Press, New York.
37. Weissmann, G., Smolen, J. E., and Korchak, H. H. (1981): The role of calcium in secretion of inflammatory mediators from neutrophils. *Sem. Arthritis Rheum.*, 11:84–86.
38. Zurier, R. B., and Weissmann, G. (1972): Effect of prostaglandins upon enzyme release from lysosomes and experimental arthritis. In: *Prostaglandins in Cellular Biology*, edited by P. W. Ramwell and B. B. Pharriss, pp. 151–172. Plenum, New York.

Leukotrienes and Other Lipoxygenase Products,
edited by B. Samuelsson and R. Paoletti.
Raven Press, New York © 1982.

Mediation of Leukocyte Components of Inflammatory Reactions by Lipoxygenase Products of Arachidonic Acid

*E. J. Goetzl, *D. W. Goldman, †P. H. Naccache, †R. I. Sha'afi, and ‡W. C. Pickett

*Howard Hughes Medical Institute Laboratories at Harvard Medical School, and Departments of Medicine, Harvard Medical School and the Brigham and Women's Hospital, Boston, Massachusetts 02115; †Departments of Pathology and Physiology, University of Connecticut Health Center, Farmington, Connecticut 06032; and ‡Lederle Laboratories, Pearl River, New York 10965

The leukocyte chemotactic activity generated by the specific lipoxygenation of arachidonic acid is attributable predominantly to 5(S),12(R)-dihydroxy-eicosa-6,14-cis-8,10-trans-tetraenoic acid (leukotriene B_4, LTB_4) and 5(S)-hydroxyeicosatetraenoic acid (5-HETE), whereas the contractile and vasoactive factors 5-hydroxy-6-sulfido-glutathionyl-eicosatetraenoic acid (LTC_4) and 5-hydroxy-6-sulfido-cysteinyl-glycyl-eicosatetraenoic acid (LTD_4) lack significant chemotactic activity for leukocytes (9,12,14). LTB_4 and the less potent 5-HETE elicit neutrophil and eosinophil chemotactic responses in vitro and in vivo and account for a substantial part of the leukotactic activity in the lesions of some human inflammatory diseases, e.g., rheumatoid arthritis and the spondyloarthritides (5,6,9,12,14,17,19). LTB_4 and C5a, the anaphylatoxic cleavage fragment of the fifth component of complement, exhibit a similarly high chemotactic potency and are the major natural leukotactic factors that have been defined structurally (8,13). However, LTB_4 and C5a possess different profiles of other functional effects on leukocytes. The unique aspects of the activation of leukocytes by LTB_4 appear to reflect in part the presence of a specific subset of receptors for LTB_4 which are not shared with other classes of chemotactic factors.

Although any consideration of the leukotrienes as a family of mediators leads to the designation of LTB_4 as the predominant leukocyte stimulus, and of LTC_4 and LTD_4 as the principal contractile and vasoactive factors, LTB_4 is a weak mediator of some humoral effects (10,21). The biochemical prerequisites and time courses of the activities of the leukotriene mediators also exhibit numerous similarities, despite significant differences in the target cell preferences of each leukotriene. The complex dependence of the chemotactic potency of LTB_4 on numerous portions of the molecular structure is analogous to the multiplicity of structural determinants of the contractile and other activities of LTC_4 and LTD_4. This chapter describes

the structural determinants of the leukocyte effects of LTB$_4$, the cellular characteristics of the activation of leukocytes by LTB$_4$ and 5-HETE, and the evidence for
unique leukocyte receptors for LTB$_4$.

STRUCTURAL DETERMINANTS OF THE CHEMOTACTIC ACTIVITY OF LTB$_4$

In a modified Boyden chamber assay, LTB$_4$ elicited a maximal human neutrophil
chemotactic response and a response that was one-half of the maximal level at
concentrations of approximately 10^{-7} M and 6×10^{-9} M, respectively, which were
similar in magnitude to those evoked by 4×10^{-6} M and 4×10^{-7} M 5-HETE
and by 1.7×10^{-8} M and 1.3×10^{-9} M C5a (8,13,21). The availability of a
series of other structurally distinct 5,12-dihydroxy-HETES (5,12-DHETEs) that had
been purified from natural sources, and of derivatives of LTB$_4$ permitted the initial
analysis of the functional role of the various molecular substituents, which can be
considered in terms of five relatively polar domains in LTB$_4$ (Fig. 1). In contrast
to the methyl ester of 5-HETE, which retains only 10% or less of the chemotactic
activity of native 5-HETE, LTB$_4$ methyl ester was as potent and active as LTB$_4$ in
stimulating human neutrophil chemotaxis (13,15). Although the carboxyl group
thus appeared not to contribute to the chemotactic activity of LTB$_4$, the other four
domains proved to be critical to chemotactic potency. The acetylation of both
hydroxyl groups reduced the chemotactic potency of LTB$_4$ by 75% but full activity
was expressed at concentrations 3- to 10-fold higher than that required to achieve
maximal stimulation by LTB$_4$ (13).

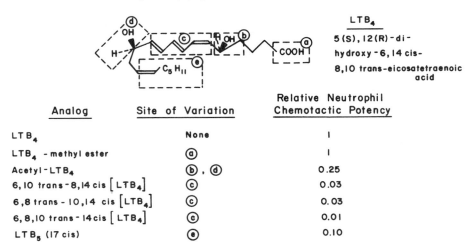

Analog	Site of Variation	Relative Neutrophil Chemotactic Potency
LTB$_4$	None	1
LTB$_4$ - methyl ester	ⓐ	1
Acetyl-LTB$_4$	ⓑ , ⓓ	0.25
6,10 trans-8,14 cis [LTB$_4$]	ⓒ	0.03
6,8 trans - 10,14 cis [LTB$_4$]	ⓒ	0.03
6,8,10 trans - 14cis [LTB$_4$]	ⓒ	0.01
LTB$_5$ (17 cis)	ⓔ	0.10

FIG. 1. Chemotactic functional domains of LTB$_4$. The chemotactic potency of highly purified
native LTB$_4$ for human neutrophils is arbitrarily assigned a value of 1. The chemotactic potencies
of derivatives of LTB$_4$ and of purified natural analogs of LTB$_4$ with differences in double bond
structure are expressed relative to that of LTB$_4$, based on the results of five to eight *in vitro*
neutrophil chemotactic studies of each principle.

When the *cis* double bond at position 6 in LTB$_4$ was placed at position 8 or was replaced by a *trans* double bond, the chemotactic potency of the resultant analogs was reduced by 90 to 99% (13) (Fig. 1). The 5,12-dihydroxyeicosapentaenoic acid chemotactic factor (LTB$_5$), which is derived from the specific lipoxygenation of eicosapentaenoic acid rather than arachidonic acid, contains one additional *cis* double bond at position 17 and is only 1/10 as potent as LTB$_4$ in stimulating neutrophil chemotaxis *in vitro*. The functional importance of the geometry of the double bonds in the conjugated triene portion of LTB$_4$ was confirmed recently in studies of synthetic analogs (21). Synthetic LTB$_4$ had the same potency and activity as the purified natural compound in neutrophil chemotactic assays carried out with an incubation period of 45 min instead of the conventional 1.5 to 2 hr. Synthetic analogs with the *cis* double bond in the 8 or 10 position, instead of the 6 position, exhibited approximately 1/100 the chemotactic potency of LTB$_4$, and analogs without a *cis* double bond in the conjugated triene had only approximately 1/300 the potency of LTB$_4$, based on the concentration required to achieve a response that was 50% of maximal neutrophil chemotaxis (21).

Utilizing guinea pig lung parenchymal strips, which represent a model for peripheral airways, LTB$_4$ elicited a contraction at concentrations as low as 0.5 ng/ml, whereas the central airway of guinea pig tracheal spirals failed to respond to LTB$_4$ at a concentration as high as 10 ng/ml (21). Thus although the potency of LTB$_4$ on the small airways of guinea pig lung was only 1/1,000 to 1/300 that of LTD$_4$ and 1/100 to 1/30 that of LTC$_4$, based on the concentration required to evoke a contraction that was 25% of the maximal response, the regional preference for peripheral airways, as contrasted with central airways, was the same for each of the leukotriene mediators. Similarly, LTB$_4$ induced dermal vasodilatation and localized wheals in the skin of normal human subjects, but the potency of LTB$_4$ again was less than that of LTD$_4$ (22). As for the leukocyte chemotactic effects of LTB$_4$, the modest vasoactive and contractile activity was reduced substantially by replacing the 6-*cis* double bond with a 6-*trans* double bond. Whereas most investigations of the structural bases of the contractile activities of LTC$_4$ and LTD$_4$ have been directed to variations in the peptide portion of the molecules (7,20), a few analogs have been synthesized with modifications in the fatty acid carboxyl or hydroxyl groups or in double bonds. The C-1 monoamide of LTD$_4$ was as potent as LTD$_4$, whereas removal of the 5-hydroxyl group and hydrogenation of the 9, 11, and 14 double bonds reduced the potency by over 1,000-fold when the compounds were assessed in terms of contraction of guinea pig lung parenchymal strips (7,20).

LEUKOCYTE RECEPTORS FOR CHEMOTACTIC FACTORS

The relationship of the induction of chemotaxis to the specificity of leukocyte receptors for chemotactic factors initially was suggested by the close correlation between binding affinity and chemotactic potency for a series of formyl-methionyl peptides (1,28,30,33). Chemotactically inactive peptide analogs inhibited the leukocyte chemotactic response to formyl-methionyl peptides but not to C5a, and the inhibitory potency of the analogs was determined directly by their affinity for the receptors (27). A similar approach to the functional definition of the interaction of LTB_4 with human neutrophils has suggested that a distinct subset of receptors recognizes LTB_4 but not C5a or formyl-methionyl peptides (13). Analogs and derivatives of LTB_4 were added to LTB_4 in the stimulus compartment of Boyden chambers at various molar ratios. The neutrophil chemotactic responses to the mixtures were corrected by subtracting the minimal activity of the modified LTB_4 and were expressed as a percentage of the response to LTB_4 alone (Fig. 2). Acetyl LTB_4, which exhibits a chemotactic potency approximately one-fourth that of native LTB_4, suppressed the neutrophil response to LTB_4 in relation to the concentration ratio of acetyl LTB_4 to LTB_4 (Fig. 2). At an equimolar ratio to LTB_4, acetyl LTB_4 inhibited chemotaxis by a mean of 47%, suggesting that the mechanism of inhibition was receptor-directed competition. A natural 5,12-DHETE analog, which differs from LTB_4 in placement of the first *cis* double bond in the 8 position instead of the 6 position, retains approximately 3 to 10% of the chemotactic potency of LTB_4. This analog inhibited the neutrophil chemotactic response to LTB_4 by 27% and 47%, respectively, at an equimolar ratio and a 10-fold molar ratio to LTB_4 (Fig. 2) and thus was less effective than acetyl LTB_4. In contrast, LTB_4 methyl ester, which is as potent a chemotactic factor as LTB_4, and 9-*cis*-11,13-*trans*-octadeca-trienoic acid, which contains the triene portion but not the hydroxyl groups of LTB_4 and lacks chemotactic activity, both failed to inhibit substantially the neutrophil chemotactic response to LTB_4. Thus the structural analogs of LTB_4 which acted as

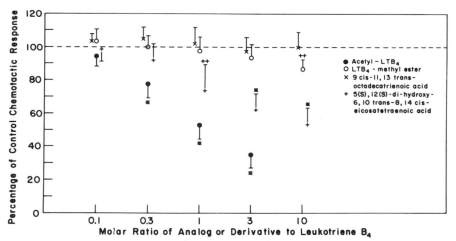

FIG. 2. Inhibition of human neutrophil chemotactic responses to LTB_4 by derivatives and structural analogs. The chemotactic response to each mixture of LTB_4 1 ng/ml and a derivative or analog of LTB_4 is expressed as a percentage of the control response (100%) of neutrophils to the same concentration of LTB_4 alone, which ranged from 23.9 to 34.2 net neutrophils/hpf. The chemotactic chambers were incubated for 45 min at 37°C. Each point and bracket is the mean ± SD of the results of three separate studies. The statistical significance of the differences between the response to a mixture and to LTB_4 alone was calculated with the Student's t-test: * = $p < 0.01$ and + + = $p < 0.05$.

competitive inhibitors of the chemotactic activity of the parent principle were agonists of lower potency than native LTB_4 and apparently retained an affinity for the specific receptors that exceeded the residual capacity to activate chemotaxis.

That neither competitive inhibitor of LTB_4 chemotactic activity suppressed the neutrophil responses to C5a or formyl-methionyl peptides confirmed other functional data which suggested the presence of separate subsets of receptors for each type of chemotactic factor. Further support for this contention was offered by the finding that maximally chemotactic concentrations of C5a and LTB_4 failed to inhibit the binding of [^{3}H]f-Met-Leu-Phe to human neutrophils. Several membrane protein constituents of neutrophil receptors for f-Met-Leu-Phe have been recovered from non-ionic detergent solutions of neutrophil membranes by f-Met-Leu-Phe-Sepharose affinity chromatography, utilizing specific elution with f-Met-Leu-Phe, and have been resolved by gel filtration in sodium dodecyl sulfate (16). The principal constituent is an intrinsic membrane protein of molecular weight 68,000, which binds [^{3}H]f-Met-Leu-Phe in equilibrium dialysis chambers with two apparent affinities. The mean high affinity was approximately $1 \cdot \times 10^9 \, M^{-1}$, which is equivalent to the affinity of receptors on intact neutrophils; the mean low affinity was only 1/30 the high affinity. This specific f-Met peptide-binding protein fails to bind [^{3}H]LTB_4 or [^{3}H]5-HETE in equilibrium dialysis chambers.

LTB_4, 5-HETE, and LTC_4 enhanced the expression of C3b receptors on neutrophils and eosinophils, as assessed by a rosetting technique (15,25). The rank order of activity, based on the magnitude of the enhancing effect, was $LTB_4 > 5$-HETE $> LTC_4$. LTB_4 was approximately 10-fold more potent than 5-HETE, whereas LTC_4 exerted an enhancing effect only in a restricted concentration range. LTC_4 and 5-HETE exhibited high-zone inhibition, as manifested by the loss of enhancement of expression of C3b receptors at concentrations 10-fold or more above the level required to achieve maximal enhancement. That a similar enhancement of C3b and possibly immunoglobulin G (IgG) Fc receptors on neutrophils and eosinophils has been achieved with other chemotactic factors suggests either that the C3b and chemotactic factor receptors are related structurally in the plasma membrane or simply that this is a capability shared by many potent stimuli of chemotaxis.

CELLULAR CHARACTERISTICS OF NEUTROPHIL ACTIVATION BY LTB_4

Although studies of the biochemical concomitants of neutrophil activation by LTB_4 are still incomplete, some alterations in specific plasma membrane functions and in calcium homeostasis have been documented and shown to be similar to the responses evoked by C5a and formyl-methionyl peptides. LTB_4 stimulated the stereospecific uptake of [^{3}H]deoxyglucose by neutrophils and eosinophils in a concentration-dependent manner with an EC_{50} of 17 nM for neutrophils, as contrasted with an EC_{50} of 110 nM for 5-HETE (3,4). As for C5a and formyl-methionyl peptides, the LTB_4 dose-response relationships for the stimulation of [^{3}H]deoxyglucose uptake by neutrophils and for the elicitation of neutrophil chemotaxis exhibited a similar EC_{50}. Lipoxygenase products of arachidonic acid and other chemotactic factors also significantly enhance the influx of calcium into neutrophils, as assessed by the uptake of ^{45}Ca after 1 min at 37°C (23,24) (Fig. 3). Monohydroperoxy and monohydroxy metabolites of the 11-lipoxygenation and 5-lipoxygenation of arachidonic acid increase neutrophil membrane permeability to calcium to a similar extent at equally chemotactic concentrations. LTB_4 again was the most active of the principles in this class (Fig. 3). The potency of LTB_4 in the enhancement of calcium

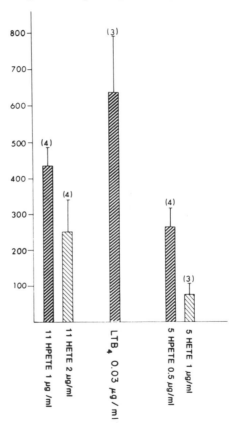

FIG. 3. Effect of lipoxygenase metabolites of arachidonic acid on the initial rate of uptake of $^{45}Ca^{++}$ by rabbit neutrophils. Purified rabbit peritoneal neutrophils were suspended in magnesium-free Hanks' solution containing 0.5 mM $CaCl_2$ and were preincubated for 20 min at 37°C prior to the addition of the stimuli or buffer alone and of $^{45}Ca^{++}$ (23,24). Neutrophil-associated radioactivity was quantified 1 min after the introduction of $^{45}Ca^{++}$ using a rapid sampling silicone oil method (24). The effect of each stimulus is expressed as the percentage increase in $^{45}Ca^{++}$ uptake relative to that observed for control neutrophils exposed to buffer alone. Each bar and bracket depicts the mean ± SEM for the number of experiments noted in parentheses. 5-HPETE = 5-hydroperoxyeicosatetraenoic acid. 11-HPETE = 11-hydroperoxyeicosatetraenoic acid.

influx exceeded that of 5-HETE by approximately 30- to 100-fold, a ratio similar to that observed for the stimulation of neutrophil chemotaxis.

LTB_4, but not mono-HETEs, evoked a release of intraneutrophil calcium from previously unexchangeable pools, as had been demonstrated previously for C5a and formyl-methionyl peptides (23,24). The net effect of LTB_4 was to increase transiently the intraneutrophil concentration of freely exchangeable calcium. The effects on neutrophil transport of hexose and calcium homeostasis attributable to LTB_4 and the potent peptide chemotactic factors are considered to be important prerequisites for the stimulation of a variety of leukocyte primary functions, including chemotaxis. However, the time course and mechanisms of integration of these and other biochemical events critical to leukocyte activation by chemotactic factors remain to be elucidated fully.

REGULATION OF LEUKOCYTE FUNCTION BY ENDOGENOUS LIPOXYGENASE PRODUCTS OF ARACHIDONIC ACID

A role for lipoxygenase products of arachidonic acid as functional intracellular constituents has been suggested by the ability of inhibitors of lipoxygenation to suppress neutrophil and eosinophil migration, hexose transport, aggregation, and lysosomal enzyme release at concentrations that substantially depleted the neutrophil content of HETEs (2,11,17,26,29,31). The possible intracellular role of some endogenous HETEs was supported directly by the finding that exogenous 5-HETE was capable of bypassing the suppression of hexose transport by lipoxygenase inhibitors (3) and of restoring the capacity of HETE-depleted neutrophils and eosinophils to migrate (11,17). Although the mechanism of the lipoxygenase-dependence of some neutrophil functions has not been identified, two possibilities have been formulated: Human neutrophils reincorporate into cellular phospholipids a substantial portion of the 5-HETE and other lipoxygenase products, which may alter the properties of plasma membranes, possibly at specific sites of prior deacylation (23). Alternatively, the reactive intermediates such as 5-hydroperoxyeicosatetraenoic acid (5-HPETE) can covalently derivatize intracellular peptides, proteins, and other constituents critical to cell function. Ethanol precipitates of human neutrophils that had been preincubated with [3H]arachidonic acid or [3H]5-HPETE contained relatively polar radioactive products that were resolved by LH-20 column chromatography into three distinct fractions (18). One such fraction, which was distinct from previously identified lipoxygenase products, reached a peak level by 30 sec at 37°C and by 10 to 15 min at 0°C. The appearance of this unique polar metabolite in neutrophils was blocked by lipoxygenase inhibitors (e.g., ETYA) and stimulated by C5a (18).

SUMMARY

The leukocyte lipoxygenase products LTB_4 and 5-HETE elicit human neutrophil and eosinophil chemotactic responses *in vitro* and *in vivo* (Fig. 4), and are present

Major Effects   **Minor Effects**

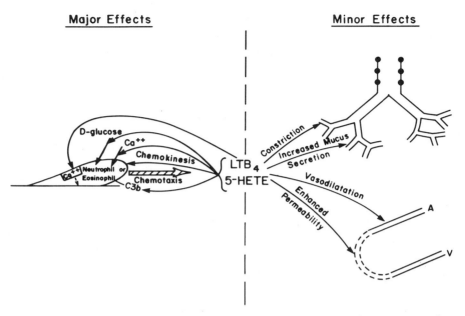

FIG. 4. The roles of LTB$_4$ and 5-HETE as mediators of allergic and inflammatory reactions. The effects of LTB$_4$ and 5-HETE on human leukocytes, pulmonary airways, and microvasculature are depicted in terms of relatively high potency (major effects) or low potency (minor effects) relative to LTC$_4$ and LTD$_4$. A = arteriole. V = venule. Ca^{++}→ = extracellular calcium. Ca^{++}--→ = release of calcium from intracellular stores, C3b = receptor for C3b. → = stimulation.

at elevated concentrations in the lesions of some human inflammatory diseases, such as rheumatoid arthritis and the spondyloarthritides. The chemotactic potency of LTB$_4$ is similar to that of the minor fragment of the fifth component of human complement, termed C5a, and is 30- to 300-fold greater than that of 5-HETE and of other natural DHETE isomers. Human neutrophils possess distinct subsets of chemotactic factor receptors that are specific for LTB$_4$ and for 5-HETE, as demonstrated by the selective competitive inhibition of the chemotactic responses to the parent stimuli by acetyl LTB$_4$ and 5-HETE methyl ester, respectively, and by the failure of the lipid chemotactic factors to bind to isolated membrane protein constituents of the human neutrophil receptors for chemotactic formyl-methionyl peptides. LTB$_4$ and 5-HETE also elicit human neutrophil and eosinophil chemokinesis, stimulate the uptake of calcium and D-glucose, and enhance the expression of C3b receptors on the leukocytes; however, they exert only a minimal effect on superoxide generation and lysosomal enzyme release. LTB$_4$, but not 5-HETE, stimulates the release of calcium from previously unexchangeable intraneutrophil pools, as has been described for potent peptide chemotactic factors. Although far less potent than LTC$_4$ and LTD$_4$, LTB$_4$ constricts peripheral airways, enhances mucous secretion in the airways of the lung, and dilates and enhances the permeability of the microvasculature in skin and other organs (Fig. 4). A variety of leukocyte functions, including chemotaxis, D-glucose uptake, and lysosomal enzyme release, are impaired in association with the depletion of endogenous lipoxygenase products. Exogenous 5-HETE reverses some of the functional deficits of HETE-depleted leukocytes. Inhibition of leukocyte lipoxygenase activity also suppresses the intracellular content of hydroperoxyeicosatetraenoic acids and of novel polar metabolites of arachidonic acid that may be critical to the activation of human neutrophils and eosinophils. Thus LTB_4 and the less potent 5-HETE are active extracellular mediators of the leukocytic components of hypersensitivity and inflammation and may also serve an important role as intracellular mediators of leukocyte function.

ACKNOWLEDGMENTS

Dr. Goldman is supported by a postdoctoral research fellowship from the Arthritis Foundation. The investigations described were supported in part by NIH grants HL-19777 and AI-13734.

REFERENCES

1. Aswanikumar, S., Corcoran, B. A., Schiffmann, E., Day, A. R., Freer, R. J., Showell, H. J., Becker, E. L., and Pert, C. B. (1977): Demonstration of a receptor on rabbit neutrophils for chemotactic peptides. *Biochem. Biophys. Res. Commun.*, 74:810–817.
2. Bass, D. A., O'Flaherty, J. T., Szejda, P., DeChatelet, L. R., and McCall, C. E. (1980): Role of arachidonic acid in stimulation of hexose transport by human polymorphonuclear leukocytes. *Proc. Natl. Acad. Sci. USA*, 77:5125–5129.
3. Bass, D. A., Thomas, T. J., Goetzl, E. J., DeChatelet, L. R., and McCall, C. E. (1981): Lipoxygenase-derived products of archidonic acid mediate stimulation of hexose uptake in human polymorphonuclear leukocytes. *Biochem. Biophys. Res. Commun.*, 100:1–7.
4. Bass, D. A., Szejda, P., Goetzl, E. J., Love, S., O'Flaherty, J. T., and McCall, C. E. (1981): Stimulation of hexose uptake of human eosinophils by chemotactic factors. *J. Infect. Dis.*, 143:719–725.
5. Bhattacherjee, P., Eakins, K. E., and Hammond, B. (1981): Chemotactic activity of arachidonic acid lipoxygenase products in the rabbit eye. *Br. J. Pharmacol. (in press)*.
6. Carr, S. C., Higgs, G. A., Salmon, J. A., and Spayne, J. A. (1981): The effects of arachidonate lipoxygenase products on leukocyte migration in rabbit skin. *Br. J. Pharmacol. (in press)*.
7. Drazen, J. M., Lewis, R. A., Austen, K. F., Toda, M., Brion, F., Marfat, A., and Corey, E. J. (1981): Contractile activities of structural analogues of leukotrienes C and D: necessity of a hydrophobic region. *Proc. Natl. Acad. Sci. USA*, 78:3195–3198.
8. Fernandez, H. N., Henson, P. M., Otani, A., and Hugli, T. E. (1978): Chemotactic response to C3a and C5a. I. Evaluation of C3a and C5a leukotaxis in vitro and under simulated in vivo conditions. *J. Immunol.*, 120:109–115.
9. Ford-Hutchinson, A. W., Bray, M. A., Doig, M. V., Shipley, M. E., and Smith, M. J. H. (1980): Leukotriene B₄, a potent chemotactic and aggregating substance released from polymorphonuclear leukocytes. *Nature*, 286:264–265.
10. Goetzl, E. J. (1980): Mediators of immediate hypersensitivity derived from arachidonic acid. *N. Engl. J. Med.*, 303:822–825.
11. Goetzl, E. J. (1980): A role for endogenous mono-hydroxy-eicosatetraenoic acids (HETEs) in the regulation of human neutrophil migration. *Immunology*, 40:709–726.
12. Goetzl, E. J., and Pickett, W. C. (1980): The human PMN leukocyte chemotactic activity of complex hydroxy-eicosatetraenoic acids (HETEs). *J. Immunol.*, 125:1789–1791.
13. Goetzl, E. J., and Pickett, W. C. (1981): Novel structural determinants of the human neutrophil chemotactic activity of leukotriene B. *J. Exp. Med.*, 153:482–487.
14. Goetzl, E. J., and Sun, F. F. (1979): Generation of unique mono-hydroxy-eicosatetraenoic acids from arachidonic acid by human neutrophils. *J. Exp. Med.*, 150:406–411.

15. Goetzl, E. J., Brash, A. R., Tauber, A. I., Oates, J. A., and Hubbard, W. C. (1980): Modulation of human neutrophil function by monohydroxy-eicosatetraenoic acids. *Immunology*, 39:491–501.
16. Goetzl, E. J., Foster, D. W., and Goldman, D. W. (1981): Structural characteristics of human neutrophil receptors for chemotactic N-formyl-methionyl peptides. *Fed. Proc.*, 40:365.
17. Goetzl, E. J., Weller, P. F., and Sun, F. F. (1980): The regulation of human eosinophil function by endogenous mono-hydroxy-eicosatetraenoic acids (HETEs). *J. Immunol.*, 124:926–933.
18. Goldman, D. W., and Goetzl, E. J. (1981): Novel metabolites of arachidonic acid in human neutrophils. *Fed. Proc.*, 40:1004.
19. Klickstein, L. B., Shapleigh, C., and Goetzl, E. J. (1980): Lipoxygenation of arachidonic acid as a source of polymorphonuclear leukocyte chemotactic factors in synovial fluid and tissue in rheumatoid arthritis and spondyloarthritis. *J. Clin. Invest.*, 66:1166–1170.
20. Lewis, R. A., Drazen, J. M., Austen, K. F., Toda, M., Brion, F., Marfat, A., and Corey, E. J. (1981): Contractile activities of structural analogues of leukotrienes C and D: role of the polar substituent. *Proc. Natl. Acad. Sci. USA*, 78:4579–4583.
21. Lewis, R. A., Goetzl, E. J., Soter, N. A., Drazen, J. M., Corey, E. J., and Austen, K. F. (1981): Leukotriene B₄: stereochemical dependency of chemotactic and spasmogenic activities. *Fed. Proc.*, 40:791.
22. Lewis, R. A., Soter, N. A., Corey, E. J., and Austen, K. F. (1981): Local effects of synthetic leukotrienes (LTs) on monkey (M) and human (H) skin. *Clin. Res.*, 29:492A.
23. Naccache, P. H., Sha'afi, R. A., Borgeat, P., and Goetzl, E. J. (1981): Mono- and di-hydroxy-eicosatetraenoic acids alter calcium homeostasis in rabbit neutrophils. *J. Clin. Invest.*, 67:1584–1587.
24. Naccache, P. H., Showell, H. J., Becker, E. L., and Sha'afi, R. I. (1977): Transport of sodium, potassium and calcium across rabbit polymorphonuclear leukocyte membranes: effect of chemotactic factor. *J. Cell. Biol.*, 73:428–444.
25. Nagy, L., Lee, T. H., Goetzl, E. J., and Kay, A. B. (1981): Complement receptor enhancement and chemotaxis of human neutrophils and eosinophils by leukotrienes and other lipoxygenase products. Presented at the Int. Symp. on Leukotrienes and Other Lipoxygenase Products, Florence.
26. O'Flaherty, J. T., Showell, H. J., Becker, E. L., and Ward, P. A. (1979): Role of arachidonic acid derivatives in neutrophil aggregation: a hypothesis. *Prostaglandins*, 17:915–927.
27. O'Flaherty, J. T., Showell, H. J., Kreutzer, D. L., Ward, P. A., and Becker, E. L. (1978): Inhibition of in vivo and in vitro neutrophil responses to chemotactic factors by a competitive antagonist. *J. Immunol.*, 120:1326–1340.
28. Pike, M. C., Fischer, D. G., Koren, H. S., and Snyderman, R. (1980): Development of specific receptors for N-formylated chemotactic peptides in a human monocyte cell line stimulated with lymphokines. *J. Exp. Med.*, 152:31–40.
29. Sha'afi, R. I., Naccache, P. H., Molski, T. F., Borgeat, P., and Goetzl, E. J. (1981): Cellular regulatory role of leukotriene B₄: its effects on cation homeostasis in rabbit neutrophils. *J. Cell Physiol.*, 108:401–408.
30. Showell, J. J., Freer, R. J., Zigmond, S. H., Schiffmann, E., Aswanikumar, S., Corcoran, B. A., and Becker, E. L. (1976): The structure-activity relations of synthetic peptides as chemotactic factors and inducers of lysosomal enzyme secretion for neutrophils. *J. Exp. Med.*, 143:1154–1169.
31. Showell, H. J., Naccache, P., Sha'afi, R. I., and Becker, E. L. (1980): Inhibition of rabbit neutrophil lysosomal enzyme secretion, nonstimulated and chemotactic factor stimulated locomotion by nordihydroguaiaretic acid. *Life Sci.*, 27:421–426.
32. Stenson, W. F., and Parker, C. W. (1979): Metabolism of arachidonic acid in ionophore stimulated neutrophils. *J. Clin. Invest.*, 64:1457–1465.
33. Williams, L. T., Snyderman, R., Pike, M. C., and Lefkowitz, R. J. (1977): Specific receptor sites for chemotactic peptides on human polymorphonuclear leukocytes. *Proc. Natl. Acad. Sci. USA*, 74:1204–1208.

Leukotrienes and Other Lipoxygenase Products,
edited by B. Samuelsson and R. Paoletti.
Raven Press, New York © 1982.

Biological Activities of Leukotriene B₄ (Isomer III)

M. J. H. Smith

*Department of Chemical Pathology, King's College Hospital Medical School,
London, SE5 8RX England*

There are a limited number of positions at which the inclusion of oxygen-containing groups may occur in the molecule of arachidonic acid following lipoxygenase action (35). At carbon 5 the first defined product is 5-monohydroperoxyeicosatetraenoic acid (5-HPETE), which is further transformed enzymatically into the unstable epoxide 5(S)-*trans*-5,6-oxido-7,9-*trans*-11,14-*cis*-eicosatetraenoic acid (leukotriene A₄, or LTA₄). One metabolic pathway of LTA₄ is its conversion to the 5,12-dihydroxy-eicosatetraenoic acid (5,12-DHETE; LTB₄), and a second pathway is the formation of the slow-reacting substances LTC₄, etc. (3,33). The particular isomer of LTB₄ which is derived enzymatically from LTA₄ has been characterized as 5(S),12(R)-dihydroxy-6-*cis*-8-*trans*-10-*trans*-14-*cis*-eicosatetraenoic acid (10). In this chapter it will be designated as LTB₄ (isomer III) in order to distinguish it from other isomers of LTB₄ obtained either by total synthesis (9), nonenzymatic hydrolysis of LTA₄ (31), or biological formation from precursors other than LTA₄. This is a matter of more than mere academic nicety, as there appears to be a wide variation in biological activities between the known isomers of LTB₄ (9,13,20).

The formation of LTB₄ (isomer III) by leukocyte cell types *in vitro* and its occurrence *in vivo* have been reviewed elsewhere (6,36). The purposes of the present chapter are to describe its biological activities, *in vitro* and *in vivo*, and to discuss their possible relevance to the development of inflammatory responses.

IN VITRO BIOLOGICAL ACTIVITIES OF LTB₄ (ISOMER III)

The *in vitro* biological activities of LTB₄ (isomer III) are listed below. The most potent, and certainly the best documented, comprise effects on the aggregation and movement (chemokinesis and chemotaxis) of a variety of leukocyte cell types.

1. Stimulation of leukocyte behavior and function
   a. Aggregation
   b. Chemokinesis
   c. Chemotaxis

d. Release of lysosomal enzymes
2. Activity in isolated tissue preparations

Leukocyte Aggregation

It was found by O'Flaherty et al. (26,27) that arachidonic acid caused a transient aggregation of human polymorphonuclear leukocytes (PMNs) *in vitro*, that the effect was enhanced by cytochalasin B, and that it was not induced by structurally similar fatty acids. The response was blocked, however, by 5,8,11,14-eicosatetraynoic acid, indicating that a metabolite of arachidonic acid was the active agent. These observations were confirmed and extended by other workers (5,15) to show, first, that the aggregating activity was associated with lipoxygenase rather than the cyclo-oxygenase pathway of arachidonic acid metabolism and, secondly, that the biological activity could be released when either human or rat PMNs were exposed to the calcium ionophore A23187. Maximal release occurred after 4 min and could be detected in dilutions of the supernatant from the leukocyte suspensions of up to 1:1,000. Fractionation of the supernatant by a sequence of solvent extraction, silicic acid chromatography, and reverse-phase high-pressure liquid chromatography (RP-HPLC) revealed that >95% of the aggregating activity could be recovered in a single peak after HPLC. The substance present in this active fraction was identified as LTB_4 (isomer III) by gas chromatography-mass spectrometry (GC-MS) (14).

The separated LTB_4 (isomer III) causes the aggregation *in vitro* of a number of leukocyte cell types. These include human, rat, and rabbit PMNs, monocytes, and macrophages, the last being either resting cells obtained by lavage of the peritoneal cavity or elicited at this site by prior injection of thioglycollic acid. The ED_{50} values for the aggregating activity of LTB_4 (isomer III) are between 0.5 and 1.0 ng ml^{-1} for human and rat PMN suspensions (14,16). The aggregating activity is measured as a change in light transmission of a stirred suspension of PMNs using a Payton aggregometer (12), and this procedure serves as an easy and sensitive bioassay for the leukotriene. LTB_4 (isomer III) does not cause aggregation of either lymphocyte or blood platelet suspensions from various species.

Leukocyte Chemokinesis

An equally convenient and sensitive bioassay for LTB_4 (isomer III) is the measurement of the chemokinesis (stimulated random migration) of leukocyte suspensions using the agarose microdroplet technique (39). The published data (14,19,29) relating to the potency of the leukotriene as a chemokinesin is in general accord, except that some authors report the concentrations showing maximal activity whereas others derive ED_{50} values. Using human peripheral PMNs as the test cells, LTB_4 (isomer III) shows maximal chemokinetic activity at concentrations ranging from 10 to 30 ng ml^{-1}. There is little variation between different leukocyte cell types, the ED_{50} values for LTB_4 (isomer III) being as follows: human peripheral PMNs 250 pg ml^{-1}, human peripheral monocytes 380 pg ml^{-1}, guinea pig eosinophils 230 pg ml^{-1}, and rat peritoneal macrophages 140 pg ml^{-1} (36). There is no significant difference between the potency of preparations of LTB$_4$ (isomer III) obtained either synthetically or biologically, i.e., from rat peritoneal PMNs exposed to calcium ionophore (16).

Leukocyte Chemotaxis

It has also become clear that, despite differences in assay methods, LTB$_4$ (isomer III) is also a potent chemotaxin *in vitro*. The leukotriene has been reported to produce a maximal stimulation of the directed migration of human peripheral leukocyte suspensions at concentrations ranging from 300 pg to 30 ng ml^{-1} (19,20,29,38). Furthermore, its potency as a chemotaxin, chemokinesin, and aggregating agent *in vitro* is equivalent to that of the established cytotaxins, C5a, and the synthetic formyl-methionyl-leucyl-phenylalanine (FMLP). Thus the respective ED$_{50}$ values for chemokinetic stimulation of human peripheral neutrophils are LTB$_4$ (isomer III) 7.6×10^{-10} M, C5a (Des Arg) 1.9×10^{-9} M, and FMLP 2.0×10^{-10} M (36). Similar data have also been obtained for the chemotactic response of the same cell type to these agents (20).

There is also general agreement that LTB$_4$ (isomer III) is considerably more potent *in vitro* than either 5-HETE, other mono-HETEs (6,19,20,29), or the natural dihydroxy-HETE (DHETE) isomers (13,20). There is also provisional evidence (9,20) that the synthetically available isomers of LTB$_4$ (isomer III), in which the geometry of the double bonds in the triene portion of the molecule are either *trans,trans,cis* or *trans,cis,trans*, are less active than the biologically derived material. Minor chemical modifications of the LTB$_4$ (isomer III) structure such as methylation of the carboxyl group or acetylation of the hydroxyls (20) cause a loss of biological activity. More exact conclusions about structure/activity requirements in this group of hydroxy eicosatetraenoic acids must await the results of more comprehensive studies. It appears, however, that LTB$_4$ (isomer III) is the most potent metabolite, derived via initial lipoxygenase attack on arachidonic acid, with respect to the *in vitro* effects on leukocyte aggregation and movement.

Release of Lysosomal Enzymes from Leukocytes

A further similarity between LTB$_4$ (isomer III) and the other cytotaxins, C5a and FMLP, is its ability to cause increased release of lysosomal enzymes from leukocytes treated with cytochalasin B (30). Here, however, the resemblance is only qualitative. The maximal level of release of lysozyme and β-glucuronidase is less, approximately one-third, of that produced by the chemotactic peptides (19). The dose-response curve for the release of the enzymes from human peripheral PMNs by LTB$_4$ (isomer III) is much flatter than that for FMLP (32). Hence there appears to be a dichotomy between the equipotent effects of LTB$_4$ (isomer III) and the other cytotaxins on leukocyte aggregation and movement and on their differing potencies with respect to lysosomal enzyme release.

Activity in Isolated Tissue Preparations

LTB_4 (isomer III), at concentrations up to 20 ng ml^{-1} (i.e., within the range causing maximal responses on leukocyte aggregation and movement) produces little or no effect on isolated smooth muscle preparations (11). These include the fundus from the rat; the ileum, vas deferens, and tracheal spiral from the guinea pig; and the aorta from the rabbit. Guinea pig ileum gives characteristic slow contractions with LTC_4 and LTD_4 at concentrations of 5 to 60 ng ml^{-1} (22), and the response of this preparation to other agonists (histamine 8 to 20 ng ml^{-1} and acetylcholine 2 to 10 ng ml^{-1}) is potentiated by LTB_4 (isomer III). Furthermore, there is an increased spontaneous activity of the ileal strip after exposure to and washing out of the LTB_4 (isomer III) from the organ bath (11).

Much larger amounts of LTB_4 (isomer III) have been reported to cause a sustained contraction of lung parenchymal strips in a manner similar to that of LTC_4 (34). A bolus of 10 μg in 5 μl of medium, in a cascade superfusion procedure, was necessary to produce the effect, although the leukotriene was three times more potent than histamine. The effect was abolished by indomethacin but not by FPL 55712, a receptor antagonist of SRS-A, and it was suggested that this action of LTB_4 (isomer III) on the isolated lung tissue is mediated by the release of endogenous prostaglandins. At the present time it is difficult to assess the significance of these interesting observations. The effective concentrations of LTB_4 (isomer III) *in vitro* with respect to leukocyte aggregation and movement are very much less than those necessary to cause effects on the lung parenchyma. It is not easy to understand why the perfusion of SRS-A and LTB_4 (isomer III) should evoke an apparently specific release of thromboxanes and prostaglandins with the implication that these products of the cyclo-oxygenase pathway are the final mediators of leukotriene action at the tissue level.

IN VIVO BIOLOGICAL ACTIONS OF LTB₄ (ISOMER III)

There is now a substantial body of evidence showing that LTB_4 (isomer III) possesses cytotactic activity in animals and man. The experimental data include the accumulation of leukocytes at local sites of administration of the leukotriene; more direct observations on preparations (e.g., hamster cheek pouch and rabbit mesentery); the production of neutropenia after intravenous injection in the rabbit; and effects on vascular permeability in the presence of a vasodilator. These actions have been elicited by single doses of LTB_4 (isomer III) within the range of the concentrations per milliliter reported to occur in human synovial fluid (23). In some of the models a dose-dependent response has been observed, and there are close parallels between the *in vivo* effects of LTB_4 (isomer III) and those of the established cytotaxins C5a and FMLP.

Accumulation of Leukocytes at Sites of Administration

Single doses of LTB_4 (isomer III), ranging from 25 to 500 ng, cause the accumulation of neutrophils at local sites of administration in a number of animal species and man. An intraperitoneal injection of 100 ng of the leukotriene in the guinea pig produced a significant increase in the total white cell count in the peritoneal fluid (38). The effect was restricted to neutrophils, with no increased migration of either mononuclear cells or eosinophils being observed. It was not apparent at time intervals up to 1 hr after the injection of LTB$_4$ (isomer III), but it was significant 5 hr after administration.

The introduction of LTB$_4$ (isomer III), in doses ranging from 25 to 400 ng, into the aqueous humor of the rabbit eye has been reported to produce a dose-dependent increase in total white cell counts (2). Significant increases in leukocyte migration into the aqueous humor, samples of which were obtained 4 hr after the intracameral injections, were observed with 200 and 400 ng doses of LTB$_4$ (isomer III) and 250 and 500 ng doses of FMLP.

A third model is rabbit skin. Two groups of workers (7,8) obtained similar results in that the intradermal injection of doses of LTB$_4$ (isomer III), varying from 25 to 100 ng, caused a dose-dependent increase in the numbers of leukocytes in the dermis at intervals of 30 min to 4 hr after the injection. The leukocytes were counted in five high power fields arranged vertically through the dermis in sections from skin samples cut at several points at each injection site. The increased leukocyte migration at 30 min was restricted to the accumulation of neutrophils, there being no change in the monocyte counts and very few eosinophils present. A further significant increase in PMN accumulation in the dermis occurred when 100 ng prostaglandin E$_2$ (PGE$_2$) was included in the intradermal injection (7).

An even more dramatic effect of LTB$_4$ (isomer III) on neutrophil accumulation occurs when the substance is placed in skin chambers (41) attached to either shaved rabbit skin or the human forearm. The chambers are placed over superficial abrasions made with a spherical dental burr and are filled with 2 ml of a suitable vehicle (control) with or without LTB$_4$ (isomer III) 100 ng ml^{-1}. After 5 hr the fluid from the chamber is withdrawn, and total and differential cell counts are performed by conventional methods. In both species, there was no accumulation of leukocytes in the control chambers, whereas in the presence of the leukotriene the total leukocyte counts were between 100 and 500 $\times$ 10^3 ml^{-1} (37). Differential staining revealed that the cells were PMNs, no eosinophils or monocytes being detected.

Effects on Adherence and Diapedesis of Leukocytes

An important aspect of many inflammatory reactions is the adherence of white cells to vascular endothelium and their emigration across the vessel walls of small vascular beds. The sticking and diapedesis of leukocytes is the dominant microscopic event and has been described in many animal preparations (21). One simple and convenient model is the intestinal mesentery of the rabbit, which may be exteriorized, immersed in a suitable bathing fluid, with or without LTB$_4$ (isomer III), observed visually, and then fixed, sectioned, stained, and examined histologically. When such preparations are exposed to a 10 ng ml^{-1} solution of LTB$_4$ (isomer III), an enhanced sticking of leukocytes to the walls of the small blood vessels is observed within a few minutes, followed by an increased escape of the white cells from the vessels. On histological examination, small venules are seen to contain an increased number of leukocytes, mainly PMNs, within the lumen, and some of these adhere to the vascular endothelium after 5 min exposure to the leukotriene. After 20 min exposure, neutrophils not only stick to the vessel walls but are observed to have migrated into the extravascular areas (7).

Similar effects occur in the cheek pouch of the hamster (1), which has been extruded and superfused with a 5 ng ml^{-1} LTB$_4$ (isomer III) solution. Using this preparation, it was consistently found that increased numbers of leukocytes adhered to, and passed through, the walls of small venules. The effect was reflected by a sharp decrease in the numbers of rolling granulocytes counted in the postcapillary venules 2 to 3 min after commencing the infusion, the peak response occurring after 8 to 10 min.

A related *in vivo* effect of leukotriene, explicable in terms of an increased adherence of circulatory leukocytes to vascular beds, occurs in the rabbit. The intravenous injection of 1 μg LTB$_4$ (isomer III) causes a transient but profound neutropenia (7). The effect is maximal 1 min after administration and returns to control levels within 3 min. No changes in either peripheral monocyte or platelet counts were observed. The dose of the leukotriene used is comparable in terms of molarity to those of C5a and FMLP necessary to produce a similar degree of neutropenia in the same species (28). However, the duration of the neutropenic response elicited by LTB$_4$ (isomer III) is much shorter than those of the other cytotaxins. Thus neutropenia induced by FMLP continues for up to 20 min. One possible explanation is that LTB$_4$ (isomer III) is more rapidly metabolized *in vivo* than are the other cytotaxins. Information about its metabolic fate is sparse except that it has been reported to be transformed into a 5,12,19 tri-HETE by rat mononuclear cells (24). The fate and persistence of LTB$_4$ (isomer III) in the circulation is a topic of some potential interest, particularly in relation to its possible role as a localized inflammatory mediator.

Effects on Vascular Permeability

The cytotactic activity of LTB$_4$ (isomer III) *in vivo* appears to be the basis of a further effect of the leukotriene on vascular permeability (4). In the rabbit the intradermal injection of LTB$_4$ (isomer III), in doses of 1 to 10 ng, has no effect on blood flow in the skin, assessed by the ^{133}Xe clearance technique; in doses up to 100 ng, it has little if any action on plasma exudation, measured by the extravasation of labeled albumin (38,43). The administration of PGE$_2$ alone, 1 to 100 ng per skin site, produces a dose-dependent response in blood flow but no change in plasma exudation. When the two substances are given together, there is a significant increase in plasma exudation, not only in the rabbit but also in the rat and guinea pig (4). Similar results have been obtained with other cytotaxins (C5a and FMLP). When mixed with PGE$_2$ these substances show highly potent permeability—increased activity—but cause little or no edema when given alone (42). These effects, as well as that with LTB$_4$ (isomer III), occur only in normal animals and not in those whose circulating PMNs have previously been depleted, e.g., by treatment with nitrogen mustard. It has been concluded (42) that mechanisms exist to regulate local fluid accumulation which are dependent on the presence of PMNs, attracted to the inflammatory site by complement activation (C5a), from microorganisms (FMLP), or from leukocytes (LTB$_4$ isomer III). The cytotaxins induce minimal edema alone, and the concomitant generation of vasodilator prostaglandins seems to be essential for the formation of a fluid exudate. It is of interest that products of the cyclo-oxygenase (PGE$_2$, etc.) and lipoxygenase (LTB$_4$ isomer III) pathways of arachidonic acid metabolism may act in concert to cause increased vascular permeability in inflammatory responses.

There are further interactions in that both PGE$_2$ and LTB$_4$ (isomer III) potentiate the effects of bradykinin on vascular permeability. The mechanisms, however, are different. The potentiation by PGE$_2$ is due to its vasodilator action, whereas that of the leukotriene is the result of the vasodilator activity of the bradykinin. The lack of vasodilator activity by LTB$_4$ (isomer III) may explain its failure to enhance swelling in the carrageenan-induced paw reaction in the rat (38).

It therefore appears that there are two types of mediators which increase vascular permeability: (a) those acting directly on vascular endothelium and also possessing some intrinsic vasodilator properties (histamine and bradykinin), and (b) the cytotaxins (C5a, FMLP, and LTB$_4$ isomer III), which have no vasodilator properties and affect vascular permeability indirectly through the rapid involvement of PMNs at the inflammatory site. What must occur in the latter instance is a yet undefined mechanism by which the leukocytes, attracted to the inflammatory site by the cytotaxins, interact with the vascular endothelium to increase permeability. This may involve secretion of enzymes by the PMNs which could act on basement membranes or may degrade substances present at the junctions between adjacent endothelial cells. An alternative idea is that some of the attracted leukocytes lodge underneath endothelial cell junctions and, after being exposed to the cytotaxins, swell and thus mechanically force open the junctions (42).

SUMMARY AND CONCLUSIONS

LTB$_4$ (isomer III) is a metabolite of arachidonic acid formed, after initial lipoxygenase action, via 5-HPETE and LTA$_4$, and its detailed structure has been confirmed by total synthesis. Its principal biological activities *in vitro* are on leukocyte aggregation and movement. In this context it is not only equipotent to C5a and FMLP but appears to be significantly more active than related hydroxy- and hydroperoxyeicosatetraenoic acids (mono-, di-, and tri-HETEs, mono-HPETEs) which are formed either naturally or nonenzymatically. LTB$_4$ (isomer III), either in single doses or at concentrations within the range of those reported to occur in human synovial fluid, exerts a range of cytotactic effects *in vivo*. These comprise the accumulation of neutrophils at local sites of administration, neutropenia, adherence to and diapedesis of leukocytes from small blood vessels, and increased vascular permeability in the presence of vasodilatory prostaglandins.

LTB₄ (isomer III) may be a natural mediator of inflammation. It is produced by the leukocytes present in developing inflammatory exudates, and its effects resemble and supplement those of the complement-derived peptide C5a, which is formed in the fluid phase. A further interaction is with vasodilator prostaglandins, the combination causing an increased vascular permeability and the formation of local edema. Thus prostaglandins, C5a, and LTB₄ (isomer III), all of which occur in the synovial fluid of patients with chronic self-destructive conditions, such as rheumatoid arthritis, may act sequentially and in combination to produce the vascular permeability and cellular infiltration which characterize inflammatory responses.

There are many potential actions of LTB₄ (isomer III) which remain to be explored. In addition to effects on cell movement, it may also express surface receptors on leukocytes, as mono-HETEs and HPETEs enhance the expression of C3b receptors on eosinophils (17,18). A similar comment applies to effects on intraleukocyte cyclic GMP, which is increased by 12-HPETE (18). The effects of hydroxyeicosatetraenoic acids on phenomena such as lymphocyte activation by mitogens and the histamine release reaction from basophils and mast cells deserve further investigation. The results of inhibitor studies (25,40) suggest the involvement of lipoxygenase-derived products of arachidonic acid which may or may not include LTB₄ (isomer III).

A specific and rapid assay for LTB₄ (isomer III) together with the availability of the labeled leukotriene are now required. These should enable the metabolism, including its incorporation into phospholipids, and the binding of LTB₄ (isomer III) to cells and plasma proteins to be studied. The very rapid but short action of the leukotriene in producing a neutropenia in the rabbit suggests that its life in the circulation is not only short but possibly of little significance *in vivo*. Another potentially rewarding area of future research could be the production of specific inhibitors of either its formation from LTA₄ or its actions at the leukocyte cell surface. Such compounds should provide more definitive evidence about the real or apparent importance of LTB₄ (isomer III) as a putative mediator of inflammation.

ACKNOWLEDGMENTS

The author wishes to acknowledge the help of his co-workers Dr. M. A. Bray, Miss L. M. Brown, and Drs. F. M. Cunningham, E. M. Davidson, M. V. Doig, A. W. Ford-Hutchinson, M. Irani, S. Rae, and M. E. Shipley, and is grateful to Mrs. P. Carter and Miss C. Davis for secretarial assistance. The work was supported by grants from the Medical Research Council, the Arthritis and Rheumatism Research Council, and the Joint Research Committee of King's College Hospital and Medical School.

REFERENCES

1. Atherton, A., and Born, G. V. R. (1972): Quantitative investigation of the adhesiveness of circulating polymorphonuclear leucocytes to blood vessel walls. *J. Physiol (Lond.)*, 222:447–474.
2. Bhattacherjee, P., Eakins, K. E., and Hammond, B. (1981): Chemotactic activity of arachidonic acid lipoxygenase products in the rabbit eye. *Br. J. Pharmacol.*, 73:254–255P.

3. Borgeat, P., and Sirois, P. (1981): Leukotrienes: a major step in understanding of immediate hypersensitivity reactions. *J. Med. Chem.*, 24:121–126.
4. Bray, M. A., Cunningham, F. M., Ford-Hutchinson, A. W., and Smith, M. J. H. (1981): Leukotriene B₄: a mediator of vascular permeability. *Br. J. Pharmacol.*, 72:483–486.
5. Bray, M. A., Ford-Hutchinson, A. W., Shipley, M. E., and Smith, M. J. H. (1980): Calcium ionophore A23187 induces release of chemokinetic and aggregating factors from polymorphonuclear leucocytes. *Br. J. Pharmacol.*, 71:507–512.
6. Bray, M. A., Ford-Hutchinson, A. W., and Smith, M. J. H. (1981): Leukotriene B₄: biosynthesis and biological activities. In: *SRS-A and Leukotrienes*, edited by P. Piper, pp. 253–270. Wiley, New York.
7. Bray, M. A., Smith, M. J. H., and Ford-Hutchinson, A. W. (1981): Leukotriene B₄ : An inflammatory meditor *in vivo*. *Prostaglandins,* 22:213–218.
8. Carr, S. C., Higgs, G. A., Salmon, J. A., and Spayne, J. A. (1981): The effects of arachidonate lipoxygenase products on leukocyte migration in rabbit skin. *Br. J. Pharmacol.*, 73:253–254P.
9. Corey, E. J., Hopkins, P. B., Munroe, J. E., Marfat, A., and Hashimoto, S. (1980): Total synthesis of 6-*trans*, 10-*cis* and (±)-6-trans, 8-cis isomer of leukotriene B. *J. Am. Chem. Soc.*, 102:7986–7987.
10. Corey, E. J., Marfat, A., Goto, G., and Brion, F. (1980): Leukotriene B: total synthesis and assignment of stereochemistry. *J. Am. Chem. Soc,* 102:7984–7985.
11. Cunningham, F. M., Brown, L. M., and Smith, M. J. H. (1981): *Br. J. Pharmacol. (in press).*
12. Cunningham, F. M., Shipley, M. E., and Smith, M. J. H. (1980): Aggregation of rat polymorphonuclear leucocytes in vitro. *J. Pharm. Pharmacol.*, 32:377–380.
13. Ford-Hutchinson, A. W., Bray, M. A., Cunningham, F. M., Davidson, E. M., and Smith, M. J. H. (1981): Isomers of leukotriene B possess different biological potencies. *Prostaglandins*, 21:143–152.
14. Ford-Hutchinson, A. W., Bray, M. A., Doig, M. V., Shipley, M. E., and Smith, M. J. H. (1980): Leukotriene B, a potent chemokinetic and aggregating substance released from polymorphonuclear leucocytes. *Nature*, 286:264–265.
15. Ford-Hutchinson, A. W., Bray, M. A., and Smith, M. J. H. (1979): The aggregation of rat neutrophils by arachidonic acid: a possible bioassay for lipoxygenase activity. *J. Pharm. Pharmacol.*, 31:868–869.
16. Ford-Hutchinson, A. W., Smith, M. J. H., and Bray, M. A. (1981): Leukotriene B₄ (isomer III): biological actions of synthetic and biologically derived preparations. *J. Pharm. Pharmacol.*, 33:332.
17. Goetzl, E. J., Brash, A. R., Tauber, A. I., Oates, J. A., and Hubbard, W. C. (1980): Modulation of human neutrophil function by monohydroxy-eicosatetraenoic acids. *Immunology*, 39:491–501.
18. Goetzl, E. J., Hill, H. R., and Gorman, R. A. (1980): Unique aspects of the modulation of human neutrophil function by 12-L-hydroperoxy-5,8,10,14-eicosatetraenoic acid. *Prostaglandins*, 19:71–85.
19. Goetzl, E. J., and Pickett, W. C. (1980): The human PMN leukocyte chemotactic activity of complex hydroxyeicosatetraenoic acids (HETEs). *J. Immunol.*, 125:1789–1791.
20. Goetzl, E. J., and Pickett, W. C. (1981): Novel structural determinants of the human neutrophil chemotactic activity of leukotriene B. *J. Exp. Med.*, 153:482–487.
21. Grant, L. (1973): The sticking and emigration of white blood cells in inflammation. In: *The Inflammatory Process*, Vol. II, edited by B. W. Zweifach, L. Grant, and R. T. McCluskey, pp. 205–249. Academic Press, New York.
22. Hedqvist, P., Dahlen, S., Gustaesson, L., Hammarström, S., and Samuelsson, B. (1980): Biological profile of leukotrienes C₄ and D₄. *Acta Physiol. Scand.*, 110:331–333.
23. Klickstein, L. B., Shapleigh, C., and Goetzl, E. J. (1980): Lipoxygenation of arachidonic acid as a source of polymorphonuclear leukocyte chemotactic factors in synovial fluid and tissue in rheumatoid arthritis and spondyloarthritis. *J. Clin. Invest.*, 66:1166–1170.
24. Maas, R. L., Brash, A. R., and Oates, J. A. (1981): Novel leukotriene and lipoxygenase product from rat mononuclear cells. In: *SRS-A and Leukotrienes*, edited by P. Piper, pp. 151–159. Wiley, New York.
25. Marone, G., Hammarström, S., and Lichtenstein, L. L. (1980): An inhibitor of lipoxygenase inhibits histamine release from human basophils. *Clin. Immunol. Immunopathol.*, 17:117–122.
26. O'Flaherty, J. T., Showell, H. J., Becker, E. L., and Ward, P. A. (1979): Neutrophil aggregation and degranulation: effect of arachidonic acid. *Am. J. Pathol.*, 95:433–444.

27. O'Flaherty, J. T., Showell, H. J., Becker, E. L., and Ward, P. A. (1979): Role of arachidonic acid derivates in neutrophil aggregation: a hypothesis. *Prostaglandins*, 17:915–927.
28. O'Flaherty, J. T., Showell, H. J., and Ward, P. A. (1977): Neutropenia induced by systemic infusion of chemotactic factors. *J. Immunol.*, 118:1586–1589.
29. Palmer, R. M. J., Stepney, R. J., Higgs, G. A., and Eakins, K. E. (1980): Chemotactic activity of arachidonic acid lipoxygenase products on leucocytes of different species. *Prostaglandins*, 20:411–418.
30. Palmer, R. M. J., and Yeats, D. A. (1981): Leukotriene B: a potent chemotactic agent for lysosomal enzyme secretion for human neutrophils (PMN). *Br. J. Pharmacol.*, 73:260–261P.
31. Rådmark, O., Malmsten, C., Samuelsson, B., Clark, D. A., Goto, G., Marfat, A., and Corey, E. J. (1980): Leukotriene A: stereochemistry and enzymatic conversion to leukotriene B. *Biochem. Biophys. Res. Commun.*, 92:954–961.
32. Rae, S., and Smith, M. J. H. (1981): The stimulation of lysosomal enzyme secretion by leukotriene B₄. *J. Pharm. Pharmacol.*, 33:616–617.
33. Samuelsson, B., Borgeat P., Hammarström, S., and Murphy, B. (1980): Leukotrienes: a new group of biologically active compounds. In: *Advances in Prostaglandin and Thromboxane Research, Vol. 6*, edited by B. Samuelsson, P. Ramwell, and R. Paoletti, pp. 1–18. Raven Press, New York.
34. Sirois, P., Borgeat, P., Jeanson, A., Roy, S., and Girard, G. (1980): The actions of leukotriene B₄ (LTB₄) on the lung. *Prostaglandins*, 5:429–444.
35. Smith, M. J. H. (1981): Leukotriene B₄. *Gen. Pharmacol.*, 12:211–216.
36. Smith, M. J. H. (1981): Biological aspects of leukotriene B₄. In: *Colloquia of Institute Pasteur.* INSERM, Paris, 100:129–145.
37. Smith, M. J. H. (1981): Leukotriene B₄. *Agents Actions (in press)*.
38. Smith, M. J. H., Ford-Hutchinson, A. W., and Bray, M. A. (1980): Leukotriene B: a potential mediator of inflammation. *J. Pharm. Pharmacol.*, 32:517–518.
39. Smith, M. J. H., and Walker, J. R. (1980): The effects of some antirheumatic drugs on an in vitro model of human polymorphonuclear leucocyte chemokinesis. *Br. J. Pharmacol.*, 69:473–478.
40. Sullivan, T. J., and Parker, C. W. (1979): Possible role of arachidonic acid and its metabolites in mediator release from rat mast cells. *J. Immunol.*, 122:431–436.
41. Walker, J. R., James, D. W., and Smith, M. J. H. (1979): Directed migration of circulating polymorphonuclear leucocytes in patients with rheumatoid arthritis: a defect in the plasma. *Ann. Rheum. Dis.*, 38:215–218.
42. Wedmore, C. V., and Williams, T. J. (1981): Control of vascular permeability by polymorphonuclear leukocytes in inflammation. *Nature*, 289:646–650.
43. Williams, T. J. (1979): Prostaglandin E₂, prostaglandin I₂ and the vascular changes in inflammation. *Br. J. Pharmacol.*, 65:517–524.

Leukotrienes and Other Lipoxygenase Products,
edited by B. Samuelsson and R. Paoletti.
Raven Press, New York © 1982.

Effects of Leukotrienes on In Vitro Neutrophil Functions

Jan Palmblad, Ingiäld Hafström, *Curt L. Malmsten, Ann-Mari
Udén, *Olof Rådmark, Lars Engstedt, and *Bengt Samuelsson

*Departments of Medicine III and IV; Karolinska Institute, Södersjukhuset, S-10064
Stockholm, Sweden; and *Department of Physiological Chemistry, Karolinska Institute,
S-104 01 Stockholm, Sweden*

After perturbation with substances formed within an inflammatory or infected site, polymorphonuclear neutrophilic granulocytes (PMNs) exhibit increased adherence, chemotaxis, degranulation, oxidative metabolism, and bacterial killing. These reactions to, for example, the C5a fragment and bacterial products such as f-Met-Leu-Phe (fMLP) are critical for host defense but also contribute to tissue damage because of extrusion of PMN lysosomal enzymes and formation of toxic oxygen radicals (2,7).

We and others have recently shown that leukotrienes, in particular leukotriene B_4 (LTB$_4$), as well as other lipoxygenase products, stimulate neutrophil migration, aggregation, and to a lesser degree enzyme release (4–6,8,13). This chapter concerns the further investigations of the effects of LTB$_4$ on several neutrophil functions *in vitro*, its nonenzymatically formed isomers compounds I and II (10), LTC$_4$ [one component of slow-reacting substances of anaphylaxis (SRS-A) (11)], and 5-hydroxyeicosatetraenoic acid (5-HETE) (5), as well as the effects of two inhibitors of the lipoxygenase and cyclo-oxygenase pathways (indomethacin and eicosatetraynoic acid, ETYA) on migration.

MATERIALS AND METHODS

Chemicals, 5-HETE, LTB$_4$, compounds I and II, and LTC$_4$ were obtained after incubation of PMNs with arachidonic acid (Nu-Chek Prep. Inc., Elysian, Minn.) and ionophore A23187 (Eli Lilly, Indianapolis, Ind.). The incubate was extracted and purified by high-pressure chromatography (10,11).

Leukocyte Preparation and PMN Assays

Twice washed, dextran sedimented human leukocytes were resuspended in HBSS containing 0.1% gelatin for the degranulation, bactericidal, and chemiluminescence assays and supplemented with 1% human albumin for adherence assays and 3.1% Hepes buffer (Sigma, St. Louis, Mo.) for the chemotaxis assays. In some migration experiments PMNS were incubated with indomethacin or ETYA for 15 min at 37°C, both dissolved in ethanol (at a final concentration of 2.5%). Cells incubated with solvent only served as controls. Neutrophils (2.5×10^9/L) were mixed with suspensions of *Staphylococcus aureus* (70×10^9 colony-forming units/L) and serum (10%). After 90 min incubation at 37°C, samples were removed for quantitation of viable bacteria. The results are given as the percentages of living bacteria of the initial counts (1).

Adherence was assayed by a 40 mg nylon fiber column to which leukocytes (preincubated with leukotrienes for 2 min at room temperature) were added. The results are given as the percentage of neutrophils adhering to the fibers (1).

Spontaneous and stimulated locomotion were assayed with an agarose method. Ten microliters of the leukocyte suspension (10×10^9 PMNs/liter) was placed in a central well. The chemotactic factors (either 10^{-7} M fMLP, 5-HETE, or the leukotrienes, dissolved in 10% ethanol in HBSS) were placed in the outer wells, and control media in the inner wells. The distance migrated by the leading front neutrophils was measured by microscopy after incubation for 0.5 to 3 hr and is given in millimeters (1). In order to assess if locomotion was directed (i.e., chemotaxis), the orientation of lamellopodia and nuclei of 600 neutrophils migrating toward the cytotaxin wells was estimated by microscopy (1). PMNs 2×10^6/ml were, in some experiments, preincubated with 2.5 μg cytochalasin B dissolved in dimethylsulfoxide (DMSO, Sigma; at a final concentration of 0.5%) for 15 min at 37°C. Leukotrienes, 5-HETE, or fMLP, dissolved in ethanol (at a final concentration of 1%) in HBSS, were added to the PMNs. After 15 min at 37°C the reaction was stopped by lowering the temperature rapidly to $+4$°C, and the cell-free supernatants were collected after centrifugation. These were assayed for β-glucuronidase (3) and lysozyme (9) activities. The stimulated net release of enzymes was expressed as a percentage of the total cellular enzyme contents, determined after lysis of PMNs with 0.2% Triton X-100 (Sigma) (12).

RESULTS

Preincubation of PMNs with indomethacin ($\geqslant 2.5 \times 10^{-5}$ M) was followed by a dose-dependent inhibition of fMLP-stimulated migration (Fig. 1). However, with indomethacin at 2.5×10^{-7} M, stimulated migration was significantly enhanced. Lower concentrations of indomethacin had no significant effect. PMNs pretreated with ETYA exhibited markedly impaired migration when the final ETYA concentration was $\geqslant 2.5 \times 10^{-5}$M (Fig. 1). At ETYA concentrations of $\geqslant 2.5 \times 10^{-6}$ M, however, spontaneous or stimulated migration was unaffected.

When 5-HETE, LTC$_4$, LTB$_4$, and compounds I and II were added to the cytotaxin well, PMN migration was significantly stimulated by LTB$_4$ and compounds I and II (Fig. 1). After incubation for 3 hr with LTB$_4$ at 10^{-5} M, PMNs were densely packed around the border of the PMN well facing the LTB-containing well; the previously spontaneously moving cells facing the control well showed stimulated movement, a pattern previously recognized as desensitization (8). With 5-HETE

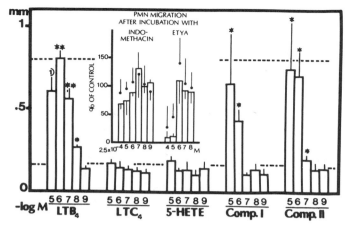

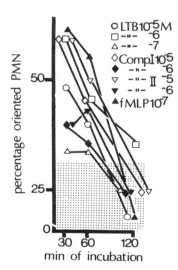

FIG. 1. Migration of neutrophils toward LTB$_4$, LTC$_4$, 5-HETE, and compounds I and II, measured as the distance to the leading front cells. Mean and SD values for triplicates performed on PMNs from four to six subjects. The mean distance migrated after stimulation with fMLP at 10^{-7} M is 0.80 ± 0.27 mm and is depicted as (------). The mean value for spontaneously migrating PMNs is 0.13 ± 0.02 mm and is marked as (·-·-·-·-·). **Inset:** Inhibition of migration of indomethacin- or ETYA-treated cells, expressed as a percentage of untreated control. Mean and SD values for triplicates with PMNs from four to six subjects. Bars = fMLP-induced migration. (●) Spontaneously moving cells. (*) $p < 0.05$ and (**) $p < 0.01$ compared with spontaneously migrating cells or untreated controls.

FIG. 2. Orientation of PMNs migrating toward the leukotriene- or fMLP-containing agarose wells. The shaded area represents the normal value (mean $\pm$ 2 SD) for spontaneously moving cells.

or LTC$_4$ migration was not significantly different from that of spontaneously moving cells (Fig. 1). When the degree of orientation of migrating cells was determined, it was found that PMNs stimulated by fMLP showed the highest degree of orientation at 30 and 60 min of incubation, but this decreased after 120 min to control values (i.e., to values characterizing spontaneously moving cells). LTB_4 at 10^{-6} M and compounds I and II at 10^{-5} M also stimulated a directed migration at 30 and 60 min, and LTB_4 still did so at 120 min (Fig. 2).

Neutrophil adherence was enhanced after 2 min of incubation with LTB_4 and compounds I and II, but not after incubation with 5-HETE and LTC_4 (Fig. 3). As found for chemotaxis, LTB_4 at 10^{-6} M was as potent a stimulator of adherence as fMLP at 10^{-7} M, whereas compounds I and II were most effective at 10^{-5} M.

LTB_4 at 10^{-5} M caused a significant net release of lysozyme and β-glucuronidase from cytochalasin-B-treated PMNs, but not at lower concentrations (Fig. 4). Even from PMNs not pretreated with cytochalasin B, LTB_4 released small amounts of lysozyme. Similarly, compound II at 10^{-5} M caused a small release of lysozyme but not of β-glucuronidase from cytochalasin-B-treated neutrophils. Neither compound I, LTC_4, or 5-HETE stimulated a net secretion of any of the enzymes. The release of both enzymes from cytochalasin-B-treated PMNs by LTB_4 at 10^{-5} M was approximately half that noted after stimulation with fMLP at 10^{-5} M. However, in contrast to LTB_4, fMLP did not induce any net secretion of enzymes from PMNs not pretreated with cytochalasin B.

Neither leukotrienes nor 5-HETE stimulated phagocytosis-associated chemiluminescence or the bactericidal capacity (Table 1). Furthermore, none of the compounds elicited a chemiluminescence when added to neutrophils in the absence of bacteria (data not shown).

DISCUSSION

This study has shown that LTB_4 and, to a lesser degree, its isomers enhance neutrophil adherence, chemotaxis, and enzyme release, but they do not affect bactericidal mechanisms or oxidative metabolism. Furthermore, LTC_4 and 5-HETE

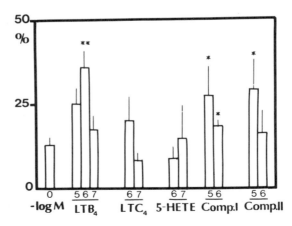

FIG. 3. Adherence of neutrophils to 40-mg nylon fibers. fMLP 10^{-7} M augmented adherence with $25 \pm 7\%$ units from untreated cells.

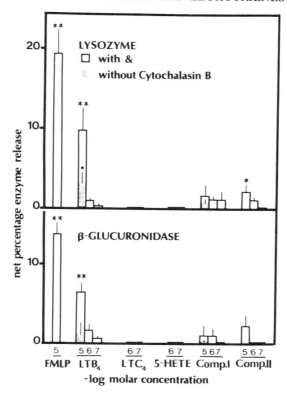

FIG. 4. Release of lysozyme and β-glucuronidase from neutrophils after stimulation with leukotrienes, 5-HETE, or fMLP, with or without pretreatment with cytochalasin B. The results are expressed as mean and SE for the net percentage of enzyme release from PMNs from duplicate determinations from three subjects. (*) $p < 0.05$. (**) $p < 0.01$.

TABLE 1. *PMN bactericidal capacity and phagocytosis-associated chemiluminescence after incubation with 5-HETE or LTs[a]*

Compound	Molar conc.	Bactericidal capacity (% CFU at 90 min)	Chemiluminescence in % of controls after		
			5 min	10 min	15 min
LTB$_4$	10^{-5}	9.3 ± 1.9	99 ± 4	114 ± 13	130 ± 19[b]
	10^{-6}	10.7 ± 2.3	98 ± 13	103 ± 9	102 ± 6
	10^{-7}	12.4 ± 3.0	95 ± 8	104 ± 6	104 ± 6
LTC$_4$	10^{-6}		83 ± 3	85 ± 4	90 ± 4
5-HETE	10^{-6}		78 ± 5	81 ± 4	84 ± 12
Compound I	10^{-5}		76 ± 5	81 ± 4	85 ± 10
Controls		10.9 ± 1.9	100	100	100

[a]Mean $\pm$ SEM values of four to six determinations in duplicates on PMNs from three or four donors.

[b]$p < 0.5$.

did not stimulate any of these PMN functions at the concentrations tested. These results suggest a role for LTB_4 in the inflammatory response.

The observation that ETYA caused a dose-dependent decrease of PMN migration [as well as of enzyme release (14)], induced by fMLP, suggests that oxygenation of arachidonic acid is essential for the chemotactic and secretagogue responses of human PMNs to the peptide, as ETYA inhibits the lipoxygenase and cyclooxygenase pathways. High concentrations of indomethacin (previously shown to inhibit mainly cyclo-oxygenase and to a lesser extent lipoxygenase) also reduced the chemotactic effect of fMLP, as well as the release of enzymes (14), whereas lower concentrations (inhibiting only the cyclo-oxygenase) stimulated migration. These findings are in accordance with the hypothesis that lipoxygenase products are important for neu-trophil locomotion and degranulation, as inhibition of cyclo-oxygenase diverts ar-achidonic acid into leukotriene synthesis. However, additional studies employing specific lipoxygenase inhibitors are required to establish a mediator role of leu-kotrienes in PMNs. It cannot be excluded at this stage that ETYA, in addition to inhibiting lipoxygenase and cyclo-oxygenase, affects other mechanisms involved in the response of PMNs to fMLP. When the lipoxygenase products 5-HETE, LTC_4, LTB_4, and compounds I and II were studied, LTB_4 was found to confer augmented adherence, directed and stimulated migration (i.e., chemotaxis), and secretion of lysozyme and β-glucuronidase. The migratory and adherence responses to LTB_4 was of the same magnitude as that to fMLP but required a 10-fold higher molar concentration. These responses also showed similar dose-response curves with an optimum at 10^{-6} M. Moreover, LTB_4 released lysosomal enzymes from untreated and cytochalasin B-treated cells, a response differing from that of fMLP, which was completely dependent on cytochalasin B. Not only was the response of cyto-chalasin-treated PMNs to LTB_4 half the magnitude of that for fMLP at the same molar concentration, it also occurred at a 10-fold higher molar concentration than the optimal one for stimulation of chemotaxis and adherence. This is in accordance with what has been noted for other chemotactic and secretagogue stimuli (5–7,12). The nonenzymatic isomers of LTB_4 (compounds I and II) were considerably less active than the enzymatically formed LTB_4, and, finally, 5-HETE and LTC_4 did not show any significant effects on any of the neutrophil functions studied.

The stereospecificity in the migration, adherence, and secretion responses sug-gests a physiological role for LTB_4. It is hypothesized that LTB_4 is generated primarily in order to recruit neutrophils to an inflammatory area. As adherence is critical for attachment to endothelial cells, the first step in the emigration of PMNs into the tissues, as well as a prerequisite for locomotion, it is noteworthy that LTB_4 affects both of these functions in a similar way. However, in contrast to C5a and fMLP, LTB_4 did not augment the production of cytotoxic oxygen radicals and caused only a marginal extrusion of lysosomal enzymes, thereby possibly avoiding some of the cytotoxic effects of stimulated PMNs on the tissues of the host.

ACKNOWLEDGMENTS

Supported in part by grants from the Swedish Medical Research Council (03X-217 and 19X-05991), the funds of the Karolinska Institute, and the Swedish Defense Research Institute. The skillful technical assistance of Mrs. I. Friberg, S. Koinberg, and N. Venizelos, M. Sci. is gratefully acknowledged. Dr. J. Pike, Upjohn Co., Kalamazoo, Mich., kindly supplied the ETYA.

REFERENCES

1. Afzelius, B. A., Ewetz, L., Palmblad, J., Udén, A.-M., and Venizelos, N. (1980): Structure and function of neutrophil leukocytes from patients with the immotile-cilia syndrome. *Acta Med. Scand.*, 208:145–154.
2. Boxer, L. A., Yoder, M., Bonsib, S., Schmidt, M., Ho, P., Jersild, R., and Baehner, R. L. (1979): Effects of a chemotactic factor, N-formyl-methionyl peptide, on adherence, superoxide anion generation, phagocytosis, and microtubule assembly of human polymorphonuclear leukocytes. *J. Lab. Clin. Med.*, 93:506–514.
3. Fishman, W. H., Kato, K., Anstiss, C. L., and Green, S. (1967): Human serum β-glucuronidase; its measurement and some of its properties. *Clin. Chim. Acta*, 15:435–447.
4. Ford-Hutchinson, A. W., Bray, M. A., Doing, M. V., Shipley, M. E., and Smith, J. H. (1980): Leukotriene B, a potent chemokinetic and aggregating substance released from polymorphonuclear leukocytes. *Nature*, 286:264–265.
5. Goetzl, E. J., Brash, A. R., Tauber, A. I., Oates, J. A., and Hubbard, W. C. (1980): Modulation of human neutrophil function by monohydroxy-eicosatetraenoic acids. *Immunology*, 39:491–501.
6. Goetzl, E. J., and Pickett, W. C. (1980): The human PMN leukocyte chemotactic activity of complex hydroxy-eicosatetraenoic acids (HETEs). *J. Immunol.*, 125:1789–1791.
7. Goldstein, I. M. (1980): Endogenous regulation of complement (C5)-derived chemotactic activity: fine-tuning of inflammation. *J. Lab. Clin. Med.*, 93:13–16.
8. Malmsten, C. L., Palmblad, J., Udén, A.-M., Rådmark, O., Engstedt, L., and Samuelsson, B. (1980): Leukotriene B$_4$: a highly potent and stereospecific factor stimulating migration of polymorphonuclear leukocytes. *Acta Physiol. Scand.*, 110:449–451.
9. Osserman, E. F., and Lawlor, D. P. (1966): Serum and urinary lysozyme (muramidase) in monocytic and nomomyelocytic leukemia. *J. Exp. Med.*, 124:921–952.
10. Rådmark, O., Malmsten, C. L., Samuelsson, B., Clark, D. A., Goto, G., Marfat, A., and Corey, E. J. (1980): Leukotriene A, stereochemistry and enzymatic conversion to leukotriene B. *Biochem. Biophys. Res. Commun.*, 92:954–961.
11. Samuelsson, B., Hammarström, S., Murphy, R. C., and Borgeat, P. (1980): Leukotrienes and slow reacting substance of anaphylaxis (SRS-A). *Allergy*, 35:375–381.
12. Showell, H. J., Freer, R. J., Zigmond, S. H., Schiffman, E., Aswanikumar, S., Corcoran, B., and Becker, E. L. (1976): The structure-activity relations of synthetic peptides as chemotactic factors and inducers of lysosomal enzyme secretion for neutrophils. *J. Exp. Med.*, 143:1154–1169.
13. Smith, M. J. H., Ford-Hutchinson, A. W., and Bray, M. A. (1980): Leukotriene B: a potential mediator of inflammation. *J. Pharm. Pharmacol.*, 32:517–518.
14. Smolen, J. E., and Weissman, G. (1980): Lysosomal enzyme release from human granulocytes is inhibited by indomethacin, ETYA, and BPB. In: *Advances in Prostaglandin and Thromboxane Research, Vol. 8*, edited by B. Samuelsson, P. W. Ramwell, and R. Paoletti, pp. 1695–1700. Raven Press, New York.

Leukotrienes and Other Lipoxygenase Products,
edited by B. Samuelsson and R. Paoletti.
Raven Press, New York © 1982.

Induction and Comparison of the Eosinophil Chemotactic Factor With Endogeneous Hydroxyeicosatetraenoic Acids: Its Inhibition by Arachidonic Acid Analogs

W. König, *H. W. Kunau, and †P. Borgeat

*Lehrstuhl für Medizinische Mikrobiologie und Immunologie, Arbeitsgruppe für Infektabwehrsmechanismen, and *Institut für physiologische Chemie, Arbeitsgruppe Bioorganische Chemie, Ruhr-Universität Bochum, Postfach, 4630 Bochum, West Germany; and †Departement d'Endocrinologie Moléculaire, Le Centre de l'Université Laval, Ste-Foy, Québec, Canada*

Recent interest has focused on the interdependence between mast cells poly-morphonuclear neutrophils (PMNs) and mononuclear cells during the inflammatory reaction (1,6,14,25,26,28). In this regard, it has been demonstrated that these cells can be stimulated to release biologically active substances derived from arachidonic acid turnover (11,12,33,34). As has been recently established, the lipoxygenase pathway induces a series of mono- and dihydroxylated eicosatetraenoic acids which have chemotactic and spasmogenic properties (1,8,13,23,29).

INDUCTION OF ECF ACTIVITY FROM VARIOUS CELLS

As early as 1975 we provided evidence that a nonpreformed mediator which specifically attracts guinea pig and human eosinophils was generated on stimulation of human PMNs, mononuclear cells, and basophils, as well as rat mononuclear and mast cells (6,7,14,15). The observation that preformed eosinophil chemotactic factor activity (termed tetrapeptide ECF-A) in purified rat mast cells was minimal compared to the ECF obtained after stimulation of the cells led us to further analyze the chemotactic factor derived from various cells with various stimuli.

Usually human PMNs were incubated with the calcium ionophore. At various times the samples were centrifuged and the supernatants were assayed for ECF in the modified Boyden chamber. In addition, the secretion of granular and microsomal enzymes was studied (Fig. 1). ECF rapidly appeared in the supernatant and decreased markedly at later times of secretion (18,20). As the biological activity was rather stable, we suggested that a cell-derived inactivator might induce the decrease in activity at later times of secretion (7,9,15,21). As for the enzyme release, it is apparent that the calcium ionophore represents a noncytotoxic stimulus, as determined by the enzyme lactate dehydroxygenase. Furthermore, the chemotactic ac-

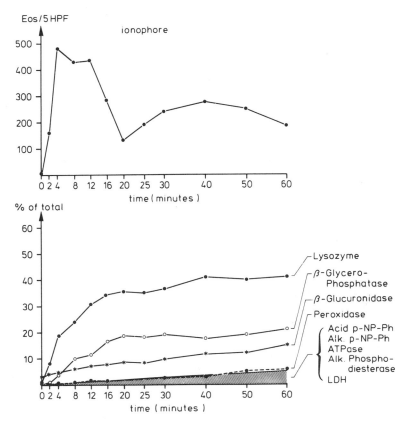

FIG. 1. Kinetics of ionophore-induced ECF release. The precision coefficient of the chemotactic assay varied between 8 and 12%. Lower part of the figure represents the pattern of enzyme release. Eos/5 hpf = eosinophils per five high power fields. (●)Lysozyme. (○) β-Glycerophosphatase. (□) β-Glucuronidase. (■) Peroxidase. Hatched area = acid-p=NP-Ph, alkaline p-NP-Ph, ATPase, alkaline phosphodiesterase, and LDH.

tivity appeared prior to the release of granular enzymes. Similar findings were observed when human PMNs or rat mononuclear cells were incubated with opsonized zymosan or antigen-antibody complexes (14,18). After a rise in chemotactic activity, a donor-dependent decline occurred.

It was previously reported that the tetrapeptide ECF-A is present within the granules of mast cells (10). The question therefore arose if these findings can be applied to the ECF as well. However, as sonicates of human PMNs did not show ECF activity (9), our studies were directed toward analyzing the subcellular distribution of ECF within PMNs. Stimulated PMNs were homogenized and the various fractions isolated by sucrose density gradient centrifugation. It became apparent that ECF was present and that it cosediments with microsomal fractions of the cells. It was therefore suggested that ECF might be derived from the plasma membrane of human PMNs (9,17). The data were confirmed when we reported in 1977 that incubation of cells such as human PMNs, mononuclear cells, or rat basophilic leukemia (RBL) cells and purified mast cells with arachidonic acid (AA) induced ECF generation, whereas human lymphocytes did not (16,18,19,30,31) (Fig. 2).

The kinetics of ECF release was similar to that induced by the above-mentioned stimuli. With the observation that AA is able to generate ECF in a kinetic pattern similar to other stimuli, the question arose as to the common link in the course of cell activation leading to ECF production. Both stimuli—the calcium ionophore and opsonized zymosan—led to cell triggering, resulting in activation of phospholipase with the subsequent turnover of AA. Furthermore, the activation of phospholipase is calcium-dependent, whereas the stimulatory effect or the turnover of AA by the cyclo-oxygenase and lipoxygenase systems does not require extracellular calcium. These results correlated with the data obtained for ECF. Calcium ions are required to trigger the cells for ECF release in the presence of the ionophore and phagocytic particles, whereas the AA-induced ECF release is not calcium-dependent.

Our experiments were therefore directed toward the question of whether the incubation of cells with various concentrations of phospholipase A, C, and D leads to ECF induction. It became apparent that only phospholipase A_2 and not C and D induced significant ECF production (32). The optimal concentration for phospholipase A_2 ranged between 1 and 18 ng/ml.

In order to analyze the membrane biological events more precisely, melittin was applied in our assay system. Melittin, the major component of honey bee venom, is an amphipathic peptide which acts as a direct lytic factor on biomembranes. The cationic peptide is unusual in that its amino acids are unequally distributed: positions 1–20 are occupied by largely hydrophobic amino acids, and positions 21–26 are

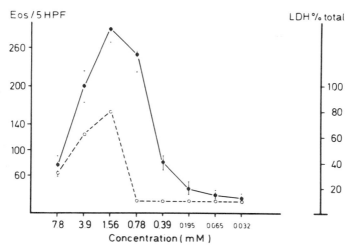

FIG. 2. Dose-dependent release of ECF from human PMNs by arachidonic acid. It occurs at toxic and nontoxic concentrations.

hydrophilic amino acids. Since the peptide is too small to accommodate itself to a globular configuration, it appears to associate as tetramer micelles in aqueous media.

Purified human PMNs (1×10^7/ml) were incubated with various concentrations of melittin; after 10 min of incubation the cells were centrifuged and the supernatant assayed for ECF activity (26). ECF-A was detected at concentrations of 2.9 μM and reached its optimum at 11.7 μM. With increasing concentrations, less ECF activity was generated. It was also demonstrated that melittin leads to the release of a granular enzyme marker such as β-glucuronidase; lactate dehydrogenase reaches only 20% of its total activity at the highest melittin concentration used in the experiments, suggesting that ECF induction is probably initiated at subtoxic concentrations of melittin. The kinetics of ECF induction are similar to those described for the calcium ionophore or the opsonized zymosan particles.

We also analyzed the effect of the various melittin fragments on the cells with regard to ECF generation. Human PMNs or rat peritoneal cells (10% mast cells) were incubated with either melittin or melittin 20–26 at various concentrations. Only the melittin portion 20–26 induced ECF release from human PMNs and rat peritoneal cells. This fragment was also able to release histamine from rat mast cells. When both fragments were combined and added to the cells, no increase in ECF activity was observed.

These data indicate that the intact molecule is required for optimal ECF induction. It should be emphasized that other membrane active substances [e.g., the mast cell degranulating peptide (MCD) from honey bee venom, compound 48/80, and polymyxin B] were not able to induce ECF generation, suggesting a different pathway of membrane activation.

These results were confirmed when purified rat peritoneal mast cells were analyzed with regard to the stimulation of anti-immunoglobulin E (IgE), compound 48/80, polymyxin B, AA, melittin, and the calcium ionophore. The cells were incubated with the various agents for 15 min at 37°C and the supernatant as well as the sonicate of the cells were analyzed for chemotactic activity. It is apparent that in the control (cells without stimuli) insignificant amounts of chemotactic activity were detected in the sonicates of the cells. High amounts of chemotactic activity were generated on incubation of the cells with AA, melittin, and the calcium ionophore, whereas no chemotactic activity was generated with anti-IgE, compound 48/80, or polymyxin B. These results clearly indicated that the chemotactic activity induced by various stimuli is generated upon stimulation of the cells, and that preformed chemotactic activity directed against eosinophils is negligible. We therefore suggested that the various stimuli lead to phospholipase activation with the subsequent turnover of phospholipids (Fig. 3). AA is then generated and is converted by a lipoxygenase-like enzyme within the cytosol to ECF (32). Our data clearly emphasize the precursor role of AA for the generation of ECF.

GENERATION OF ECF FROM SOLUBLE COMPONENTS

Based on this hypothesis, we devised experiments to generate ECF from soluble components in the absence of intact cells (22,24). To prove this assumption, human

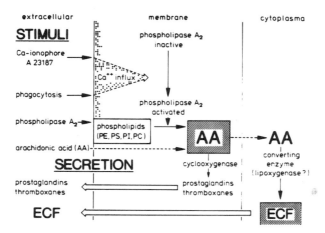

FIG. 3. Hypothetical scheme of ECF generation by various stimuli. AA = Arachidonic acid. PE = Phosphatidyl ethanolamine. PS = Phosphatidyl serine. PI = Phosphatidyl inositol. PC = Phosphorylcholine.

PMNs were homogenized and the various fractions recovered at 400, 3,000, 20,000, and 200,000 $\times$ *g* with a final supernatant fraction. AA and AA derivatives were then added to the various fractions. The results indicated that the 200,000$\times$ *g* supernatant fraction contains enzymatic activity which specifically transforms free AA into biologically active compounds such as ECF.

Since we suggested that phospholipase A_2 activation is required to induce AA release from phospholipids within the membrane of intact cells, an experimental approach was devised to analyze the biochemical cascade of ECF generation from subcellular components, e.g., in the absence of intact cells. Purified bee venom phospholipase A_2 and the 200,000 $\times$ *g* supernatant were incubated with various concentrations of either phosphatidyl inositol, phosphatidyl ethanolamine, or lysophosphatidyl choline for 15 min at 37°C. Standard ECF in the presence of the various phospholipids or incubation of the fraction at 200,000 $\times$ *g* and phospholipase A_2 in the absence of phospholipids were the controls. The latter incubation induced an ECF activity showing 250 eosinophils/5 high power fields. The mixture of phospholipase A_2 and the 200,000 $\times$ *g* supernatant in the presence of phosphatidyl inositol led to a 230% increase of ECF release at a phosphatidyl inositol concentration of 0.13 μg/ml. At this concentration phosphatidyl inositol induced a 38.1% inhibition of the eosinophil chemotaxis as assessed with a standard ECF. With phosphatidyl ethanolamine in the incubation mixture, a 260% increase was observed at this concentration. Similar results were obtained with lysophosphatidyl choline (Fig. 4).

We previously demonstrated that the ECFs obtained by the various stimuli are of low molecular weight and eluted after vitamin B_{12} by gel filtration on Sephadex G 25. When ECF-containing supernatants were extracted with chloroform, almost 50 to 80% of the original chemotactic activity was recovered with the organic phase.

phospholipid	concentration (μg/ml)	+ fraction V	+ fraction V + PLA$_2$	+ ECF (inhibition) %
phosphatidyl-	3.33	0	25	60.0
inositol	0.67	0	110	41.5
	0.13	0	610	38.1
	0.03	0	490	10.2
phosphatidyl-	3.33	0	0	99.8
ethanolamine	0.67	0	150	66.4
	0.13	0	680	30.1
lysophospha-	3.33	0	0	88.2
tidylcholine	0.67	0	410	44.5
	0.13	0	680	29.5

FIG. 4. Generation of ECF by phospholipase A$_2$ in the presence of the ECF converting enzyme (fraction 5 = 200,000 × g supernatant) and phospholipids.

The ECF-containing material after chloroform extraction was further analyzed by silicic acid chromotography. Sequential elution with hexane, ethyl acetate, acetone, diethyl ether, and finally with ethanol:concentrated ammonia:water was carried out. ECF was recovered in the ethyl acetate fraction.

COMPARISON OF ECF WITH HYDROXYEICOSATETRAENOIC ACIDS

Borgeat, Hammarström, Samuelsson and co-workers analyzed AA metabolism in leukocytes with regard to the products of the lipoxygenase pathway (2–5,29). They identified the following compounds: 5-D-hydroxy-6,8,11,14-eicosatetraenoic acid, 15-L-hydroxy-5,8,11,13-eicosatetraenoic acid, and 5-D,12D-dihydroxy-6,8,10,14-eicosatetraenoic acid. It was also reported that the two hydroxy acids formed from AA and microsomal platelet fraction [i.e., 12-L-hydroxy-5,8,10,14-eicosatetraenoic acid (HETE) and 12L-hydroxy-5,8,10-heptadecatrienoic acid (HHT)] lead to chemokinesis and chemotaxis of PMNs (11,12). A comparison of the melittin- or AA-induced ECF with standard preparations of 15-HETE, LTB$_4$, 5-HETE, 12-HETE, LTC$_4$, LTD$_4$, and 15-HPETE clearly indicated that a number of spots were identified by thin-layer chromatography (Fig. 5). These results were recently confirmed by high-pressure liquid chromatography (HPLC) analysis. When these compounds were analyzed with regard to their ECF activity, it became apparent that only LTB$_4$ and to a lesser degree 12-HETE and 5-HPETE revealed ECF activity as assessed by the Boyden chamber technique (Fig. 6). Injection of the various endogeneous HETEs into the peritoneal cavity of rats led to a substantial change

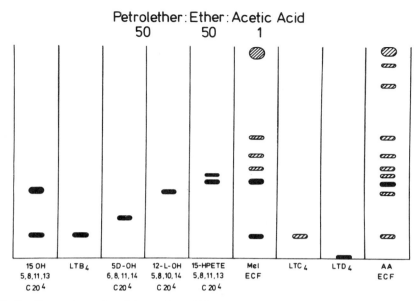

Petrolether:Ether:Acetic Acid
50   50   1

| 15 OH 5,8,11,13 C 20 4 | LTB4 | 5D-OH 6,8,11,14 C 20 4 | 12-L-OH 5,8,10,14 C 20 4 | 15-HPETE 5,8,11,13 C 20 4 | Mel ECF | LTC4 | LTD4 | AA ECF |

FIG. 5. Comparison of melittin- and arachidonic acid-induced ECF with various mono- and di-HETEs by thin-layer chromatography.

	μg / ml				
	9.3	1.33	0.186	0.093	0.008
Leukotriene B,	18	174	349	123	31
Leukotriene C,	—	—	—	—	—
Leukotriene D,	—	—	—	—	—
12-HETE	12	113	28	16	—
5-HPETE	—	45	32	21	—

FIG. 6. Analysis of mono- and di-HETEs with regard to their eosinophil chemotactic properties.

in the distribution of the various cells. The peritoneal cells of control animals are comprised of 5 to 10% mast cells, 4 to 8% eosinophils, 0 to 2% PMNs and 75 to 80% peritoneal cells.

Injections of 1 μg 12-HETE induced an influx of PMNs-increased by 50 to 53% after 30 min. LTB$_4$ led to a 20 to 25% increase of eosinophils. Higher concentrations of LTB$_4$ (more than 3 μg) under the experimental conditions also induced a significant number of PMNs. LTC$_4$ led to an increase of PMN influx, and LTD$_4$ slightly increased the number of eosinophils. 5-HETE, 15-HETE, and 15-HPETE induced a significant increase of neutrophils, ranging between 30 to 53%. With 5-HETE a slight increase in the number of eosinophils was observed as well. These data indicate that under *in vivo* conditions the cellular distribution is markedly modulated in the presence of mono- and dihydroxylated eicosatetraenoic acids (Fig. 7). They also suggest that the biological activity which had been termed ECF operationally is most likely due to the presence of LTB_4 or isomers of this compound. Indeed, quite recently we compared the eosinophil chemotactic properties of LTB_4 and $LTBX_4$. It was observed that both compounds are active in a similar dose-response range, although $LTBX_4$ also induced aggregation of the cells. LTB_4 was then injected into the skin of guinea pigs, and the influx of cells was examined during various times. Ten minutes after injection, a marked increase of granulocytes (eosinophils, neutrophils) were observed in the blood vessels; after 30 min these cells were distributed in the surrounding tissue; and 2 hr after injection the cells slowly disappeared.

INTERACTION OF HETES WITH C5a

Endogenous HETEs are not only mediators of inflammation, they also modulate the chemotactic responsiveness of eosinophil granulocytes in the presence of C5a (Fig. 8). For this purpose, eosinophils in the presence or absence of HETEs were stimulated with C5a (applied at a suboptimal biological concentration).

When 15-HETE was analyzed, it was observed that the presence of this compound in the upper and lower compartments led to an increased chemotactic responsiveness which exceeded that obtained when 15-HETE was present in the upper compartment alone or in the lower compartment with C5a.

A modest chemotactic response was obtained when LTB_4 was present in the upper as well as the lower compartment, whereas a marked increase in chemotactic responsiveness toward C5a was observed when LTB_4 was present in the upper or lower compartment. The *trans,trans*,tr. isomer of LTB_4 increased the C5a-induced responsiveness when it was present in the upper compartment; the activity obtained was less when it was present in the lower compartment and intermediate when this agent was present in the upper and lower compartment as well.

	PMN	Eos	MC	PC
CONTROL	0 - 2	4 - 8	5 - 10	75 - 80
12-HETE	50 - 53	3 - 8	3 - 7	35 - 40
LEUKOTRIENE B_4	0 - 2	25 - 35	15 - 20	50 - 55
LEUKOTRIENE C_4	60 - 75	4 - 12	5 - 10	15 - 20
LEUKOTRIENE D_4	0 - 1	12 - 18	10 - 15	75 - 85
5-HETE	50 - 53	12 - 15	3 - 5	28 - 35
15-HETE	40 - 50	4 - 8	3 - 6	35 - 45
15-HPETE	30 - 40	8 - 10	4 - 6	40 - 50

FIG. 7.   Influx of cells into the peritoneal cavity after injection of mono- and di-HETEs.

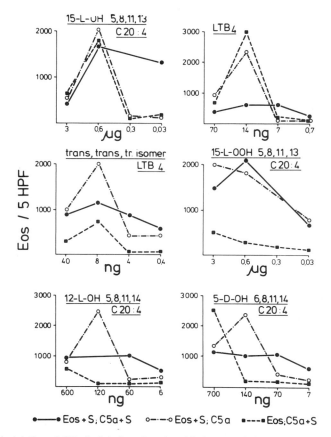

FIG. 8. Modulation of C5a (rat)-induced eosinophil chemotaxis in the presence of mono- and di-HETEs.

The presence of 15-hydroperoxyeicosatetraenoic acid (15-HPETE) led to a modest increase in the chemotactic response when it was present in the lower compartment together with C5a; a marked increase was observed when it was present in the upper compartment together with eosinophils or was applied in the upper and lower compartment.

12-HETE led to a significant increase in chemotactic activity when it was present with eosinophils in the upper compartment; an intermediate response was observed when the compound was present in the upper and lower compartments; a negligible chemotactic response was observed when 12-HETE was present with C5a in the lower compartment.

A significant increase was demonstrated when 5-HETE was present with eosinophils in the upper compartment; and an intermediate response was observed when the compound was present in the upper and lower compartments. A significant increase was observed only when the compound in the lower compartment was added at concentrations above 140 ng.

These results indicate that addition of endogeneous HETEs to the target cells induces significant chemokinetic activation of the cells in the presence of C5a. The fact that an intermediate response is observed with LTB$_4$, the *trans,trans,tr.* isomer of LTB$_4$, 12-HETE, and 5-HETE present in the upper and lower compartments can be attributed to the fact that these compounds deactivate the cells for a further response toward C5a. This also holds true for the *trans,trans,*tr. isomer of LTB$_4$, which, however, revealed no chemotactic activity toward the eosinophils by itself but apparently has receptor sites on the cells to deactivate the cells for a further response toward C5a. In general, addition of mono- and di-HETEs to the target cells in the upper compartment induces more activity than their addition with C5a in the lower compartment. This may be partially due to the fact that the volumes in the lower compartment do not correspond to the volume in the upper compartment, and it partially emphasizes that the mono- and di-HETEs exert profound chemokinetic properties and thus modulate the responsiveness of the cells toward other stimuli. So far, our results clearly indicate that the biologically defined ECF is a lipoxygenase product.

IMMUNOPHARMACOLOGICAL MODULATION OF ECF GENERATION

Further confirmation was obtained when human PMNs were stimulated with the calcium ionophore in the presence of various concentrations of indomethacin or 5,8,11,14-eicosatetraynoic acid (ETYA). With indomethacin in the incubation mixture, an increase in ECF was observed at concentrations which inhibit the cyclooxygenase pathway. With ETYA in the incubation mixture, a marked decrease in the production of ECF was detected (24).

The question therefore arose whether polyynoic or polienoic acids may differ with respect to their ECF release by various stimuli from PMNs, purified rat mast cells, and subcellular components. Such an analysis could be helpful in elucidating the possibilities of inhibiting the cyclo oxygenase and/or lipoxygenase pathway or even individual HETEs selectively.

Human PMNs (1×10^7/ml) were incubated with the various ECF-generating stimuli—e.g., AA 109 μM, the calcium ionophore 0.33, μM, and melittin 11.7 μM in the presence of various AA analogs. The inhibitory activities of 5,8,11-eicosatriynoic, 4,7,10,13-eicosatetraynoic, 5,8,11-eicosatetraenoic, and 4,7,10,13-eicosatetraenoic acid were assayed at different concentrations (27). Cells prepared in the presence of the various stimuli but without the inhibitors served as controls.

In the presence of the polyynoic acids ($\sim$110 μM), the inhibition ranged from 78 to 93% when the various stimuli were used for ECF induction. For the polienoic acids, a decrease in ECF release by 40 to 60% was observed. In addition, we studied the effect of the polienoic and polyynoic acids on the cells alone (Fig. 9). These compounds, by themselves, did not stimulate cells. Our results also dem-

	5,8,11-Eicosatriynoic acid (111.0 µM)	4,7,10,13-Eicosatetraynoic acid (112.5 µM)	5,8,11-Eicosatrienoic acid (108.8 µM)	4,7,10,13-Eicosatetraenoic acid (109.5 µM)	TCM control
Arachidonic acid (109.5 µM)	39 ± 4.3	16 ± 2.7	114 ± 11.3	96 ± 8.5	240 ± 15.3
Ionophore (0.33 µM)	26 ± 5.2	13 ± 3.0	67 ± 7.9	58 ± 6.2	164 ± 16.1
Melittin (11.7 µM)	61 ± 5.0	47 ± 14.2	139 ± 14.2	105 ± 7.6	273 ± 19.3
Zymosan (x) (4.5 mg/ml)	25 ± 2.9	22 ± 2.3	76 ± 5.4	71 ± 4.7	130 ± 12.7

FIG. 9. Comparison of the degree of inhibition with regard to the various inhibitors. ECF-inducing stimuli: AA, the calcium ionophore, melittin, and opsonized zymosan (Zx). TCM control: incubation of the PMNs with the various stimuli in the absence of the inhibitors.

onstrated that among the various AA analogs, 4,7,10,13-eicosatetraynoic acid proved to be the most potent inhibitor.

As our previous results demonstrated that stimulation of PMNs leads to a time-dependent generation of ECF, experiments were carried out to study the inhibitory properties in a kinetic analysis. Human PMNs (1×10^7) were incubated with either melittin (~11.7 µM) or AA (~109.5 µM) in the presence of the inhibitors. In the absence of the inhibitors, potent ECF was generated after 15 min of incubation. In the presence of 5,8,11-eicosatriynoic acid or 4,7,10,13-eicosatetraynoic acid, a marked reduction in ECF activity was observed which was more pronounced with the 4,7,10,13-eicosatetraynoic acid as compared to the 5,8,11-eicosatriynoic acid. Although the inhibition is remarkable, the pattern of ECF release resembles in each case that of the control: ECF generation occurs rapidly during the first 15 min of incubation and then decreases with time. Similar experiments were performed with AA as the ECF-generating stimulus, and the results were similar.

We then addressed our studies to the question of whether the generation of ECF from purified rat mast cells is also affected by AA analogs. Such an analysis was of particular interest as it has been reported that the tetrapeptides ECF-A are present as preformed mediators within the cells (10). However, our data had strongly suggested that the most potent chemotactic factors for granulocytes were of lipid origin and were not preformed but were generated in the course of stimulation from rat mast cells, human basophils, and polymorphonuclear and mononuclear cells.

In the experiments, rat mast cells (5×10^5/ ml) were incubated with either AA (~109.5 µM), the calcium ionophore (~0.33 µM), or melittin (~11.7 µM) in the presence or absence of the various inhibitors. It was shown that the polynoic acids are more active than the polienoic acids. It was also observed that 4,7,10,13-eicosatetraynoic acid is a more potent inhibitor than 5,8,11-eicosatriynoic acid.

These data indicate that the ECF obtained from purified rat mast cells can be blocked by inhibitors which interfere with the lipoxygenase pathway of AA transformation. The results support our previous notion that the most potent chemotactic factor from rat mast cells is a lipid product.

Experiments were then performed to analyze whether these inhibitors also affect the generation of ECF from subcellular components. We previously demonstrated that the $200,000 \times g$ supernatant fraction of homogenized cells releases ECF on incubation with AA but not with the calcium ionophore. When melittin was used as the stimulus, the $200,000 \times g$ pellet also proved to be a potent source of ECF. There is evidence that melittin leads to phospholipid turnover in cells and in subcellular components with the subsequent generation of AA.

From these data we suggest that a soluble component (most likely a lipoxygenase) is recovered from homogenized cells which is able to transform free AA into biologically active ECF. In the described experiments, subcellular fractions of human PMN or rat mast cells were obtained and incubated with either AA ($\sim$547.5 μM) or melittin ($\sim$58.5 M) in the absence or presence of polynoic acids. It was confirmed that 4,7,10,13-ETYA is the most potent inhibitor for ECF obtained from subcellular components. These results clearly indicate that AA analogs may serve as a potential tool for modulating the lipoxygenase and/or cyclo-oxygenase pathway of AA transformation and may thus represent a novel immunopharmacological approach to modulating the generation of mono- and di-HETEs in the inflammatory disease process.

ACKNOWLEDGMENTS

This work was supported by the Deutsche Forschungsgemeinschaft (DFG Kö 427/3). The authors appreciate the careful typing of the manuscript by Mrs. Weber.

REFERENCES

1. Bach, M. K., and Brashler, J. R. (1978): Ionophore A23187-induced production of slow reacting substance of anaphylaxis (SRS-A) by rat peritoneal cells in vitro: evidence for production by mononuclear cells. *J. Immunol.*, 120:998–1005.
2. Borgeat, P., and Samuelsson, B. (1979): Arachidonic acid metabolism in polymorphonuclear leukocytes: effects of ionophore A23187. *Proc. Natl. Acad. Sci. USA*, 76:2148–2152.
3. Borgeat, P., and Samuelsson, B. (1979): Arachidonic acid metabolism in polymorphonuclear leukocytes: unstable intermediate in formation of dihydroxy acids. *Proc. Natl. Acad. Sci. USA*, 76:3213–3217.
4. Borgeat, P., and Samuelsson, B. (1979): Transformation of arachidonic acid by rabbit polymorphonuclear leukocytes. *J. Biol. Chem.*, 254:2643–2646.
5. Borgeat, P., and Samuelsson, B. (1979): Metabolism of arachidonic acid in polymorphonuclear leucocytes: structural analysis of novel hydroxylated compounds. *J. Biol. Chem.*, 254:7865–7869.
6. Czarnetzki, B. M., König, W., and Lichtenstein, L. M. (1976): Eosinophil chemotactic factor (ECF). I. Release from polymorphonuclear leukocytes by the calcium ionophore A23187. *J. Immunol.*, 117:229–234.
7. Czarnetzki, B. M., König, W., and Lichtenstein, L. M. (1978): Eosinophil chemotactic factor (ECF). IV. Inhibitors in human peripheral leukocytes. *Int. Arch. Allergy Appl. Immunol.*, 56:398–407.
8. Ford-Hutchinson, A. W., Bray, M. A., Cunningham, F. M., Davidson, E. M., and Smith, M. J. H. (1981): Isomers of leukotriene B$_4$ possess different biological potencies. *Prostaglandins*, 21:143–152.

9. Frickhofen, N., and König, W. (1979): Subcellular localization of the eosinophil chemotactic factor (ECF) and its inactivator in human PMNs. *Immunology*, 37:111–122.
10. Goetzl, E. J., and Austen, K. F. (1970): Purification and synthesis of eosinophilotactic tetrapeptides of human lung tissue: identification as eosinophil chemotactic factor of anaphylaxis. *Proc. Natl. Acad. Sci. USA*, 72:4123–4127.
11. Goetzl, E. J., Woods, J. M., and Forman, R. R. (1977): Stimulation of human eosinophil and neutrophil polymorphonuclear leukocytes chemotaxis and random migration by 12-L-hydroxy-5,8,10,14-eicosatetraenoic acid. *J. Clin. Invest.*, 59:179–183.
12. Goetzl, E. J., and Forman, R. R. (1977): Chemotactic and chemokinetic stimulation of human eosinophil and neutrophil polymorphonuclear leukocytes by 12-L-hydroxy 5,8,10-heptadecatrienoic acid (HHT). *J. Immunol.*, 120:526–531.
13. Goetzl, E. J., Derian, C., and Valone, F. H. (1980): The extracellular and intracellular roles of hydroxy-eicosatetraenoic acids in the modulation of polymorphonuclear leukocyte and macrophage function. *J. Reticuloendothel. Soc.*, 28:1059–1119.
14. König, W., Czarnetzki, B. M., and Lichtenstein, L. M. (1976): Eosinophil chemotactic factor (ECF). II. Release during phagocytosis of human polymorphonuclear leukocytes. *J. Immunol.*, 117:235–245.
15. König, W., Czarnetzki, B. M., and Lichtenstein, L. M. (1978): Eosinophil chemotactic factor (ECF). III. Generation in human peripheral leukocytes. *Int. Arch. Allergy Appl. Immunol.*, 56:364–375.
16. König, W., Tesch, H., and Frickhofen, N. (1978): Generation and release of eosinophil chemotactic factor from human polymorphonuclear neutrophils by arachidonic acid. *Eur. J. Immunol.*, 8:434–437.
17. König, W., Tesch, H., and Frickhofen, N. (1978): Initiation and modulation of ECF generation and secretion from human PMNs. *Naunyn Schmiedebergs Arch. Pharmacol. (Suppl.)*, 302:R48. (abstract).
18. König, W., Frickhofen, N., and Tesch, H. (1979): Generation and secretion of eosinophilotactic activity from polymorphonuclear neutrophils by various mechanisms of cell activation. *Immunology*, 36:733–742.
19. König, W., Tesch, H., and Kroegel, C. (1979): On the nature of the eosinophil chemotactic factor (ECF) from human PMNs. *J. Allergy Clin. Immunol.*, 63:178 (abstract).
20. König, W., Frickhofen, N., and Tesch, H. (1979): Generation of an eosinophilotactic factor from human PMNs by various mechanisms of cell activation in function and structure of the immune system. *Adv. Exp. Med. Biol.*, 114:521–526.
21. König, W., Tesch, H., and Frickhofen, N. (1979): On the mechanism of ECF generation, secretion and inactivation. *Monogr. Allergy*, 14:203–208.
22. König, W., Kroegel, C., and Tesch, H. (1980): Generation and modulation of the eosinophil chemotactic factor (ECF) from human PMN and rat mononuclear cells by phospholipids. *Fed. Proc.*, 39:691.
23. König, W., Kroegel, C., Kunau, H. W., and Borgeat, P. (1981): Comparison of the eosinophil-chemotactic factor (ECF) with endogenous hydroxy-eicosatetraenoic acids. *Int. Arch. Allergy Appl. Immunol.*, 66:168–171.
24. König, W., Kroegel, C., Pfeiffer, P., and Tesch, H. (1981): Modulation of the eosinophil chemotactic factor (ECF) release from various cells and their subcellular components by phospholipids. *Int. Arch. Allergy Appl. Immunol.*, 65:417–431.
25. König, W., Pfeiffer, P., and Kunau, H. W. (1981): Effect of arachidonic acid metabolites on the histamine release from human basophils and rat mast cells. *Int. Arch. Allergy Immunol.*, 66:149–151.
26. Kroegel, C., König, W., Mollay, C., and Kreil, G. (1980): Generation of the eosinophil chemotactic factor (ECF) from various cell types by melittin. *Mol. Immunol.*, 18:227–236.
27. Kroegel, C., Kunau, H. W., and König, W. (1981): Inhibition of the eosinophil chemotactic factor (ECF) release from human PMNs and rat mast cells by arachidonic acid analogues. *Cell. Immunol.*, 60:480–488.
28. Marone, G., Kagey-Sobotka, A., and Lichtenstein, L. M. (1979): Effects of arachidonic acid and its metabolites of antigen induced histamine release from human basophils in vitro. *J. Immunol.*, 123:1669–1677.
29. Samuelsson, B., Borgeat, P., Hammarström, S., Murphy, R. S. (1980): Leukotrienes: a new group of biologically active compounds. In: *Advances in Prostaglandin and Thromboxane Re-*

search, Vol. 6, edited by B. Samuelsson, P. W., Ramwell, and R. Paoletti, pp. 1–18. Raven Press, new York.

30. Tesch, H., and König, W. (1978): Generation of eosinophil chemotactic factor from human neutrophils by arachidonic acid. *J. Allergy Clin. Immunol.,* 61:129 (abstract).

31. Tesch, H., König, W., and Frickhofen, N. (1979): Eosinophil chemotactic factor: release from human polymorphonuclear neutrophils by arachidonic acid. *Int. Arch. Allergy Appl. Immunol.,* 58:436–448.

32. Tesch, H., and König, W. (1980): Phospholipase A_2 and arachidonic acid: a common link in the generation of the eosinophil chemotactic factor (ECF) from human PMN by various stimuli. *Scand. J. Immunol.,* 11:409–418.

33. Turner, S. R., Campbell, J. A., and Lynn, W. S. (1975): Polymorphonuclear leukocyte chemotaxis toward oxidized lipoid components of cell membranes. *J. Exp. Med.,* 141:1437–1441.

34. Turner, S. R., Tainer, J. A., and Lynn, W. (1975): Biogenesis of chemotactic molecules by the arachidonic lipoxygenase system of platelets. *Nature,* 257:680–681.

Leukotrienes and Other Lipoxygenase Products,
edited by B. Samuelsson and R. Paoletti.
Raven Press, New York © 1982.

Lipoxygenase Products of Arachidonic Acid: Role in Modulation of IgE-Induced Histamine Release

Stephen P. Peters, Robert P. Schleimer, Gianni Marone, Anne Kagey-Sobotka, Marvin I. Siegel, and Lawrence M. Lichtenstein

Clinical Immunology Divison, Department of Medicine, The Johns Hopkins University School of Medicine at The Good Samaritan Hospital, Baltimore, Maryland 21239; and Burroughs-Wellcome Company, Research Triangle Park, North Carolina

Most of the interest in arachidonic acid metabolism via lipoxygenase pathways has centered on the effect of these metabolites as extracellular mediators of inflammatory processes, particularly as chemotactic molecules (5,6,27) or spasmogens (3,9). Our interest is different: For the last several years, we have concentrated on pharmacologic studies which focus on the role of endogenous products of the lipoxygenase pathways in two processes: (a) the secretory release of the mediators of inflammation, and (b) the specific reversal of the inhibition of that release by endogenous hormones (19,20). Our data suggest that a product of lipoxygenase pathway(s)—perhaps 5-hydroperoxyeicosatetraenoic acid (5-HPETE)—is necessary for histamine release from human basophils (28). Moreover, the same product specifically antagonizes the effects of a variety of hormones that act via adenylate cyclase to down-modulate release.

These studies were carried out with human basophils, in which the release mechanism has been well characterized by several decades of study (14,16,17,23,25). The cell preparations used for almost all studies were dextran-sedimented leukocytes or Percoll-separated mononuclear cells: In both, the concentration of basophils did not exceed a few percent and was usually < 1%. On several occasions, however, experiments were carried out with basophils purified to the extent of 40 to 80% (18). By decreasing the concentration of other cell types by approximately 100-fold, we have provided strong evidence that the effects to be described are due to the direct interaction between the agents used and the basophils, rather than involving another cell type. The mediator studied in the release reaction is histamine (28,29,31). Finally, release was usually triggered by an immunoglobulin E (IgE) antibody-related mechanism, i.e., release by either appropriate antigens or by anti-IgE.

The first observation which suggested that the metabolism of arachidonic acid was critical for IgE-mediated histamine release was that eicosatetraynoic acid (ETYA), a competitive antagonist of arachidonic acid, caused complete, dose-dependent inhibition of release (20). This result was essentially similar to that described by Sullivan and Parker using rat mast cells (33). However, although in this respect the two systems are similar, there are features of the release mechanisms which appear to be quite different. We were led to the hypothesis that it was the lipoxygenase pathways which were important in release by the observation that the cyclooxygenase inhibitors (i.e., indomethacin, meclofenamic acid, and aspirin) potentiated rather than inhibited IgE-mediated release (20). This result is shown in Fig. 1: Nanomolar concentrations of indomethacin caused a clearly discernible enhancement of release, and this effect was optimal at micromolar concentrations. The enhancement caused by indomethacin was variable depending on the cells of different donors and, of course, the intensity of the stimulus for release. The effect is extremely consistent, however, having been noted in essentially all of several dozen experiments. The fact that ETYA, an inhibitor of arachidonic acid metabolism, ablated release whereas indomethacin enhanced the process suggested that arachidonic acid itself would enhance release. Experiments designed to test this suggestion provided positive results: Arachidonic acid in submicromolar concentrations consistently enhanced IgE-mediated histamine release (Fig. 1). This is different from the rat mast cell system in which arachidonic acid and indomethacin had minimal effects on histamine release (33).

The most direct way of demonstrating that a product(s) of the lipoxygenase pathways is required for histamine release would be to use a specific inhibitor of this pathway. Unfortunately, no such inhibitor is available to date. However, to-

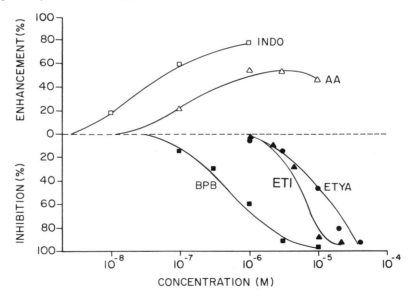

FIG. 1. Effect of indomethacin (INDO), arachidonic acid (AA), bromophenacyl bromide (BPB), eicosatriynoic acid (ETI), and eicosatetraynoic acid (ETYA) on antigen-induced histamine release from human basophils.

gether with Hammarström, we have used a relatively selective inhibitor, the triynoic derivative of arachidonic acid (eicosatriynoic acid, ETI) (19). In platelets and cultured human skin cells, ETI is 40 to 100 times more active against the lipoxy-genase enzymes than against those in the cyclo-oxygenase pathway (8). As shown in Fig. 1, ETI was also a potent inhibitor of IgE-mediated release, causing a 50% effect at micromolar concentrations. Although we have been unable, for technical reasons, to study the effects of this inhibitor on the relevant basophil enzymes, in skin cells these concentrations of ETI are sufficient to abolish lipoxygenase activity without significantly influencing the cyclo-oxygenase pathways. The inhibition observed with ETI (and ETYA) occurs only after antigen-induced activation of the cells; that is, before antigen exposure, the basophils can be incubated with mi-cromolar concentrations of ETI and, if washed, are fully able to release to subsequent immunologic stimuli. It should be noted that ETI (and ETYA) are also potent inhibitors of the release induced by all of the active stimuli we have studied; these include the calcium ionophore A23187, f-Met peptides and 12-O-tetradecanoyl-phorbol-13-acetate (TPA), as well as antigen and anti-IgE.

These pharmacologic data, taken together, strongly suggest that a product of arachidonic acid generated by a lipoxygenase pathway represents a critical step in the release mechanism.

Before describing experiments which further define the nature of such a product, it is necessary to describe another activity which is shared by the nonsteroidal anti-inflammatory drugs and arachidonic acid. As shown in Table 1, indomethacin produced a reversal of the inhibition of histamine release caused by a histamine-like agonist, dimaprit, which activates H_2 receptors, and by prostaglandin E_2 (20). This reversal is dose-dependent and is more than the result of a balance between an inhibiting agent and an enhancing agent; the reversal is essentially complete. Additional experiments showed that each of the several endogenous hormones which act through cell surface receptors to activate adenylate cyclase, increase cyclic AMP, and inhibit histamine release was similarly affected by indomethacin. These include H_2 agonists, beta-adrenergic agonists, prostaglandins E_1 and E_2, and aden-osine. The effects noted with indomethacin are also observed with aspirin and meclofenamic acid at appropriately higher concentrations. Arachidonic acid itself had a similar effect, causing the reversal of inhibition at micromolar concentrations.

TABLE 1. *Reversal of the inhibition of histamine release by 1 μM indomethacin*

Agonist	% Inhibition	% Reversal	No.
PGE$_2$ 1 μM	53 ± 5	94 ± 5	7
Dimaprit 10 μM	67 ± 8	102 ± 7	4
IBMX 0.5 mM	85 ± 5	7 ± 5	4
Dibutyryl cAMP 3 mM	71 ± 8	2 ± 2	6

As also shown in Table 1, the ability to reverse inhibition is strictly limited to agents which inhibit histamine release by activating adenylate cyclase. Indomethacin, aspirin, and arachidonic acid have no effect on the inhibition caused by agents which increase cyclic AMP by other means; for example, they do not reverse the inhibition caused by dibutyryl cyclic AMP, the phosphodiesterase inhibitors, iso-butylmethylxanthine, or theophylline. We cannot answer the interesting question of whether indomethacin's ability to block cyclase agonists is due to a disassociation of the receptor-cyclase link and a failure to increase cyclic AMP levels or if different pools of cAMP account for the differences observed. Measurements of cyclic AMP levels after treatment of mononuclear cells or whole leukocyte preparations with prostaglandin E_2 showed that the rise in cyclic AMP in these crude cell preparations is not appreciably altered by indomethacin (20). However, the measurements of cAMP in appropriately treated, purified basophil preparations have not been completed.

The experiments thus far described indicate that the postulated product(s) of the lipoxygenase pathway has two activities, each with a proinflammatory result; one is necessary for the release process, whereas the other diminishes the activity of endogenous hormones which negatively modulate release. The next step was to ascertain which products of the lipoxygenase pathway were carrying out these functions. For these experiments, we were fortunate to have a number of collaborators who supplied us with purified and characterized lipoxygenase products. We first examined the leukotrienes. Neither leukotriene C or D (LTC, LTD), synthetic or natural, at concentrations ranging from 10^{-10} to 10^{-5} M enhanced or inhibited histamine release. Although we have not carried out an exhaustive series of experiments, a variety of approaches designed to find either direct releasing activity or some sort of modulating activity by the leukotrienes have been completely unsuccessful.

We then turned our attention to more proximal products of the 5-lipoxygenase pathway: 5-hydroperoxyeicosatetraenoic acid (5-HPETE) and 5-hydroxyeicosatetraenoic acid (5-HETE). Figure 2 shows that both had enhancing activity. 5-HPETE enhanced release at nanomolar concentrations, with a peak effect at 1 μM. 5-HETE was consistently less active than the parent molecule, with an average potency 3- to 10-fold lower than that of 5-HPETE. In a large series of simultaneous experiments, the maximal enhancing efficacy of 5-HPETE was essentially equal to that of indomethacin. Further, when one agent caused maximal enhancement, there was no additional effect with the addition of the other. This lack of an additive effect supports the concept that 5-HPETE and indomethacin are acting on the same release pathway. Control experiments showed that release enhanced by 5-HPETE or indomethacin was further stimulated by enhancing agents, such as D_2O, which act on separate pathways.

Preliminary experiments have been carried out with other hydroperoxy products of lipoxygenase pathways. 11-HPETE, 15-HPETE, and the stable linoleic acid derivative 13-hydroperoxylinoleic acid (13-HPLA) are approximately 100-fold less active than the 5-hydroperoxy product. Indeed, enhancing effects were consistently

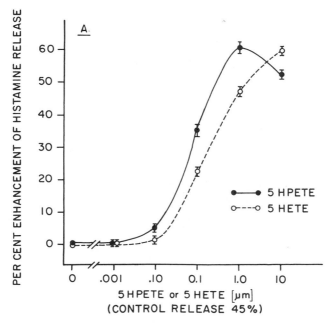

FIG. 2. Enhancement of antigen-induced histamine release by 5-HPETE and 5-HETE. The experiment involved the addition of cells to tubes to which the drugs had been added in ether, and dried under nitrogen.

seen only at concentrations of 10 μM, which raises questions regarding the specificity of this activity. High concentrations of H_2O_2 had no effect on histamine release, suggesting that the enhancement seen with 5-HPETE was not due to nonspecific effects related only to the peroxide moiety.

The basophil histamine release system has been functionally divided into two stages (15). The first is a calcium-independent stage during which cyclic AMP-dependent events occur. The second stage is calcium-dependent and involves a series of steps which are blocked by a variety of agents such as metabolic antagonists. 5-HPETE has little enhancing effect if present in the first stage of release; if added in the second stage of the release process, however, the enhancement is about equal to that seen in the whole reaction. These experiments are not absolutely clear as there is some residual enhancing effect when 5-HPETE is present in the first stage, presumably due to the fact that it cannot be completely washed away. In a series of six experiments, however, the effects of 5-HPETE in the first stage were always less than those in the second stage. Therefore, the enhancing events caused by 5-HPETE occur relatively late in the release process, after the cAMP-dependent events.

It was therefore surprising to find that the same lipoxygenase product, 5-HPETE, also had activity in reversing the inhibition caused by agonists which act through adenylate cyclase. These two activities appeared to be quite different in nature and therefore likely to be mediated by separate products. However, as shown in Fig. 3, 5-HPETE also caused a reversal of the prostaglandin E_2 inhibition of histamine release; other experiments have shown that 5-HPETE and 5-HETE both reverse the inhibition caused by dimaprit and other cyclase-active agonists in a dose-dependent fashion. Like the effects noted earlier with indomethacin and arachidonic acid, the activity of 5-HPETE is limited to reversing this particular set of agonists, having no effect on the inhibition caused by dibutyryl cyclic AMP or isobutylmethylxanthine (Fig. 3). Again, this activity is relatively selective for the 5-lipoxygenase products. Preliminary studies with 11- and 15-HPETE and with 13-HPLA show that they have either no action or reverse inhibition only marginally at 10 μM concentrations.

These results demonstrate that 5-HPETE (and 5-HETE) have both of the activities we have attributed to potentiation of the lipoxygenase pathway by indomethacin or arachidonic acid. Although it is, perhaps, not unique in this respect, these activities are not shared equally by the several other hydroperoxy derivatives of arachidonic acid which were tested. However, we have not studied all of the products of lipoxygenase pathways. We are particularly interested in whether or not LTB (5,12-dihydroxyeicosatetraenoic acid), the very potent chemotactic agent (5,6,27), is also active. These experiments will be carried out when material becomes available to us.

Finally, we have been interested in the mechanism by which arachidonic acid is generated after immunologic stimulation. This is generally thought to be due to the activation of cell-membrane-bound phospholipase A_2 (4,10,24), although in the rat mast cell there is evidence that other pathways may be involved (12,32). We can comment only indirectly on this matter from our studies with human basophils. The relatively specific inhibitor of phospholipase A_2, *p*-bromophenacyl bromide (BPB) (34,35), completely blocks histamine release at micromolar concentrations (21,22). The characteristics of this inhibition suggest that BPB is indeed blocking phospholipase A_2, but no direct biochemical measurements have been carried out. The

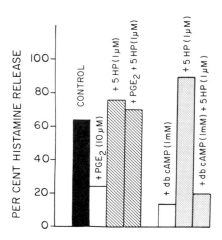

FIG. 3. Inhibition of histamine release by PGE_2 and its reversal by 5-HPETE. The experiment also shows that inhibition by dibutyryl cAMP is not reversed by 5-HPETE.

observations of Flower and Blackwell (4) and more recently of Hirata and co-workers (10) that anti-inflammatory steroids induced the production of a protein that inhibits the action of phospholipase A_2 suggested that those steroids should inhibit histamine release from basophils. Previous experiments with corticosteroids, however, using basophils and rat mast cells have shown absolutely no effects even at concentrations as high as 100 μM (13,26,30; E. Gillespie, *personal communication*). Notable in the studies of Hirata and co-workers (10) was the kinetics of inhibition. Rather long periods of time (16 hr) were required to see these effects. We have recently found that for steroids to have an effect in the human basophil system it also requires an appreciable amount of time and, therefore, cell culture (29). Figure 4 demonstrates the dose-dependent inhibition of basophil histamine release by hydrocortisone and several other steroid derivatives. In these experiments the basophils were incubated with the steroids for 24 hr before anti-Ige stimulation, since we had observed that the steroids had no significant effect at 2 hr and only a partial effect at 8 hr.

A number of comments can be made about these experiments. First, if one compares the ability of the several steroids to inhibit histamine release with their reported *in vivo* anti-inflammatory effects, the agreement is excellent. A variety of other steroid derivatives without glucocorticoid activity have been studied and found to have no inhibitory ability. Second, unlike the effect of steroids in guinea pig lung (4), the inhibitory activity is not rapidly reversible. [However, the effect of steroids in guinea pig lung also can be seen during short time periods, i.e., 40 to 60 min (4).] Third, in this case the effect is limited to IgE-mediated release, as the calcium ionophore A23187, phorbol diesters, and f-Met peptides cause release normally from cells incubated with hydrocortisone for 24 hr. Fourth, it should be noted that the inhibition by hydrocortisone occurs at levels which might be expected

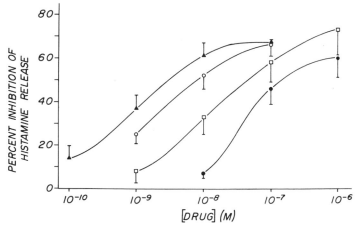

FIG. 4. Inhibition of histamine release by corticosteroids. The experiment involved a 24-hr culture of cells and steroid, washing, and then challenge with anti-Ige. (▲) Triamcinolone acetonide. (○) Dexamethasone acetate. (□) 9α-Flourocortisone. (●) Hydrocortisone.

to obtain *in vivo*. Whether the *in vivo* events are adequately reflected in these *in vitro* experiments or if there are "antisteroid" components in the *in vivo* situation remains to be determined. Finally, these results do not prove that the steroids inhibit release by inhibiting the activity of phospholipase A_2; they are simply concordant with what has been reported in other systems (10).

CONCLUSION

We have described a series of experiments which suggest that the lipoxygenase pathways, perhaps by the generation of 5-HPETE, play a critical role in the release of inflammatory mediators. They also reverse the inhibition of release caused by certain hormones. This latter effect may have interesting *in vivo* consequences during situations in which arachidonic acid or certain lipoxygenase products are elevated, e.g., the activation of platelets, neutrophils, or monocytes (1,7,8,11). The basophils may, in this situation, become insensitive to regulation by local or circulating hormones that act via elevation of cAMP (e.g., prostaglandins or catecholamines). This deregulation of the basophil may lead in turn to a greater sensitivity to antigen-induced inflammation. Regulation of leukocyte function by circulating hormones and autacoids active in the adenylate cyclase system has been proposed to be central to the control of the immune response (2). In this context, it seems important to have mechanisms by which discrete areas of inflammation might be released from systemic regulation in order to allow a vigorous response to proceed where necessary. It may also be that leukocytes other than the basophil are "deregulated" by lipoxygenase products.

Lipoxygenase products of arachidonic acid may therefore have three important roles in inflammation: first, as effector molecules (such as chemotaxins or stimulants of end-organ responses); second, as essential intracellular biochemical intermediates in the cellular mechanism of mediator release; finally, as local hormones (autacoids) which deregulate an inflammatory response.

ACKNOWLEDGMENTS

Supported by grants AI 07290 and HL 23586 from the National Institutes of Health. This is publication No. 461 of the O'Neill Laboratories, The Good Samaritan Hospital. The biologically derived LTC and LTD were gifts of Dr. Parker. The synthetic LTC and LTD were gifts from Dr. Rokach, Merck.

REFERENCES

1. Bills, T. K., Smith, J. B., and Silver, M. J. (1977): Selective release of arachidonic acid from the phospholipids of human platelets in response to thrombin. *J. Clin. Invest.*, 60:1–6.
2. Bourne, H. R., Lichtenstein, L. M., Melmon, K. L., Henney, C. S., Weinstein, Y., and Shearer, G. M. (1974): Modulation of inflammation and immunity by cyclic AMP. *Science*, 184:19–28.
3. Drazen, J. M., Austen, K. F., Lewis, R. A., Clark, D. A., Goto, G., Marfat, A., and Corey, E. J. (1980): Comparative airway and vascular activities of leukotrienes C-1 and D in vivo and in vitro. *Proc. Natl. Acad. Sci. USA*, 77:4354–4358.
4. Flower, R. J., and Blackwell, G. J. (1979): Anti-inflammatory steroids induce biosynthesis of a phospholipase A_2 inhibitor which prevents prostaglandin generation. *Nature*, 278:456–459.

5. Ford-Hutchinson, A. W., Bray, M. A., Doig, M. V., Shipley, M. E., and Smith, M. J. H. (1980): Leukotriene B, a potent chemotactic and aggregating substance released from polymorphonuclear leukocytes. *Nature*, 286:264–265.

6. Goetzl, E. J., and Pickett, W. C. (1980): The human PMN leukocyte chemotactic activity of complex hydroxy-eicosatetraenoic acids (HETES). *J. Immunol.*, 125:1789–1791.

7. Goetzl, E. J., and Sun, F. F. (1979): Generation of unique mono-hydroxy-eicosatetraenoic acids from arachidonic acid by human neutrophils. *J. Exp. Med.*, 150:406–411.

8. Hammarström, S., Lindgren, J. A., Marulo, C., Duvall, E. A., Anderson, T. F., and Voorhees, J. J. (1979): Arachidonic acid transformations in normal and psoriatic skin. *J. Invest. Dermatol.*, 73:180–183.

9. Hanna, C. J., Bach, M. K., Pape, P. D., and Schellenberg, R. R. (1981): Slow-reacting substances (leukotrienes) contract human airway and pulmonary vascular smooth muscle in vitro. *Nature*, 290:343–344.

10. Hirata, F., Schiffmann, E., Venkatasubramanian, K., Salomon, D., and Axelrod, J. (1980): A phospholipase A_2 inhibitory protein in rabbit neutrophils induced by glucocorticoids. *Proc. Natl. Acad. Sci. USA*, 77:2533–2536.

11. Humes, J. L., Bonney, R. J., Pelus, L., Dahlgren, M. E., Sadowski, S. J., Kuehl, F. R., and Davies, P. (1977): Macrophages synthesize and release prostaglandins in respect to inflammatory stimuli. *Nature*, 269:149–151.

12. Kennerly, D. A., Sullivan, T. J., Sylvester, P., and Parker, C. W. (1979): Diacylglycerol metabolism in mast cells: a potential role in membrane fusion and arachidonic acid release. *J. Exp. Med.*, 150:1039–1044.

13. Lewis, G. P., and Whittle, B. J. R. (1977): The inhibition of histamine release from rat peritoneal mast cells by non-steroid anti-inflammatory drugs and its reversal by calcium. *Br. J. Pharmacol.*, 61:224–235.

14. Lichtenstein, L. M. (1968): Mechanism of allergic histamine release from leukocytes. In: *The Biochemistry of Acute Allergic Reactions*, edited by E. L. Becker and K. F. Austen, pp. 153–174. Blackwell Scientific, London.

15. Lichtenstein, L. M., and DeBernardo, R. L. (1971): IgE-mediated histamine release: in vitro separation into two phases. *Int. Arch. Allergy Appl. Immunol.*, 41:56–71.

16. Lichtenstein, L. M., and Margolis, S. (1968): Histamine release in vitro: inhibition by catecholamines and methylxanthines. *Science*, 161:902–903.

17. Lichtenstein, L. M., and Osler, A. G. (1964): Studies on the mechanism of hypersensitivity phenomena. IX. Histamine release from human leukocytes by ragweed pollen antigen. *J. Exp. Med.*, 120:507–530.

18. MacGlashan, D. W., Jr., and Lichtenstein, L. M. (1980): The purification of human basophils. *J. Immunol.*, 124:2519–2521.

19. Marone, G., Hammarström, S., and Lichtenstein, L. M. (1980): An inhibitor of lipoxygenase inhibits histamine release from human basophils. *Clin. Immunol. Immunopathol.*, 17:117–122.

20. Marone, G., Kagey-Sobotka, A., and Lichtenstein, L. M. (1979): Effects of arachidonic acid and its metabolites on antigen-induced histamine from human basophils in vitro. *J. Immunol.*, 123:1669–1677.

21. Marone, G., Kagey-Sobotka, A., and Lichtenstein, L. M. (1981): Control mechanisms of histamine release from human basophils in vitro: the role of phospholipase A_2 and of lipoxygenase metabolism. *Int. Arch. Allergy Appl. Immunol. (Suppl.)*, 66:144–148.

22. Marone, G., Kagey-Sobotka, A., and Lichtenstein, L. M. (1981): Possible role of phospholipase A_2 in triggering histamine secretion from human basophils in vitro. *Clin. Immunol. Immunopathol.* 20:231–239.

23. Middleton, E., Sherman, W. B., Fleming, W., and Van Arsdale, P. P. (1960): Some biochemical characteristics of allergic histamine release from leukocytes of ragweed sensitive subjects. *J. Allergy*, 31:448.

24. Nakao, A., Buchanan, A. M., and Potokur, D. S. (1980): Possible involvement of phospholipase A_2 in A23187-induced histamine release from purified rat mast cells. *Int. Arch. Allergy Appl. Immunol.*, 63:30–43.

25. Noah, J. W., and Brand, A. (1954): Release of histamine in the blood of ragweed sensitive individuals. *J. Allergy*, 25:210.

26. Norn, S. (1965): Influence of antirheumatic agents on the release of histamine from rat peripheral mast cells after an antigen-antibody reaction. *Acta Pharmacol. Toxicol.*, 22:369–378.

27. Palmer, R. M. J., Stepney, R. J., Higgs, G. A., and Eakins, K. E. (1980): Chemokinetic activity of arachidonic acid lipoxygenase products on leukocytes of different species. *Prostaglandins*, 20:411–418.
28. Peters, S. P., Siegel, M. I., Kagey-Sobotka, A., and Lichtenstein, L. M. (1981): Lipoxygenase products modulate histamine release in human basophils. *Nature*, 292:455–457 .
29. Schleimer, R. P., Lichtenstein, L. M., and Gillespie, E. (1981): Inhibition of basophil histamine release by antiinflammatory steroids. *Nature*, 292:454–455.
30. Schmutzler, W., and Freundt, G. P. (1975): The effect of glucocorticoids and catecholamines on cyclic AMP and allergic histamine release in guinea pig lung. *Int. Arch. Allergy Appl. Immunol.*, 49:209–212.
31. Siriganian, R. P. (1974): An automated continuous flow system for the extraction and fluorometric analysis of histamine. *Anal. Biochem.*, 57:383–394.
32. Sullivan, T. J. (1981): Studies of the role of diacylglycerol metabolism in the formation and release of mediators from mast cells. In: *Biochemistry of the Acute Allergic Reaction*, edited by E. L. Becker, A. S. Simon, and K. F. Austen. Alan R. Liss, New York.
33. Sullivan, T. J., and Parker, C. W. (1979): Possible role of arachidonic acid and its metabolites in mediator release from rat mast cells. *J. Immunol.*, 122:431–436.
34. Vallee, E., Gougat, J., Navarro, J., and Delahayes, J. F. (1979): Anti-inflammatory and platelet anti-aggregant activity of phospholipase A_2 inhibitors. *J. Pharmacol.*, 31:588–592.
35. Varafty, B. B., Fouque, F., and Chighard, M. (1980): Interference of bromophenacyl bromide with platelet phospholipase A_2 activity induced by thrombin and by the ionophore A23187. *Thrombosis Res.*, 17:91–102.

Leukotrienes and Other Lipoxygenase Products,
edited by B. Samuelsson and R. Paoletti.
Raven Press, New York © 1982.

Effect of Lipoxygenase Products on Leukocyte Accumulation in the Rabbit Eye

Parimal Bhattacherjee, Brian Hammond, John A. Salmon, and Kenneth E. Eakins

The Wellcome Research Laboratories, Beckenham, Kent BR3 3BS, England

Leukocytes are a source of chemotactic lipoxygenase products in inflammation. Polymorphonuclear leukocytes (PMNs) synthesize monohydroxy and dihydroxy acids of the lipoxygenase pathway (1,3,7,10), many of which possess chemotactic activity for human PMNs (5); of these, leukotriene B_4 (LTB_4) is the most potent so far described (4,8).

In the present study we have examined the activity of LTB_4, LTC_4, and LTD_4 and other products of the lipoxygenase pathway, the synthetic chemotactic peptide formyl-methionyl-leucyl-phenylalanine (f-Met-Leu-Phe), and bradykinin (BK) on leukocyte migration following injection into the anterior chamber of the rabbit eye. In addition, we have measured the effects of these compounds on intraocular pressure (IOP) as an indication of their actions on vascular permeability.

METHODS

New Zealand White rabbits weighing 2.2 to 2.5 kg were anesthetized with urethane (1 to 2 g/kg) injected into a marginal ear vein as a 25% solution in 0.9% sodium chloride. A femoral artery was cannulated, and mean arterial blood pressure was measured with an Elcomatic strain gauge transducer. All recordings were made using a Sanborn 7700 Series recorder. Each animal received heparin, 1,000 IU/kg intravenously before cannulation of the eyes.

A 23-gauge hypodermic needle was then introduced through the cornea into the anterior chamber of each eye. Each needle was designed with a hole on either side of the central crimp as described by Hammond (6). Each end of the needle was connected by polyethylene tubing to a pressure transducer and reservoir via a closed-circuit perfusion system. Perfusion of the eye with physiological saline was maintained at 25 µl/min by means of a Watson-Marlow delta pump. Test compounds were dissolved in 25 µl saline and loaded into one-half of the perfusion system. Thus the pressure could be measured in the anterior chamber, and the drug solution could be infused by manipulating the perfusion system. All drugs were infused for 1 min after at least 10 min of steady-state recording of IOP. The pump was then turned off and the IOP recorded for the duration of the experiment. Each infusion of a test compound into one eye was accompanied by a simultaneous infusion of an equal volume of saline into the contralateral control eye. At the end of a given time period, the animal was killed with an overdose of sodium pentobarbitone, the aqueous humor removed from each eye, and leukocytes counted in an improved Neubauer chamber.

SYNTHESIS OF LIPOXYGENASE PRODUCTS

5-Hydroxyeicosatetraenoic acid (5-HPETE), 12-HPETE, and 15-HPETE, were synthesized by oxygenation of arachidonic acid (9). The HETEs were prepared by the reduction of the corresponding hydroperoxyeicosatetraenoic acids (HPETEs) with triphenyl phospine followed by high-pressure liquid chromatography (HPLC) purification. LTB_4 was prepared biosynthetically by incubating arachidonic acid with rabbit PMNs (2); the product was purified by silicic acid column chromatography followed by reverse-phase (RP) HPLC. Purified LTB_4 was quantified by ultraviolet (UV) spectrophotometry (2). LTC_4 and LTD_4 were obtained from Prof. E. J. Corey, Harvard University.

RESULTS AND DISCUSSION

The injection of either LTB_4 or the f-Met peptide into the anterior chamber of rabbit eyes resulted in the appearance of leukocytes in the aqueous humor. An equal volume of saline injected under similar experimental conditions did not result in the appearance of significant numbers of leukocytes in the aqueous humor at any time up to 8 hr after the injection. Experiments with both compounds were then carried out to determine the time course of the appearance of leukocytes following the intracameral injections. The increase in leukocytes occurred between 2 and 3 hr and was maximal by 4 hr after the injection of the f-Met peptide (500 ng). A similar time course was observed with LTB_4 (200 ng). On the basis of these results, aqueous humor samples were removed at 4 hr in all subsequent experiments.

Varying amounts of either f-Met-Leu-Phe or LTB_4 resulted in a dose-dependent accumulation of leukocytes in the aqueous humor 4 hr later (Fig. 1). LTB_4 and the f-Met peptide were approximately equipotent, the threshold doses required to increase the leukocyte counts being approximately 25 to 100 ng for each compound. LTC_4 and LTD_4 (up to 500 ng) had no effect on cell accumulation.

Doses of up to 25 μg arachidonic acid and 5 μg 12-HETE failed to induce the appearance of leukocytes in the aqueous humor. Similarly, 500 ng each of prostaglandins E_1, E_2, $F_{2α}$, and I_1 (a stable analog of prostacyclin), 5-HPETE, and 5-HETE failed to elevate leukocyte numbers.

Differential counting of leukocytes in smears of aqueous humor from the eyes 4 hr after injection of LTB_4 or the f-Met peptide demonstrated that 85 to 90% of the cells were PMNs. Furthermore, histological examination of sections of the iris-ciliary processes stained with hematoxylin and eosin showed heavy infiltration of the tissue by PMN leukocytes following the injection of LTB_4 or the f-Met peptide but not saline.

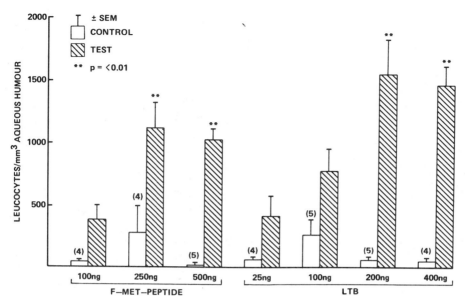

FIG. 1. Leukocyte accumulation in the anterior chamber 4 hr after injection of various doses of f-Met peptide and LTB₄ in a constant volume of 25 μl. Contralateral eyes received an equal volume of saline or an appropriate vehicle.

TABLE 1. *Effect of various substances on the intraocular pressure of the rabbit eye following their injection into the anterior chamber*

Drug[a]	Dose (ng)	IOP Test-control (mm Hg) Mean ± SEM	No.	Statistical significance t-test (p)
LTB₄	400	−0.25 ± 2.25	4	N.S.
5-HETE	500	−0.25 ± 0.8	4	N.S.
5-HPETE	500	1.00 ± 0.9	3	N.S.
12-HETE	1,000	1.00 ± 0.9	5	N.S.
FMLP	500	4.2 ± 1.4	6	< 0.05
PGE₂	500	27.5 ± 0.3	4	< 0.001
Arachidonic acid	25,000	30.4 ± 2.9	4	< 0.002
PGI₁	500	18.0 ± 4.0	4	< 0.05
PGF₂α	500	10.0 ± 4.5	5	< 0.1
BK	500	38.0 ± 6.0	4	< 0.01

[a]All drugs injected in a dose of 25 μl.

Although arachidonic acid and the metabolites of the cyclo-oxygenase pathway did not result in leukocyte accumulation, all these compounds increased IOP (Table 1). This rise in IOP was maximal between 10 and 30 min after the injection and slowly returned to the initial value within 40 to 120 min depending on the compound injected. None of the lipoxygenase products tested affected IOP at the highest doses used; the f-Met peptide had a very small but significant effect. BK was the most potent of all the compounds tested, 10 ng resulting in a profound rise in IOP.

Injection of a mixture of LTB_4 and BK did not result in an alteration in the leukocyte accumulation produced by LTB_4, whereas an additive response was found with mixtures of (FMLP) and LTB_4 (Table 2). In contrast, the prostaglandins (E_2, $F_{2\alpha}$, and I_1) inhibited leukocyte accumulation obtained in response to LTB_4 (Fig. 2). In order to investigate whether the decreased leukocyte accumulation in the aqueous humor in the presence of the prostaglandins was due to increased adherence in the vasculature of the iris-ciliary processes, tissues taken from eyes treated with the mixture of LTB_4 + PGE_2 and LTB_4 alone were examined histologically. Leukocyte infiltration of the tissues in response to the LTB_4 + E_2 mixture appeared to be less than that with LTB_4 alone, suggesting that PGE_2 had indeed decreased the chemotactic response to LTB_4.

The rabbit eye model described here appears to be a useful model for studying putative chemotactic factors *in vivo*, particularly as the aqueous humor is normally free of white blood cells. Using this model we demonstrated that LTB_4 produces a dose-dependent increase in the number of leukocytes (predominantly PMNs) in the aqueous humor. LTB_4 was more active than monohydroxy or monohydroperoxy acids of the lipoxygenase pathway, and it was approximately equipotent with the synthetic chemotactic peptide FMLP.

None of the lipoxygenase products tested affected IOP in contrast to the profound increase observed following the introduction of arachidonic acid, PGE_2, or bradykinin into the anterior chamber. This rise in IOP is thought to result from a breakdown of the blood-aqueous barrier with the consequent sudden influx of protein and fluid into the anterior chamber, indicating that the products of the lipoxygenase pathway of arachidonic acid metabolism do not have a direct effect on ocular vascular permeability. The cyclo-oxygenase and lipoxygenase products of arachi-

TABLE 2. *Effect of bradykinin and f-Met peptide on leukocyte accumulation induced by LTB_4*

Drug[a]	Dose (ng)	Leukocytes/mm³ aqueous humor	
		Mean ± SEM	No.
LTB_4	200	1,694 ± 194	19
	100	800 ± 173	5
BK	500	135 ± 138	5
BK + LTB_4	500 + 200	1,453 ± 547	4
FMLP	100	990 ± 114	4
FMLP + LTB_4	100 + 100	1,559 ± 204	4

[a]Drug solutions either alone or as a mixture were injected into the anterior chamber, and 4 hr later leukocytes were counted in the aqueous humor.

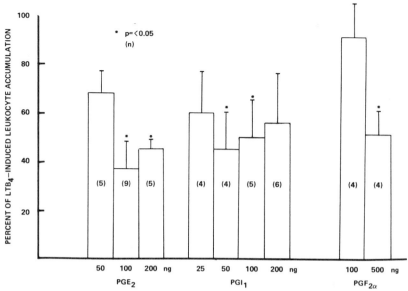

FIG. 2.   Effect of prostaglandins on leukocyte accumulation induced by LTB₄. The mixture of a prostaglandin and LTB₄ was infused simultaneously into the anterior chamber; leukocytes in aqueous humor were counted 4 hr later and expressed as the percent of those seen with LTB₄ alone.

donic metabolism may be important in the development of both acute and chronic ocular inflammation.

REFERENCES

1. Borgeat, P., Hamberg, M., and Samuelsson, B. (1976): Transformation of arachidonic acid and dihomo-Y-linolenic acid by polymorphonuclear leukocytes. *J. Biol. Chem.*, 251:7816–7820.
2. Borgeat, P., and Samuelsson, B. (1979): Arachidonic acid metabolism in polymorphonuclear leukocytes—effects of ionophore A23187. *Proc. Natl. Acad. Sci. USA*, 76:2148–2152.
3. Bragt, P. C., and Bonta, I. L. (1979): In vivo metabolism of [1-¹⁴C]-arachidonic acid during different phases of granuloma development in the rat. *Biochem. Pharmacol.*, 28:1581–1586.
4. Ford-Hutchinson, A. W., Bray, M. A., Doig, M. V., Shipley, M. E., and Smith, M. J. H. (1980): Leukotriene B: a potent chemokinetic and aggregating substance released from polymorphonuclear leukocytes. *Nature*, 286:264–265.
5. Goetzl, E. J., Valone, F. H., Reinholt, U. N., and Gorman, R. R. (1979): Specific inhibition of the polymorphonuclear leukocyte chemotactic response to hydroxy-fatty acid metabolites of arachidonic acid by methyl ester derivatives. *J. Clin. Invest.*, 63:1181–1186.
6. Hammond, B. R. (1977): Perfusion of the rabbit anterior chamber. *Exp. Eye Res.*, 24:533–534.
7. Hammarström, S., Hamberg, M., Samuelsson, B., Duell, E. A., Stawiski, M., and Voorhees, J. J. (1975): Increased concentrations of free arachidonic acid prostaglandin E₂ and F₂α and of 12L-hydroxy-5,8,10,14,eicosatetraenoic acid (HETE) in epidermis of psoriasis: evidence for perturbed regulations of arachidonic acid levels in psoriasis. *Proc. Natl. Acad. Sci. USA*, 72:5130–5134.
8. Palmer, R. J., Stepney, R., Higgs, G. A., and Eakins, K. E. (1980): Chemokinetic activity of arachidonic acid lipoxygenase products on leukocytes from different species. *Prostaglandins*, 20:411–418.

9. Porter, N. A., Wolf, R., Yarbro, E., and Weenan, H. (1979): The auto-oxidation of arachidonic acid: formation of a proposed SRS-A intermediate. *Biochem. Biophys. Res. Commun.*, 89:1058–1064.
10. Turner, S. R., Tainer, J. A., and Lynn, W. S. (1975): Biogenesis of chemotactic molecules by the arachidonic lipoxygenase system of platelets. *Nature*, 257:680–681.

Leukotrienes and Other Lipoxygenase Products,
edited by B. Samuelsson and R. Paoletti.
Raven Press, New York © 1982.

Inflammatory Properties of Lipoxygenase Products and the Effects of Indomethacin and BW755C on Prostaglandin Production, Leukocyte Migration, and Plasma Exudation in Rabbit Skin

G. A. Higgs, C. M. R. Bax, and S. Moncada

Departments of Prostaglandin Research and Pharmacology,
The Wellcome Research Laboratories, Beckenham, Kent BR3 3BS, England

It has been proposed that the oxygenation of arachidonic acid by lipoxygenase leads to the formation of important inflammatory mediators. Lipoxygenase activity was first demonstrated in platelets (8,15), and it was subsequently shown that the product of platelet lipoxygenase, 12-hydroxyeicosatetraenoic acid (12-HETE), was chemotactic for polymorphonuclear leukocytes (PMNs) (20). The theory that lipoxygenase activity contributes to leukocyte migration was supported by the observation that a lipoxygenase inhibitor reduced the accumulation of PMNs in experimental inflammation (10).

Leukocytes themselves contain a lipoxygenase which controls peroxidation of arachidonic acid in the C-5 position (2), and this leads to the production of 5-HETE and the dihydroxy acid 5,12-DHETE (leukotriene B_4, LTB_4) (18). We now know that LTB_4 is one of the most potent endogenous chemotactic factors yet discovered (6,7,16). LTB_4 is at least 100 times more active than any of the mono-HETEs tested and is equipotent as a chemotactic agent with the synthetic peptide N-formyl-methionyl-leucyl-phenylalanine (FMLP) (16). LTB_4 also induces leukocyte accumulation *in vivo* following intraperitoneal (19), intradermal (4,13), or intraocular injection (1).

Initially it was thought that lipoxygenase products were not involved in vascular changes in inflammation. LTB_4 did not cause or enhance edema formation in the rat (19), and 5-HETE or LTB_4 did not increase vascular permeability in guinea pig skin (25). It has been reported, however, that LTB_4 is equipotent with prostaglandin E_2 (PGE_2) and prostacyclin in enhancing bradykinin-induced plasma exudation in rabbit skin (5,13). This observation has been confirmed (3), and it has also been shown that vasodilator prostaglandins synergize with LTB_4 to produce plasma exudation (3,13,21). LTB_4 is not a vasodilator (3), and its effects on vascular permeability appear to depend on circulating PMNs. LTB_4, in common with other chemotactic agents such as FMLP and the C5a fraction of complement, does not cause plasma exudation in neutropenic animals (21).

Evidence is accumulating therefore that arachidonate lipoxygenase, as well as cyclo-oxygenase, is involved in the generation of inflammatory mediators. Cyclo-oxygenase products such as PGE_2 and prostacyclin mediate the vasodilatation and hyperalgesia which are characteristic of acute inflammation (for review see ref. 12), whereas the synthesis of chemotactic lipoxygenase products contributes to increased vascular permeability and leukocyte accumulation. These findings support the theory that dual inhibitors of cyclo-oxygenase and lipoxygenase offer a more comprehensive anti-inflammatory activity than drugs such as aspirin or indomethacin, which are selective inhibitors of prostaglandin synthesis (10,12).

METHODOLOGY

Leukocyte Infiltration in Rabbit Skin

Leukocyte infiltration was estimated after intradermal injection of test substances (in 100 μl 50 mM tris buffer, pH 7.5) into the shaved backs of male New Zealand White rabbits (2 to 3 kg). The animals were killed after 1 to 4 hr, the skin removed, and injection sites separated using a ⅝-inch steel punch. Skin samples were processed histologically and sections (4 to 5 μm) cut at four points of each injection site. The sections were stained with hematoxylin and eosin before microscopic examination at low power (× 160) to locate the injection site. Leukocyte numbers were counted in five high-power (× 1,000) fields (hpf) arranged vertically through the dermis. Each treatment was given at two to six injection sites in at least three animals.

Plasma Exudation in Rabbit Skin

Plasma exudation was measured following intravenous injection of 50 μCi [^{125}I] human serum albumin (Radiochemical Centre, Amersham). Test substances were injected intradermally into five or six sites, each animal receiving six treatments. After 30 min the animals were killed and the skins removed. Radioactivity in injection sites and blood samples was measured using an automatic γ-counter, and the plasma volume at each site was calculated (23). Drugs were administered intraperitoneally 1 to 2 hr before the animals were killed.

Arachidonic Acid Metabolism in Rabbit Skin

The products of arachidonic acid metabolism in rabbit skin were measured 15 min after the intradermal injection of 5.5 μg [1-^{14}C]arachidonic acid (176 μCi/mg; Radiochemical Centre, Amersham). Injection sites were chopped in 1 ml Tris buffer and mixed with 2 ml cold acetone acidified to pH 4 with 1% 5 N HCl and extracted into chloroform. The chloroform was evaporated to dryness, the residue taken up in 75 μl chloroform:methanol (2:1) and applied quantitatively to silica gel thin-layer chromatography plates. The plates were developed in ether:hexane:acetic acid (60:40:1), and the separated products were located by autoradiography. Zones of radioactivity were identified by comparison with authentic standards and quantitated by standard liquid scintillation counting procedures.

Prostaglandin Synthesis and Leukocyte Migration

Inflammatory exudates were collected 24 hr after the subcutaneous implantation of sterile polyester sponges (4.0 × 1.5 × 0.5 cm) impregnated with 2% carrageenin (Viscarin, Marine colloids; w/v in sterile saline), in the shaved backs of rabbits. Sponge exudates were squeezed into 5 ml heparinized (5 units/ml) saline, and total leukocyte numbers were estimated on a blind basis using "Improved Neubauer" counting chambers and phase contrast microscopy. Prostaglandin concentrations in exudates were determined by bioassay following acid-lipid extraction. Drugs were administered intraperitoneally at the time of sponge implantation, 5 to 8 hr later, and 3 hr before sponge removal (11).

Materials

LTB$_4$ was supplied by E. J. Corey, Department of Chemistry, Harvard University, and 5-HPETE was synthesized by autoxidation of arachidonic acid (17) by R. J. Stepney at the Wellcome Laboratories. BW755C was synthesized by F. C. Copp.

RESULTS

Inflammatory Effects of LTB$_4$, Arachidonic Acid, and 5-HPETE

LTB$_4$ (25 to 100 ng) caused a dose-dependent increase in the number of leukocytes in the dermis 1 to 4 hr after intradermal injection which was significantly greater than the response caused by injection of tris buffer alone. Arachidonic acid or 5-HPETE, which are precursors of LTB$_4$ (2,18), were relatively inactive (Fig. 1). Higher doses of arachidonic acid (10 to 50 μg) caused significant leukocyte infiltration.

Intradermal injections of LTB$_4$ (0.1 to 1.0 μg) caused small but significant increases in plasma exudation, and combinations of LTB$_4$ (1 to 200 ng) with bradykinin (0.5 μg) or PGE$_2$ (0.1 μg) resulted in a dose-dependent enhancement of plasma exudation which could not be accounted for by an additive effect. Arachidonic acid or 5-HPETE was relatively inactive in potentiating bradykinin-induced plasma exudation (Fig. 2).

High doses of arachidonic acid (10 to 200 μg) caused a dose-dependent increase in plasma exudation in rabbit skin (Fig. 3). Oleic acid (10 to 200 μg), an 18-C fatty acid which cannot be converted to prostaglandins or leukotrienes, also caused a dose-dependent increase in plasma exudation but was significantly less active than arachidonic acid. A dose of 200 μg arachidonic acid caused a plasma leakage of

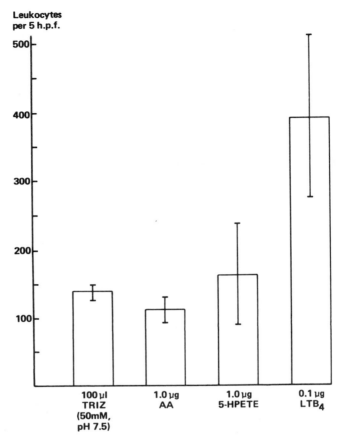

FIG. 1. Effects of intradermal injection of tris buffer, arachidonic acid, 5-HPETE, or LTB$_4$ on the total number of leukocytes in the dermis of rabbit skin. Skin samples were taken 4 hr after injection, and leukocytes in five high-power fields (hpf) were counted. Each histogram is the mean of two to six injection sites in at least three animals. The bars represent ± 1 SEM.

50.4 ± 4.2 μl (mean ± SE mn), whereas the same dose of oleic acid induced a leakage of 16.7 ± 3.4 μl.

Metabolism of Arachidonic Acid

Approximately 10% of the total [1-^{14}C]arachidonic acid injected into rabbit skin was converted to a variety of products. In skin receiving [1-^{14}C]arachidonic acid alone, metabolites which cochromatographed with authentic 5-HETE and 5,12-DHETE accounted for 23.6 ± 2.7% of the total products generated. When [1-^{14}C] arachidonic acid was injected into a site which had received an intradermal injection of LTB$_4$ (100 ng) 1 hr previously, the generation of 5-lipoxygenase products was increased to 34.9 ± 3.2%. This was accompanied by an increase in leukocyte infiltration in the dermis.

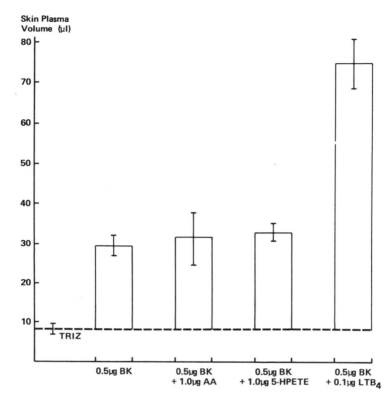

FIG. 2. Effects of arachidonic acid, 5-HPETE, or LTB_4 on bradykinin-induced plasma exudation in rabbit skin. Each histogram is the mean of at least six values. The bars represent $\pm$ 1 SEM.

Effects of Indomethacin and BW755C on Prostaglandin Synthesis, Leukocyte Migration, and Plasma Exudation

Inflammatory exudates at 24 hr contained 36.4 ± 6.1 ng PGE_2-equivalent/ml and $54.1 \times 10^6 \pm 9.2 \times 10^6$ leukocytes/ml. Indomethacin (0.1 to 50 mg/kg) caused a dose-dependent inhibition of prostaglandin synthesis ($ED_{50} = 1.1$ mg/kg) but had a differential effect on leukocyte migration. Low doses of indomethacin (0.1 to 2.0 mg/kg) elevated the number of leukocytes in sponge exudates but higher doses (5 to 50 mg/kg) caused a dose-dependent inhibition ($ED_{50} = 7.5$ mg/kg). Indomethacin 1 mg/kg caused significant enhancement of leukocyte migration and significant depression of prostaglandin synthesis (Fig. 3). BW755C (50 mg/kg) was equiactive in decreasing the total leukocyte number and prostaglandin concentration (Fig. 3).

BW755C (50 mg/kg) significantly reduced arachidonic acid-induced plasma exudation at every dose (Fig. 3), but indomethacin (1 mg/kg) had little or no effect, producing small potentiation at the highest dose of arachidonic acid (Fig. 3). Higher doses of indomethacin (10 to 50 mg/kg) caused dose-dependent inhibition of plasma

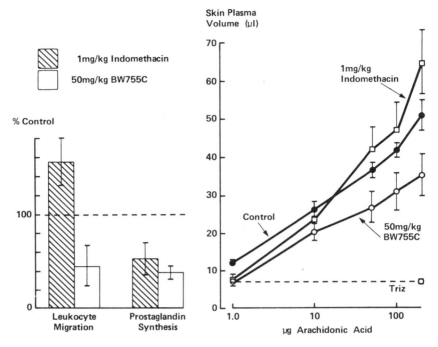

FIG. 3. Effects of indomethacin (1 mg/kg) or BW755C (50 mg/kg) on leukocyte migration, prostaglandin synthesis, and arachidonic-acid induced plasma exudation in rabbit skin. The histograms show the effects of the drugs on total leukocyte numbers and prostaglandin concentrations in 24-hr inflammatory exudates. Results are expressed as percent of control values obtained in untreated animals. The histograms represent the means of three to five experiments. The bars are ± 1 SEM. The dose-response curves show the effect of indomethacin (1 mg/kg) or BW755C (50 mg/kg) on arachidonic acid-induced plasma exudation. The filled circles (control) show responses to arachidonic acid (1 to 200 µg) in untreated animals. Each point is the mean of 5 to 11 experiments in separate animals. The bars represent ± SEM.

exudation. Indomethacin 50 mg/kg reduced the plasma exudation caused by 200 µg arachidonic acid to 34.0 ± 4.4 µl, and this was not significantly different from the reduction in exudation caused by BW755C 50 mg/kg (Fig. 3).

DISCUSSION

These results confirm that LTB_4 has potent inflammatory properties *in vivo* (1,3–5,13,19,21), and it is possible that the local generation of LTB_4 is an important mechanism in the control of vascular permeability and leukocyte accumulation in injured tissues. The relative inactivity of arachidonic acid or 5-HPETE, compared to LTB_4, in causing leukocyte infiltration or plasma exudation suggests that the enzymes necessary for leukotriene synthesis are not active in normal skin. It is likely that the source of leukotrienes in inflammation is the leukocytes themselves, and this is supported by the increase in 5-lipoxygenase metabolism in skin with elevated leukocyte infiltration. Arachidonic acid metabolism by leukocytes could therefore represent a central mechanism in the development and amplification of the acute inflammatory response.

It has been suggested that a component of arachidonic acid-induced plasma exudation in rabbit skin is due to histamine release (14). The response, however, is only partially blocked by histamine antagonists, and it has been concluded that at least some of the increased permeability is the result of arachidonic acid metabolism (14). This is supported by the observation that oleic acid, which cannot be converted to prostaglandins or leukotrienes, is considerably less active than arachidonic acid.

Prostaglandins and other cyclo-oxygenase products are weak inducers of plasma exudation (14,23), and PGE_2 has similar potency to arachidonic acid itself (14). The effects of arachidonic acid can therefore be accounted for only by cyclo-oxygenase metabolism if there is a high conversion in the skin. However, in our experiments less than 10% of injected arachidonic acid was converted, and cyclo-oxygenase products accounted for only $1.7 \pm 0.4\%$ of the total activity injected. Furthermore, indomethacin 1 to 5 mg/kg decreased prostaglandin concentrations in inflammatory exudates by 50 to 75% of control values, but in separate experiments the same doses of indomethacin did not reduce arachidonate-induced plasma exudation (Fig. 3).

The effects of indomethacin and BW755C on prostaglandin synthesis and leukocyte migration confirm similar observations in the rat (9). Indomethacin is clearly a selective inhibitor of prostaglandin synthesis, but the inhibition of leukocyte migration could indicate lipoxygenase inhibition at higher doses. The enhancement of leukocyte accumulation by indomethacin 1 mg/kg (Fig. 3) could be explained by a diversion of substrate toward the production of chemotactic lipoxygenase products. This may also account for the tendency of indomethacin 1 mg/kg to potentiate arachidonate-induced plasma exudation (Fig. 3). The results presented in this chapter also confirm that BW755C is equiactive in inhibiting prostaglandin synthesis and leukotyce migration, and it is possible that this activity is due to dual cyclo-oxygenase and lipoxygenase inhibition (10).

Arachidonic acid-induced plasma exudation is therefore probably caused by the synthesis of vasodilator prostaglandins and LTB_4, which separately act to increase blood flow and vascular permeability, thereby enhancing plasma exudation (21,24). If the LTB_4-induced increase in permeability is dependent on PMNs (21), this explains why indomethacin and BW755C reduce plasma exudation only at doses which inhibit cyclo-oxygenase activity and leukocyte migration. The disadvantage with selective cyclo-oxygenase inhibitors such as indomethacin and aspirin is that there is an onset of acute gastric toxicity, which is related to inhibition of gastric cyclo-oxygenase (22), before the dose that reduces leukocyte migration is reached. BW755C, on the other hand, does not cause gastric erosions at anti-inflammatory doses (22). It is possible that dual cyclo-oxygenase and lipoxygenase inhibitors will prove more effective than present nonsteroidal anti-inflammatory drugs, and that they will be better tolerated.

REFERENCES

1. Bhattacherjee, P., Eakins, K. E., and Hammond, B. (1981): Chemotactic activity of arachidonic acid lipoxygenase products in the rabbit eye. *Br. J. Pharmacol.*, 73:254–255P.
2. Borgeat, P., and Samuelsson, B. (1979): Transformations of arachidonic acid by rabbit polymorphonuclear leukocytes: formation of a novel dihydroxy eicosatetraenoic acid. *J. Biol. Chem.*, 254:2643–2646.
3. Bray, M. A., Cunningham, F. M., Ford-Hutchinson, A. W., and Smith, M. J. H. (1981): Leukotriene B_4: a mediator of vascular permeability. *Br. J. Pharmacol.*, 72:483–486.
4. Carr, S. C., Higgs, G. A., Salmon, J. A., and Spayne, J. A. (1981): The effects of arachidonate lipoxygenase products on leukocyte migration in rabbit skin. *Br. J. Pharmacol.*, 73:253–254P.
5. Eakins, K. E., Higgs, G. A., Moncada, S., Salmon, J. A., and Spayne, J. A. (1980): The effects of arachidonate lipoxygenase products on plasma exudation in rabbit skin. *J. Physiol. (Lond.)*, 307:71P.
6. Ford-Hutchinson, A. W., Bray, M. A., Doig, M. V., Shipley, M. E., and Smith, M. J. H. (1980): Leukotriene B, a potent chemokinetic and aggregating substance released from polymorphonuclear leukocytes. *Nature*, 286:264–265.
7. Goetzl, E. J., and Pickett, W. C. (1980): The human PMN leukocyte chemotactic activity of complex hydroxy-eicosatetraenoic acids (HETEs). *J. Immunol.*, 125:1789–1791.
8. Hamberg, M., and Samuelsson, B. (1974): Prostaglandin endoperoxides: novel transformations of arachidonic acid in human platelets. *Proc. Natl. Acad. Sci. USA*, 71:3400–3404.
9. Higgs, G. A., Eakins, K. E., Mugridge, K. G., Moncada, S., and Vane, J. R. (1980): The effects of non-steroid anti-inflammatory drugs on leukocyte migration in carrageenin-induced inflammation. *Eur. J. Pharmacol.*, 66:81–86.
10. Higgs, G. A., Flower, R. J., and Vane, J. R. (1979): A new approach to anti-inflammatory drugs. *Biochem. Pharmacol.*, 28:1959–1961.
11. Higgs, G. A., Harvey, E. A., Ferreira, S. H., and Vane, J. R. (1976): The effects of anti-inflammatory drugs on the production of prostaglandins in vivo. In: *Advances in Prostaglandin and Thromboxane Research, Vol. 1*, edited by B. Samuelsson, and R. Paoletti, pp. 105–110. Raven Press, New York.
12. Higgs, G. A., Moncada, S., and Vane, J. R. (1980): The mode of action of anti-inflammatory drugs which inhibit the peroxidation of arachidonic acid. *Clin. Rheum. Dis.*, 6:675–693.
13. Higgs, G. A., Salmon, J. A., and Spayne, J. A. (1981): The inflammatory effects of hydroperoxy and hydroxy acid products of arachidonate lipoxygenase in rabbit skin. *Br. J. Pharmacol.*, 74:429–433.
14. Ikeda, K., Tanaka, K., and Katori, M. (1975): Potentiation of bradykinin-induced vascular permeability increase by prostaglandin E_2 and arachidonic acid in rabbit skin. *Prostaglandins*, 10:747–758.
15. Nugteren, D. H. (1975): Arachidonate lipoxygenase in blood platelets. *Biochim. Biophys. Acta*, 380:299–307.
16. Palmer, R. M. J., Stepney, R. J., Higgs, G. A., and Eakins, K. E. (1980): Chemokinetic activities of arachidonic acid lipoxygenase products on leukocytes of different species. *Prostaglandins*, 20:411–418.
17. Porter, W. A., Wolf, R. A., Yarbro, E. M., and Weenan, H. (1979): The autoxidation of arachidonic acid: formation of a proposed SRS-A intermediate. *Biochem. Biophys. Res. Commun.*, 89:1058–1064.
18. Samuelsson, B., Hammarström, S., Murphy, R. C., and Borgeat, P. (1980): Leukotrienes and slow reacting substance of anaphylaxis (SRS-A). *Allergy*, 35:375–381.
19. Smith, M. J. H., Ford-Hutchinson, A. W., and Bray, M. A. (1980): Leukotriene B: a potential mediator of inflammation. *J. Pharm. Pharmacol.*, 32:517–518.
20. Turner, S. R., Tainer, J. A., and Lynn, W. S. (1975): Biogenesis of chemotactic molecules by the arachidonate lipoxygenase system of platelets. *Nature*, 257:680–681.
21. Wedmore, C. V., and Williams, T. J. (1981): Control of vascular permeability by polymorphonuclear leukocytes in inflammation. *Nature*, 289:646–650.
22. Whittle, B. J. R., Higgs, G. A., Eakins, K. E., Moncada, S., and Vane, J. R. (1980): Selective inhibition of prostaglandin production in inflammatory exudates and gastric mucosa. *Nature*, 284:271–273.
23. Williams, T. J. (1979): Prostaglandin E_2, prostaglandin I_2 and the vascular changes of inflammation. *Br. J. Pharmacol.*, 65:517–524.

24. Williams, T. J., and Peck, M. J. (1977): Role of prostaglandin-mediated vasodilatation in inflammation. *Nature*, 270:530–532.
25. Williams, T. J., and Piper, P. J. (1980): The actions of chemically pure SRS-A on the microcirculation in vivo. *Prostaglandins*, 19:779–789.

Leukotrienes and Other Lipoxygenase Products,
edited by B. Samuelsson and R. Paoletti.
Raven Press, New York © 1982.

Role of Lipoxygenases in Regulation of PHA and Phorbol Ester-Induced Mitogenesis

J. M. Bailey, R. W. Bryant, C. E. Low, M. B. Pupillo, and J. Y. Vanderhoek

Department of Biochemistry, George Washington University Medical School, Washington, D.C. 20037

There are numerous reports that cyclo-oxygenase metabolites of arachidonic acid are involved in the regulation of lymphocyte differentiation (10). In mixed spleen lymphocyte cultures these compounds have been shown to regulate the initiation of mitogenesis by plant lectins, probably by controlling the production of inhibitory lymphokines by suppressor T-lymphocytes (12). Much of the information on these systems has come from the use of inhibitors of cyclo-oxygenase such as indomethacin and eicosatetraynoic acid (ETYA). Some of these inhibitors unfortunately are not completely specific for cyclo-oxygenase, also inhibiting the cellular lipoxygenase pathways. Spleen cell cultures have been reported to have very active lipoxygenase systems, and products of these pathways have been shown to be involved in a number of important immunological functions including chemotaxis and mitogenesis (5). Much of this evidence has also come from the use of inhibitors of the lipoxygenase pathway, e.g., ETYA, which also inhibits cyclo-oxygenase.

We recently isolated the compound 15-hydroxyeicosatetraenoic acid (15-HETE), which is a potent and specific inhibitor of cellular lipoxygenases, from rabbit polymorphonuclear neutrophils (PMNs) (11). In platelets incubated with ^{14}C-arachidonic acid, this compound, used at a concentration of 25 μM, completely inhibited the synthesis of 12-HETE via the lipoxygenase pathway without affecting the synthesis of either thromboxane B_2 (TXB_2) or heptadecatrienoic acid (HHT) (Fig. 1). Over a range of concentrations, 6 to 18 μM, 12-HETE synthesis was progressively inhibited, and at the lower concentrations a concomitant small increase in the formation of the cyclo-oxygenase products was observed, presumably due to the sparing effect on the ^{14}C-arachidonic acid substrate (Fig. 2).

In similar experiments, rabbit PMNs incubated with ^{14}C-arachidonic acid were shown to synthesize considerable amounts of 5-HETE and a compound with chromatographic characteristics of a dihydroxy acid (Fig. 3). The latter compound was isolated by high-pressure liquid chromatography (HPLC), converted to the mono-trimethyl-silyl-methyl-ester derivative, and its structure confirmed as 5,12-dihydroxyeicosatetraenoic acid (5,12-DHETE; leukotriene B_4, LTB_4) by mass spectrometry (Fig. 4). Addition of 15-HETE (8 μM) completely inhibited synthesis of both of these compounds, as shown in the right-hand panels of Fig. 3.

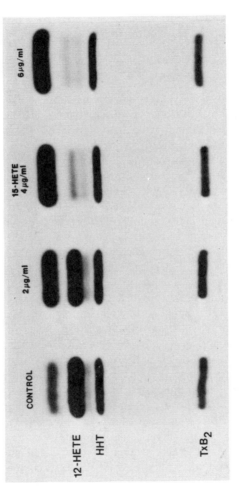

FIG. 1. Differential inhibition of human platelet lipoxygenase versus cyclo-oxygenase by 15-HETE. Human platelets were incubated for 5 min with ^{14}C-arachidonic acid and the products isolated, separated by thin-layer chromatography (TLC), and visualized by radioautography as described in ref. 4. As shown here, preincubation of platelets with increasing concentrations of 15-HETE progressively inhibited formation of 12-HETE without appreciably affecting the concentrations of the cyclo-oxygenase products thromboxane B_2 and HHT.

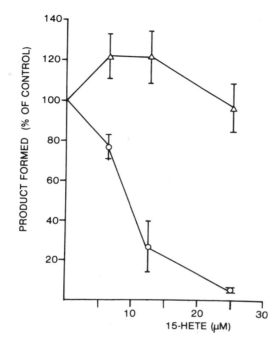

FIG. 2. Comparative inhibition of thromboxane and 12-HETE formation in human platelets by 15-HETE. For details see legend to Fig. 1.

Mouse spleen lymphocyte cultures incubated with ^{14}C-arachidonic acid produced a somewhat different pattern of lipoxygenase metabolites, including 12-HETE and a compound with chromatographic characteristics of a dihydroxy arachidonic acid (Fig. 5, left panel), similar to those of the 5,12-DHETE isolated from rabbit PMNs. Synthesis of these compounds was inhibited by added 15-HETE (Fig. 5, right panel) in a manner which suggested the possibility that it may be a product of the 12-lipoxygenase pathway. Incubation of spleen lymphocytes with either ^{14}C-12-HPETE or ^{14}C-12-HETE gave a similar product, suggesting that the compound may be a 5,12-DHETE isomeric with LTB$_4$ and formed by a different enzymatic sequence of events (3). Synthesis of 12-HETE and the presumed isomeric 5,12-DHETE was prevented by adding 15-HETE to the cultures.

The possible role of these compounds in lymphocyte mitogenesis was investigated in spleen cell cultures treated with the T-lymphocyte mitogen phytohemagglutinin (PHA) or the B-lymphocyte mitogen lipopolysaccharide (LPS). Mitogenesis was measured 72 hr after exposure to mitogen by adding ^{3}H-thymidine and measuring ^{3}H incorporation after a further 18 hr incubation. Increasing concentrations of 15-HETE progressively blocked the T-lymphocyte response without significantly inhibiting mitogenesis of B-lymphocytes (Table 1). Inhibition of the T-cell mitogenic response was half-maximal at concentrations ranging from 5 to 28 μM and was not accompanied by significant decreases in cell viability as measured by the trypan blue dye exclusion test (Table 2). It should be pointed out that the mitogenesis

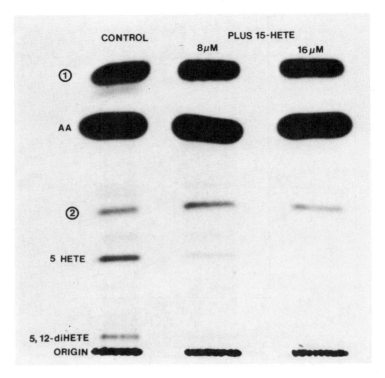

FIG. 3. Inhibition of LTB formation in rabbit peritoneal neutrophils by 15-HETE. Rabbit peritoneal neutrophils elicited with glycogen were washed and incubated for 5 min with ^{14}C-arachidonic acid and the products isolated, separated by TLC, and visualized by radioautography as described in ref. 4. Synthesis of 5-HETE and 5,12-DHETE (LTB$_4$) was confirmed by conversion to the TMS-methyl ester derivatives and analysis by gas chromatography-mass spectrometry (GC-MS) as shown in Fig. 4. Note that addition of exogenous 15-HETE progressively blocked synthesis of products of the 5-lipoxygenase pathway.

experiments are carried out in 5% serum medium which, because of binding by the serum albumin, may decrease the effective concentration of 15-HETE somewhat. In similar experiments, the cyclo-oxygenase inhibitor aspirin was shown to produce marked stimulation in the T-lymphocyte response to PHA (Fig. 6). This is in agreement with previous observations (6) that suppressor T-cell activity in spleen cell populations is dependent on prostaglandin synthesis. It is therefore of particular interest to note that the stimulation by aspirin can be progressively and completely reversed by concentrations of 15-HETE similar to those which inhibit cellular lipoxygenase pathways. Indeed, in most experiments aspirin-enhanced mitogenesis was more sensitive to 15-HETE than was the basal mitogenic response to PHA (Fig. 7), suggesting that aspirin may selectively stimulate a more sensitive T-cell subpopulation in mixed spleen lymphocyte cultures.

In additional experiments, a number of analogs of 15-HETE, including 15-OH-20:3, 15-OH-20:2, and 15-OH-20:0, were synthesized and tested for inhibition of

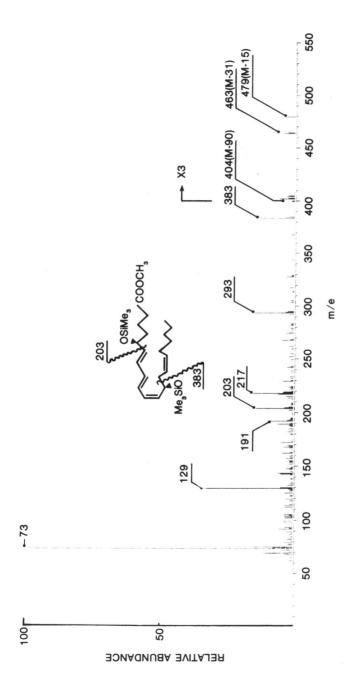

FIG. 4. Mass spectrum of Me₃Si derivative of LTB methyl ester (peak A; C 23.7) di-*trans*, di-*cis* isomer. Products from the incubation of rabbit PMNs incubated with ^{14}C-arachidonic acid were separated by HPLC. A major peak in the dihydroxy fatty acid region (peak A; C 23.7) was isolated, converted to the trimethyl-silyl-methyl-ester derivative, and analyzed by GC-MS. The mass spectrum displays ions characteristic of the structure of authentic LTB₄.

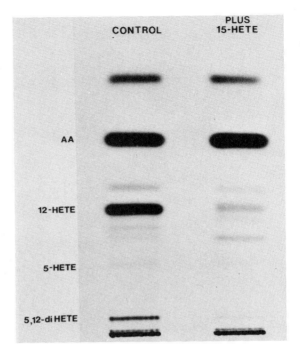

FIG. 5. Inhibition of arachidonic acid metabolism in mouse spleen lymphocytes by 15-HETE. Products of ^{14}C-arachidonic acid metabolism by mouse spleen lymphocyte suspensions were isolated and analyzed by TLC. The identification of products is by relative R_f values only. The compound migrating in the dihydroxy fatty acid region tentatively labeled 5,12-diHETE has different migration characteristics to the authentic LTB$_4$ isolated from rabbit PMNs.

TABLE 1. *Selective inhibition of T-lymphocyte versus B-lymphocyte
mitogenesis by 15-HETE*

Conc. of 15-HETE added to culture (μM)	^{3}H-thymidine incorporated (dpm)		
	Control	Plus PHA	Plus LPS
0	853 ± 80	41,495 ± 2,208	2,720 ± 122
10	846 ± 72	21,758 ± 1,121	2,908 ± 207
30	949 ± 109	9,338 ± 141	2,280 ± 103

Mouse spleen lymphocyte cultures containing 5 × 10^5 cells prepared from spleens of C57 BRG male mice 4 to 6 weeks old were established in RPMI medium containing 5% serum, together with 15-HETE at the indicated concentrations. The T-cell or B-cell mitogens phytohemagglutinin (PHA) or lipopolysaccharide (LPS), respectively, were added as indicated. After 72 hr of incubation, ^{3}H-thymidine (1 μCi) was added to all cultures and ^{3}H-thymidine incorporation measured at 90 hr on the washed harvested cells.

TABLE 2. *Trypan blue cell-viability measurements in*
15-HETE-inhibited spleen lymphocyte cultures

Sample	³H-thymidine incorporation (dpm)	Viability (% of control)
Control	153,673 ± 13,557	100 ± 12.1
15-HETE (10 μM)	70,509 ± 5,684	101.8 ± 3.6

Spleen lymphocyte cultures were incubated with PHA in the presence or absence of 15-HETE (10 μM). Mitogenesis was measured by addition of 1 μCi ³H-thymidine to cultures at 72 hr and measuring ³H incorporation at 90 hr. Viability was measured by the trypan blue exclusion procedure on all cultures at 90 hr.

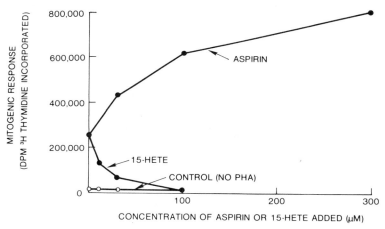

FIG. 6. Opposing effects of 15-HETE and aspirin on PHA-induced mitogenesis in spleen lymphocytes. Mouse spleen lymphocyte cultures were incubated in 5% serum RPMI medium for 72 hr in the presence of PHA (1 μg/ml) with the indicated additions of aspirin or 15-HETE. Mitogenesis was measured by adding ³H-thymidine (1 μCi). Cells were harvested 18 hr later, and incorporation of radioactivity was measured.

the cellular lipoxygenases (Fig. 8) and mitogenesis. In general there was a correlation between the two systems, so that those fatty acids which were most effective in inhibiting lipoxygenase products were also most effective in inhibiting mitogenesis (Fig. 9).

It is known that induction of mitogenesis requires a period of several hours exposure to the mitogen before cells become committed to undergoing division some 72 hr later (1). The time course of sensitivity of the lymphocyte cultures both to the lipoxygenase inhibitor 15-HETE and to the cyclo-oxygenase inhibitor aspirin was therefore studied by adding these compounds at varying intervals after PHA exposure was begun. The marked stimulation produced by aspirin was almost completely dependent on events occurring within the first 8 to 10 hr following PHA exposure (Fig. 10). In contrast, although the effects of 15-HETE were maximal if

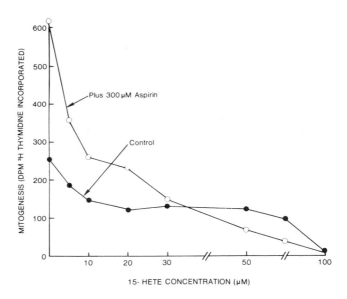

FIG. 7. Influence of 15-HETE on PHA-induced mitogenesis in spleen lymphocytes: control versus aspirin-treated cells. Mouse spleen lymphocyte cultures were stimulated with PHA (1 μg/ml) together with the indicated additions. Note that the aspirin-stimulated mitogenesis was more sensitive to inhibition by 15-HETE than were the control cultures.

added during the first 8 hr, considerable inhibition of events occurring later in the blastogenic process was also produced by this agent (Fig. 11).

Induction of mitogenesis by tumor promoters, including the phorbol diesters, has been reported to be accompanied by arachidonic acid release and prostaglandin (PG) synthesis (9). Inhibitors of PG synthesis, however, do not inhibit the mitogenic effects of the phorbol esters (7), which suggested that alternative products of arachidonic acid metabolism (e.g., the HETEs or leukotrienes) may be more important in this phenomenon.

To test this, spleen cell cultures were incubated with the tumor promoter phorbol myristate acetate (PMA), and mitogenesis was assayed after 72 hr in the usual manner. This compound considerably enhanced basal ^{3}H-thymidine incorporation, although the response was only about 20% of that produced by PHA. Addition of 15-HETE produced progressive inhibition of the mitogenic response to the phorbol diester at both of the PMA concentrations tested (Fig. 12). It seems probable that a different cell subpopulation may be responding to PMA, as in contrast to PHA, the response was not enhanced by aspirin. Furthermore, the pattern of inhibition by different fatty acids was much broader than that observed for PHA, suggesting that secondary inhibition of the epoxidase enzyme similar to that observed in PMNs (2) occurs in this system. The experiments described here provide suggestive evidence that lipoxygenase products of arachidonic acid may be involved in the responses of certain populations of lymphocytes to plant lectins and tumor promoters.

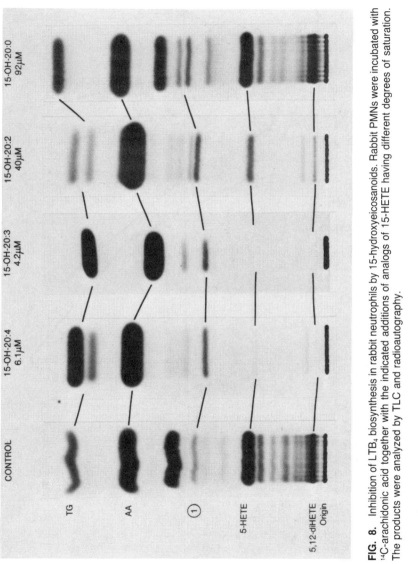

FIG. 8. Inhibition of LTB$_4$ biosynthesis in rabbit neutrophils by 15-hydroxyeicosanoids. Rabbit PMNs were incubated with ^{14}C-arachidonic acid together with the indicated additions of analogs of 15-HETE having different degrees of saturation. The products were analyzed by TLC and radioautography.

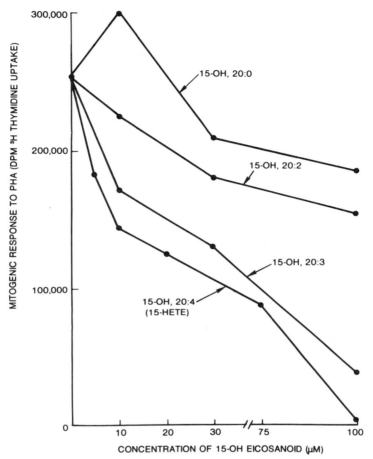

FIG. 9. Influence of various 15-hydroxyeicosanoids on PHA-induced mitogenesis in spleen lymphocyte cultures. Mouse spleen lymphocyte cultures stimulated with PHA (1 μg/ml) were incubated for 72 hr with various analogs of 15-HETE having different degrees of saturation. After 72 hr mitogenesis was measured by adding ³H-thymidine.

There is considerable supporting evidence that arachidonic acid release is a key event in the induction of mitogenesis, a possibility which we have recently confirmed using specific inhibitors of arachidonic acid release from cellular phospholipids (8). Furthermore, it has been reported that the activation of guanyl cyclase, which is induced by mitogens, can be duplicated by certain HETEs (4). Addition of 15-HETE to PHA-treated cells has also been shown to block this process.

SUMMARY

The experiments described here confirm the critical involvement of products of arachidonic acid metabolism in immune cell function. The results suggest opposing

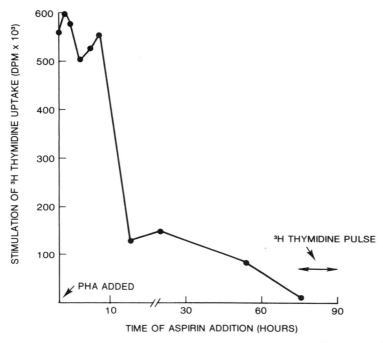

FIG. 10. Influence of the time of aspirin addition in aspirin stimulation of PHA-induced mitogenesis. Mouse spleen lymphocyte cultures were incubated with PHA (1 μg/ml) together with aspirin (300 μM) added at the indicated intervals after PHA addition. Note that stimulation by aspirin is critically dependent on events occurring during the first 8 to 10 hr of incubation with PHA.

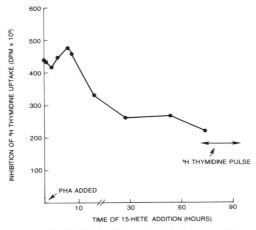

FIG. 11. Influence of time of 15-HETE addition in inhibition of PHA-induced mitogenesis. Mouse spleen lymphocyte cultures were incubated with PHA, and 15-HETE (30 μM) was added at various intervals following PHA addition. Note that mitogenesis was maximally inhibited by 15-HETE added during the first 10 hr, but that events occurring later in the mitogenic process were also sensitive to 15-HETE.

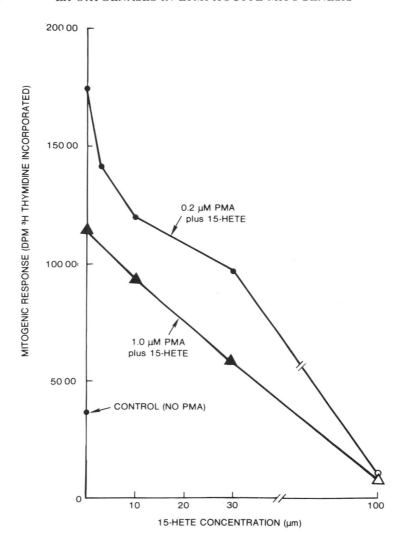

FIG. 12. Influence of phorbol-ester (PMA) induced mitogenesis by the leukotriene inhibitor 15-HETE. Mouse spleen lymphocyte cultures were incubated with the tumor promoter phorbol myristate acetate (PMA) at two levels (0.2 and 1 μM). 15-HETE at the indicated concentrations was added and the mitogenesis measured by adding ^{3}H-thymidine at 72 hr.

activities of the cyclo-oxygenase versus the lipoxygenase products in these systems. However, in contrast to the situation with the prostaglandins, where addition of the pure compounds has reproduced the effects predicted from the inhibitor studies, identification of the specific lipoxygenase products involved and the induction of the predicted biological response by addition of these compounds remains to be accomplished.

REFERENCES

1. Allan, D., and Crumpton, M. J. (1973): Phytohemagglutinin-lymphocyte interaction: characterization of binding sites on pig lymphocytes for ^{125}I-labelled phytohemagglutinin. *Exp. Cell Res.*, 78:271–278.
2. Bokoch, G. M., and Reed, P. W. (1981): Evidence for inhibition of leukotriene A_4 synthesis by 5,8,11,14-eicosatetraynoic acid in guinea pig polymorphonuclear leukocytes. *J. Biol. Chem.*, 256:4156–4159.
3. Borgeat, P. (1981): In: *Proceedings: Winter Prostaglandin Conference*, Clearwater, Florida.
4. Coffey, R. G., and Hadden, J. W. (1981): Arachidonic acid metabolites in mitogen activation of lymphocyte guanylate cyclase. In: *Advances in Immunopharmacology*, edited by J. Hadden, L. Chedid, P. Mullen and F. Spreafico, pp. 365–373. Pergamon Press, New York.
5. Goetzl, E. J., and Pickett, W. C. (1980): The human PMN leukocyte chemotactic activity of complex hydroxy-eicosatetraenoic acids (HETEs). *J. Immunol.*, 125:1789–1791.
6. Leung, K. H., and Mihich, E. (1980): Prostaglandin modulation of development of cell-mediated immunity in culture. *Nature*, 288:597–600.
7. Levine, L., and Ohuchi, K. (1978): Stimulation by carcinogens and promoters of prostaglandin production by dog kidney (MDCK) cells in culture. *Cancer Res.*, 38:4142–4146.
8. Makheja, A. N., Low, C. E., Salata, K., Pupillo, M. B., and Bailey, J. M. (1981): In preparation.
9. Ohuchi, K., and Levine, L. (1978): Tumor promoting phorbol diesters stimulate release of radioactivity from (^{3}H)-arachidonic acid labeled- but not (^{14}C) linoleic acid-labeled cells: indomethacin inhibits the stimulated release from (^{3}H) arachidonate labeled cells. *Prostaglandins Med.*, 1:421–431.
10. Parker, C. W. (1979): Prostaglandins and slow-reacting substance. *J. Allergy Clin. Immunol.*, 63:1–14.
11. Vanderhoek, J. Y., Bryant, R. W., and Bailey, J. M. (1980): Inhibition of leukotriene biosynthesis by the leukocyte product 15-hydroxy-5,8,11,13-eicosatetraenoic acid. *J. Biol. Chem.*, 255:10064–10066.
12. Webb, D. R., Wieder, K. J., and Nowowiejski, I. (1981): Prostaglandins in lymphocyte suppressor mechanism. In: *Advances in Immunopharmacology*, edited by J. Hadden, L. Chedid, P. Mullen, and F. Spreafico, pp. 383–388. Pergamon Press, New York.

Subject Index

Acetone, 15-lipoxygenase purification, 77, 82
Acetylation, LTB$_4$, 274–275, 280
Acetylcholine blockade, 191
Acetylenic acids, 20–27, 127–134
Acidic lipids, SRS cleavage, 116, 120–121
Adenylate cyclase system, 159, 322
Aerosol administration
    FPL 55712, airways, 231–232, 238–241
    LTC$_4$, bronchoconstriction, 189–190
    LTD$_4$, pulmonary effects, baboons, 206–209
Air flow, 237–241
Airway resistance, LTC$_4$, 147, 190
Airway smooth muscle, 169–180, 243–250
Allergic response, FPL 55712 effect, 232–234
Alveolar macrophages, sputum, 256
p-Amino hippuric acid inhibition, 111
Analogs, *see* Synthetic leukotrienes
Anaphylatoxins, 105; *see also* specific toxins
Anaphylaxis
    and eicosanoid system activation, 165
    leukotrienes and SRS, 11–14, 215–222
    prostaglandins and leukotrienes, 111
Angiotensin II, 196
Antagonists, SRS, 229–235; *see also* specific
    drugs
Antigen challenge, anaphylaxis, 215–221; *see
    also* IgE mediation
Anti-inflammatory steroids, 1, 13, 321
Anti-thrombotic agents, 25
Aorta
    LTC$_4$ metabolism, 98
    LTD$_4$ injection, 213
    prostaglandin formation, LTC$_4$ challenge,
        155–160, 163
Aqueous humor, chemotaxic factors, 328–329
Arachidonate-15-lipoxygenase, 77–82
Arachidonic acid
    activation by LTC$_4$, 154–155, 164
    analogs, ECF inhibition, 301–312
    as calcium ionophore, 264–265
    calcium translocation, 266
    conversion to LTC and LTD, 106–107
    conversion to LTC$_4$, 84
    conversion to 8,11,12 triols, 71–76
    ECF generation, 303–312
    histamine enhancement, IgE, 316–317
    after immunologic stimulation, 320
    inflammatory effects, rabbit skin, 333–337
    inhibition by 15-HETE, 346

and intraocular pressure, 326–327
metabolism, leukocytes, 1–6
    and acetylenic acid, RBLs, 127–134
    double dioxygenation, 45–50, 53–59
    oxidation, neutrophils, 259–270
mitogenesis induction, 350
novel products, 29–43
steroid inhibition, 13
selective inhibitors, platelets, 17–27
Arrhythmias, 216
Arsenazo III, neutrophils, 264–267
Arterial pressure, *see* Systemic arterial
    pressure
Arteriole contraction, 194–195
Aryl sulfatase, SRS inactivation, 119
Aspirin
    histamine release enhancement, 317
    inhibition by 15-HETE, 344, 348–351
    and LTC$_4$ challenge, 154–155, 162–165
    T-lymphocyte response, 344, 347–348
Asthma, 122, 232; *see also* Bronchoconstriction
Atopic status, SRS activity, 252, 255
Atropine and LTC$_4$, 159–161, 163–165

B-lymphocyte mitogenesis, 346
Baboons, LTD$_4$ effects, 204–209
Bactericidal capacity, neutrophils, 294, 296–299
Basophilic leukemia cells, *see* Rat basophilic
    leukemia cells
Basophils, 105, 111, 301, 315–322
Beta-adrenoceptor activation, LTC$_4$
    antagonism, 159–160
Beta-glucuronidase, 261, 296–298, 302
Beta-glycero-phosphatase, ECF release, 302
Bicarbonate, LTD$_4$ effect, 204
Bile, LTC$_4$ metabolism, 97
Bilirubin inhibition, LTC and LTD, 109–111
Bisamide LTD$_4$, structure and function, 143
Blood flow, 174–175, 212–214
Blood measures, LTC$_3$ metabolism, 98–100
Blood pressure; *see also* specific types
    after LTC$_4$, 162–163, 175–176, 188
    after LTD$_4$, 175–176, 202–209
Bovine coronary artery, 155–157, 161–162
Bradykinin effects
    leukocyte accumulation, rabbit eye, 325–329
    LTB$_4$ potentiation, 289
    versus LTC$_4$, prostaglandin release, 154–165

Science   Leukotrienes and other lipoxygenase
QP   products / editors, Bengt
801   Samuelsson, Rodolfo Paoletti. -- New
P68   York : Raven Press, c1982.
L48   xviii, 365 p. : ill. ; 24 cm. --
1982   (Advances in prostaglandin,
            thromboxane, and leukotriene research
            series ; v. 9)

            Contains most of the papers
        presented at an international
        conference on leukotrienes and other
        lipoxygenase products held in
        Florence, Italy, June 10-12, 1981.
            Includes bibliographical references
        and index.
            ISBN 0-89004-741-3
                              (Cont'd on next card)

MUNION ME   821116   821110 CStoC-S
C000467   KW /JW   A*   82-B9880
                                   81-40607

Science   Leukotrienes and other lipoxygenase
QP   products... c1982.
801   (CARD   2)
P68
L48
1982

1. Prostaglandins--Congresses.   I. Samuelsson,
Bengt.   II. Paoletti, Rodolfo.   III.
International Conference on Leukotrienes and
Other Lipoxygenase Products (1981 : Florence,
Italy)   IV. Series.

MUNION ME   821116   821110 CStoC-S
C000468   KW /JW   A*   82-B9880
                                   81-40607